MÜNCHNER GEOGRAPHISCHE ABHANDLUNGEN

Reihe B

in

MÜNCHNER UNIVERSITÄTSSCHRIFTEN

FAKULTÄT FÜR GEOWISSENSCHAFTEN

Again and again the scholar starts to attain a certain goal to arrive at a completely different one.

Münchener Universitätsschriften

Fakultät für Geowissenschaften

MÜNCHENER GEOGRAPHISCHE ABHANDLUNGEN

REIHE B

Herausgegeben von
Prof. Dr. H.-G. Gierloff-Emden und Prof. Dr. F. Wilhelm
Schriftleitung: Dr. K.R. Dietz und Dr. H.-J. Mette

Band B 11

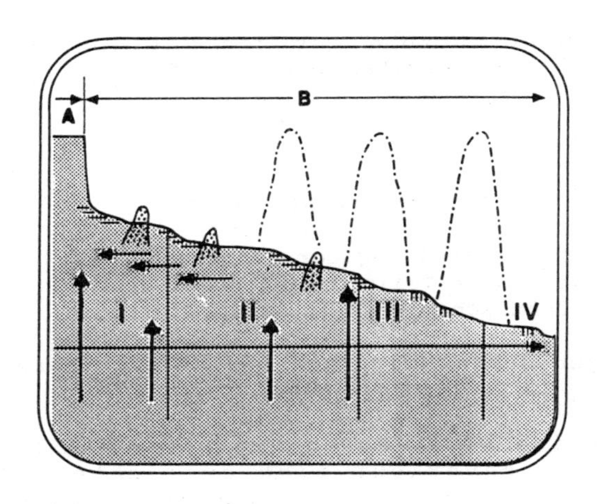

JOSEF BIRKENHAUER

The Great Escarpment of Southern Africa and its Coastal Forelands - A Re-Appraisal

Mit 55 Fotos, 50 Abbildungen und 19 Tabellen

1991
Institut für Geographie der Universität München
Kommissionsverlag: GEOBUCH-Verlag, München

Gedruckt mit Unterstützung aus den Mitteln der Münchner Universitätsgesellschaft

Die Ausführungen geben Meinungen und Korrekturstand des Autors wieder.

Anfragen bezüglich Drucklegung von wissenschaftlichen Arbeiten und Tauschverkehr sind zu richten an die Herausgeber im Institut für Geographie der Universität München, Luisenstraße 37, 8000 München 2

Druck: Fotodruck Frank GmbH, Gabelsbergerstr. 15, 8000 München 2

Kommissionsverlag: GEOBUCH-Verlag, Rosental 2, 8000 München 2

Zu beziehen durch den Buchhandel

ISBN 3-925308-96-2
ISSN 0932-3147

FOREWORD

In the years between 1981 and 1985 I was fortunate to undertake five more or less extensive journeys to **southern Africa**. Because of these journeys I was enabled to come to know my working area in a suitably thorough way. Three of the journeys were funded by the German Research Society. I am very grateful indeed to the society and its referees.

Beginning with Natal, the working area was steadily widened. The same happened to questions and problems (cf. motto on the title-page). Fig. 0.1 allows a view over the extent of the working area. Research in the field was mainly guided by these broad attempts:

(1) An attempt to gather witnesses for the **morphodynamic events** and processes going on recently and in the past (i.e. soils, duricrusts, slope development and the like) and, moreover, to gather witnesses for the **stratigraphy** of the events (i.e. sediments - though these become very scarce above the altitude of 400m).

(2) An attempt to find out **planations** in several altitudes, to bring their occurrence into connection with each other and the witnesses mentioned, and, finally, to map the planations (using the 1 : 250 000 map of South Africa and South West Africa).
(In doing so, **cross sections** were laid in more or less regular distances, covering the whole working area (Fig. 0.1), with each section running from the interior towards the sea (or vice versa). The cross sections were connected with each other by profiles running parallel to the coast, both in the neighbourhood of the sea and in the interior.

(3) An attempt, finally, to define the nature of the **Great Escarpment** and to gain a more secure knowledge about its evolution, dealing particularly with the concepts presented by ROGERS, KAYSER, KING and OLLIER. In doing so, it was thought advisable not to start from any **Mega-geomorphology**, but rather to end up with it. Prior to it, it was thought necessary to gather as many details by patient fieldwork as was possible. These **details** will be presented in the following text **within their own frame** (though this will perhaps be boring for the reader).

Another thing turning out to be boring (or even judged to be not necessary to the one or the other reader) may be my decision to include the **history of research** in this area. Yet I feel justified in doing so,

- partly out of respect to earlier scholars,
- partly in helping towards a better understanding of the research tradition (especially since German workers seem to have been prominent in this field during the wars, their findings and ideas however not known to the English-speaking community),
- partly because one's own findings and ideas can thereby be put into a stronger relief.

As will be seen, some older theories and notions will be particularly questioned - not from any private wish to do so, but from the conclusions I was forced to draw from the details of my own fieldwork. A particular theory in question is the **piedmont-treppen theory** by Walther PENCK and a particular notion is that of **parallel slope retreat** and back-wearing, so often put forward by KING. Though W. PENCK's theory has already been refuted by BÜDEL conclusively enough at that, I think it is necessary to deal with it again because this theory presents a frame of reference for both KING (knick-points) and KAYSER (gradual widening of area of **uplift**).

Though, as will be seen, **plate tectonics**, marine geology and geophysics do play their parts on the stage that forms the frame of my own research, no particular research was done by me within these fields mentioned. I dedicated myself to fieldwork wholly on land and there only to geomorphology. Such restrictions cannot be avoided, though even within such **restrictions**, one has to cross the borders between one earth science and the other more often than once.

The gradual **widening of the working area** beyond the originally narrow nucleus region proved to be exceptionally fruit- and useful, because of the resulting change in the scale of perception (cf. motto).

In spite of the continuous **mapping** of the planations while in the field I do not think it advisable (and worth while) to produce a chart of the planations for the whole working area because such a chart could be done only on a very small scale, making the chart more or less meaningless. Instead, profiles as well as charts on a larger scale and fotos are presented for particular regions. In looking at these one should bear in mind that all phenomena pertaining to planations are principally the same all over the working area.

All materials used in the treatise are presented in Table 0.1 in the way of a synopsis. The cross sections in Fig. 0.1 may be used as references.

J. Birkenhauer
Munich

Note

Native speakers of the English-speaking community who set about to read this book are kindly asked to forebear with the many semantic inadequacies, tortuous phrasings and the like, which the author himself is painfully aware of. Nevertheless the author hopes that it is to their benefit, despite all the drawbacks.

VORWORT

In den Jahren 1981 - 1985 konnte der Verfasser in fünf ausgedehnteren Reisen das Arbeitsgebiet im **südlichen Afrika** einigermaßen gründlich kennenlernen. Drei dieser Reisen wurden von der DFG unterstützt. Dafür sei dieser und den Gutachtern herzlich gedankt.

Vom ersten Ansetzen in Natal an ausgehend wurde das Arbeitsgebiet (vgl. dazu Fig. 0.1) aus sachimmanenten Gründen ständig erweitert. Gleiches widerfuhr den Fragestellungen (vgl. Motto auf der Titelseite).

Die Arbeit im Gelände war im wesentlichen von drei Hauptabsichten geleitet.
1) wurde auf Zeugen von **morphodynamischen Vorgängen** (Böden, Hartkrusten, Hangentwicklung z.B.) geachtet und auf Zeugen ihrer zeitlichen Einordnung (wobei allerdings Sedimente mit zunehmender Höhe über 400m NN leider nur äußerst spärlich vorhanden sind);

2) wurden die Zeugen wie auch die festgestellten **Verebnungen** im Zusammenhang mit der Karte 1 : 250 000 flächenhaft auskartiert. Dafür wurden in regelmäßigen Abständen **Querprofile** vom Hochland zur Küste gelegt (bzw. umgekehrt); vgl. Fig. 0.1.
Die 38 dort ausgewiesenen Profile wurden sowohl im Binnenland als auch im Küstenbereich durch Längsprofile miteinander verknüpft.

3) sollte die Natur der **Großen Randstufe** bestimmt und eine klarere Vorstellung ihrer Genese entwickelt werden, wobei besonders die vorliegenden Konzepte von ROGERS, KAYSER, KING und OLLIER zu berücksichtigen waren.
Auch wenn die Große Randstufe in den Bereich der **Makro-Geomorphologie** (oder gar Mega-Geomorphologie) gehört, sollte ein solcher Referenzrahmen jedoch nicht den Anfang bilden, sondern eher das Ende. Zuvor war es notwendig, soviel relevante Details wie möglich beizubringen. Daher werden in der folgenden Arbeit zuerst diese **Geländebefunde in dem ihnen angemessenen Rahmen** dargestellt (auch wenn dies dem Leser als langweilig erscheint).

Eine weitere ausdrückliche Absicht bei der Abfassung des Textes (die eventuell ebenfalls langweilig oder gar völlig unnötig erscheinen mag) war es, die **Forschungsgeschichte** zu integrieren. Dafür führe ich folgende Gründe an:

- den Respekt vor früheren Gelehrten;
- die Berücksichtigung der Forschungstradition (mit Berücksichtigung der deutschen geomorphologischen Forschung, besonders zwischen den beiden Weltkriegen, die im Ausland weitgehend unbekannt ist);
- die Einbettung der eigenen Arbeit in diese Tradition.

Einige ältere Vorstellungen werden besonders attackiert (man könnte sagen „geschlachtet"), doch nicht aus irgend einer privaten Einstellung heraus, sondern weil die eigenen Forschungsbefunde dazu zu zwingen schienen. Es handelt sich besonders um die **Rumpftreppentheorie** Walther PENCKs und um das **Rückwandern von Stufen** parallel zu sich selbst, wie sie im südlichen Afrika von KING erfolgreich propagiert wurde. Wenn auch PENCKs Theorie bereits durch BÜDEL in überzeugender Weise als falsch erwiesen wurde, so ist es doch notwendig, sich mit ihr zu befassen, da sie einen gewissen Referenzrahmen bildet sowohl für die Vorstellungen KINGs (Gefällsknicke) als auch KAYSERs (allmähliche Ausweitung des Hebungsgebietes).

Wie man sehen wird, spielen im Rahmen der Darstellung **Plattentektonik**, marine Geologie und Geophysik eine nicht zu unterschätzende Rolle. Doch habe ich selbst im Bereich von Meer und Schelf keine Untersuchungen durchführen können. Ich mußte mich im Rahmen von Zeit und Mitteln auf die Geomorphologie des Festlandes beschränken. Solche **Beschränkungen** sind unvermeidlich und unumgänglich. Doch auch innerhalb solcher Beschränkung hat man, mehr als einem unter Umständen lieb ist, die Grenzen verschiedener Erdwissenschaften zu überqueren.

Die **Erweiterung über das ursprünglich enge Arbeitsgebiet** hinaus in den umfassenden und übergreifenden Großraum hinein erwies sich wegen der perzeptiven Maßstabveränderung (siehe Motto) als außerordentlich sinnvoll und fruchtbar.

Es erscheint dem Verfasser jedoch nicht angebracht zu sein, alle **Kartierungsergebnisse** in einer einzigen, notwendigerweise sehr kleinmaßstäbigen Karte zusammenzufassen. An die Stelle einer einzigen solchen Karte treten eine Reihe von Abbildungen (Fotos, Profile, Karten) für ausgewählte Gebiete. Es sei schon hier vorsorglich darauf aufmerksam gemacht, daß die in solchen Abbildungen flächenhaft auftretenden Erscheinungen prinzipiell ähnlich im gesamten Arbeitsgebiet als vorhanden angesehen werden dürfen.

Die in der folgenden Abhandlung beigebrachten Belege sind im Sinne einer Gesamtdokumentation in der folgenden Synopse (= Tab. 0.1) im Zusammenhang mit den Querprofilen in Fig. 0.1 zusammengestellt.

J. Birkenhauer
München

Table 0.1: Synopsis of cross sections and related presentations as used in this book
Synopse der Arbeitsprofile und darauf bezogener Abbildungen in diesem Buch

Cross section	Specification	Fotos	Charts, profiles		
1	eastern Transvaal 1	54			
2	eastern Transvaal 2				
3	Swaziland	53			
4	Pongolo 1	40	5.4, 15.1, 15.2		)
5	Pongolo 2	16	5.4, 15.1, 15.2		)
6	Eshowe		3.1, 5.3	)	)
7	Stanger		6.3	)	)
8	Kwa Mashu		15.4	)	)
9	Pietermaritzburg			) 7.1	) 5.1
10	Richmond	5, 32		)	)
11	Port Shepstone	21, 36, 38		)	)
12	Harding		6.3	)	)
13	Margate	6, 55	14.6	)	)
14	Transkei 1		6.3		)
15	Transkei 2	19	6.3		
16	Kingwilliamstown				)
17	Grahamstown	7, 52			)
18	Algoa Bay	3			) 5.2
19	Kirkwood	23	7.1, 15.3		)
20	Port Elizabeth		7.1		)
21	Willowmore	1, (10), 20, 37, 44, 45	)		
22	Beaufort West	24	)		
23	Oudtshoorn	22, 47	)6.1, 6.2, 12.2		
24	Laingsburg	30, 31	)		
24	Barrydale		)		
25	Franschhoek		)		
26	Ceres	11, 12	)		
27	Middeldrift				
29	Calvinia	2, 9, 18, 46			
30	Kliprand	29, 50			
31	Springbok				
32	Upington	17, 26, 27	8.1, 12.4		
33	Grünau		12.4		
34	Fish river	25	12.4		
35	Keetmannshoop		12.4		
36	Maltahöhe	14, 28, 43			
37	Windhoek	34	9.1		
38	Omaruru	13, 15, 33 35			

Fig. 0.1: Research area in southern Africa and run of cross-sections taken during field-work
Untersuchungsgebiet im südlichen Afrika - Arbeitsprofile -

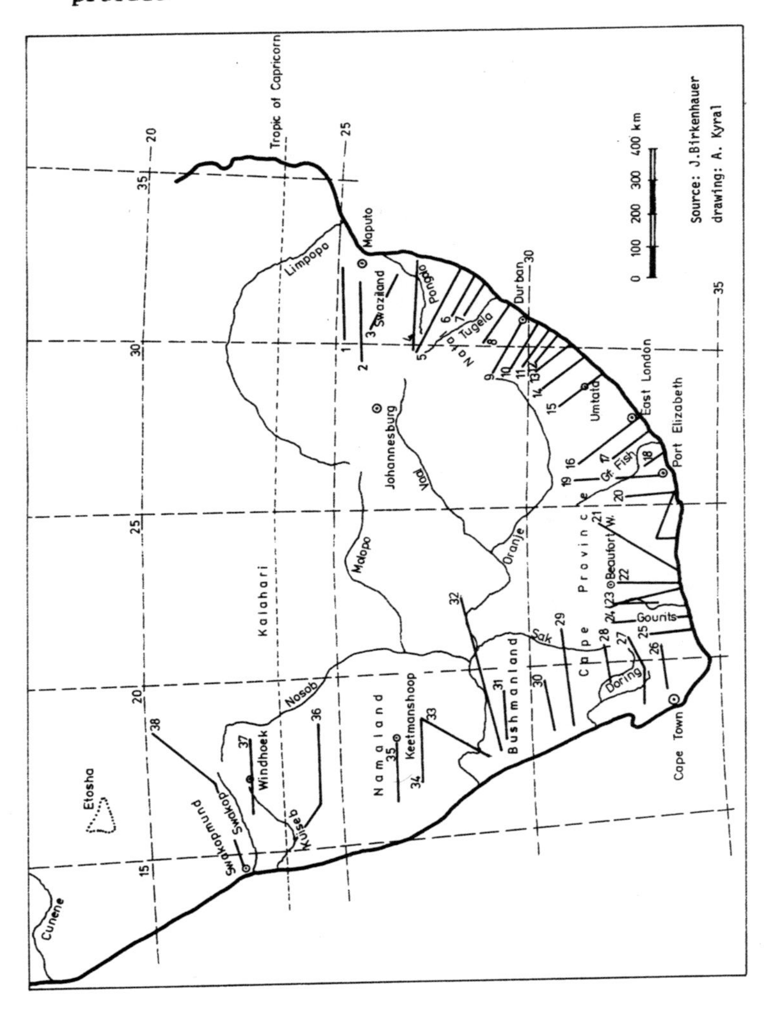

Table of contents/Inhaltsverzeichnis

LIST OF FIGURES / ABBILDUNGSVERZEICHNIS

LIST OF TABLES / VERZEICHNIS DER TABELLEN

FIRST PART: BASIC SUPPOSITIONS

1. PRINCIPAL OUTLINES

a) The geomorphic context

A swing can recently be noticed in geomorphology, a swing back from the study of geomorphic details forth to the study of larger features of the earth's morphographic face. A name has even been given to the study of large landforms: Megageomorphology (GARDNER, SCOGING 1983).

Certainly, some of the largest landforms are the continental margins, or escarpments; the Great Escarpment in southern Africa is one example of these continental margins. These continental margins received their first systematic treatment as 'continental rises' (or 'Randschwellen') about forty years ago by JESSEN. The study of these continental rises has recently been taken up by OLLIER and his co-workers (1985) in the broad frame of what OLLIER calls 'morphotectonics'. In using this term OLLIER suggests the genetic connection - union even - of tectonic processes (especially those of plate-tectonics) and major landforms. According to OLLIER the term 'morphotectonics' was first used by HILLS (1961).

As will be shown later on in some detail the Great Escarpment of southern Africa has indeed to be looked upon in this wider tectonic context. For OLLIER (1985, 1) great escarpments are the result of what he calls 'mountain-building'; 'mountain-building' is the process that results from vertical uplift, whereby plateaus are formed, and of isostatic response to erosion. Following OLLIER, it is because of this that great escarpments run parallel to the coast and separate a high interior plateau from a coastal plain by what looks like a high mountain range when viewed from the coastal plain.

OLLIER has gone on to suggest a general model for the nature and origin of the great escarpments. Whether this model can be applied to southern Africa cannot be made out without having described before the present author's findings. OLLIER, together with SWEETING (1985), gives an outline of how the Great Escarpment of southern Africa may have come into existence. At this stage of the present treatise it is however not intended to deal with the views of OLLIER and SWEETING, because it is necessary to unfold a broader context in some more detail.

b) The topography of the Great Escarpment

The Great Escarpment of southern Africa is the very marked feature of the continental margin - or 'Randschwelle' (JESSEN 1948) - that swings south in a vast garland of about 4800km if one starts south of the Sambezi basin, goes down to the Cape ranges and then runs back to the north to where the highlands of Angola find their end in the neighbourhood of the Congo basin. Whether the Great Escarpment - all along this huge crescent - indeed however has the same nature - both geomorphologically and geologically - remains open to question. The present writer's work should be seen as an attempt to clarify this question by way of his own research and inquiry. His working area, however, covers only part of the continental margin of southern Africa as described above in following JESSEN. The present writer's working area extends from the southern borders of Swaziland (with some excursions into it and into the eastern Transvaal) down to the Cape ranges and then north again up to the Erongo Mountains in South West Africa.

The Great Escarpment wins its greatest height in the highlands of Lesotho, the so-called Drakensberg (or Drakensberge). The rock formations of this particular region are so towering and overwhelming that one easily understands why the mountains were called after the dragons by the first white settlers. (Fotos: 36, 37.)

The first scholar to have seen the nature of the Drakensberge as part of a larger landform super-unit seems to have been the great South African geologist ROGERS who was the first to call this super-unit the Great Escarpment.

By the work of OBST and KAYSER (1949) and by that of KNETSCH (1940) the English term became fully adopted into the German literature, being literally translated as 'Große Randstufe'. Eventually, this term supplanted the older one used by JESSEN, namely the term 'Randschwelle' (marginal swell or rise). It should be noted, however, that JESSEN's term still seems to be more appropiate for comparative purposes on the wider scale of continents as is shown by the remarks of OLLIER (1985) and BREMER (1985).

A definition of the term Great Escarpment can at this stage be only a preliminary one. Such a preliminary definition seems to be well advised if it uses morphographic features only - the particular feature being that of the more or less sudden rise from the coastal forelands to the interior plateau. Nevertheless it would be wrong to visualize the escarpment and the forelands to be always of the same nature.

Looking at the forelands first: they differ in heights from one area to another; sometimes they are nothing but low-lying plains as the Low Veld in the eastern Transvaal; sometimes they form a high basement in front of the escarpment, at altidudes of 1000 - 2000m, as is the case for example in the Natal Drakensberge.

Going then on to have a look at the escarpment, it is not at all what low scale maps suggest, namely a simple and straight wall with a simple marked ascent along a straight line, forming just one vast crescent. Instead, it is often stepped and rugged, sinuous, as OLLIER and MARKER (1985) point out, consisting as it were of single garlands and bow-line thrusts against the coast, then swinging back again, basin-like, into the interior. Sometimes the escarpment possesses bastions and out-liers of vast dimensions far in front of it, nearly as high as the escarpment itself; sometimes the escarpment forks into separate branches as in the Cape Province to the west of Calvinia. Fig. 1.1 gives an idea of this and moreover shows that the thrusts and bays of the escarpment are connected in some way with the river drainage basins. It is these basins rather that structure and as it were articulate the escarpment. Similarly, OLLIER and MARKER (1985) remark that the sinuous nature is "in response to headway erosion by coastal rivers" (41).

From the description given it follows that the escarpment, on first sight, is nothing but a feature of erosion, connected with the erosion and the effects of rock removal happening within these basins and directed by the erosive energy of rivers. Whether this is the case, however, and to what extent such a hypothesis is true has to be discussed later on.

Table 1.1 gives an overview over the subareas of the Great Escarpment together with the heights in these sub-areas and their main outcropping rocks. Both rocks and heights show how varying indeed the escarpment is.

Fig. 1.1 Geological structures and the Great Escarpment

Geologische Strukturen und die Große Randstufe

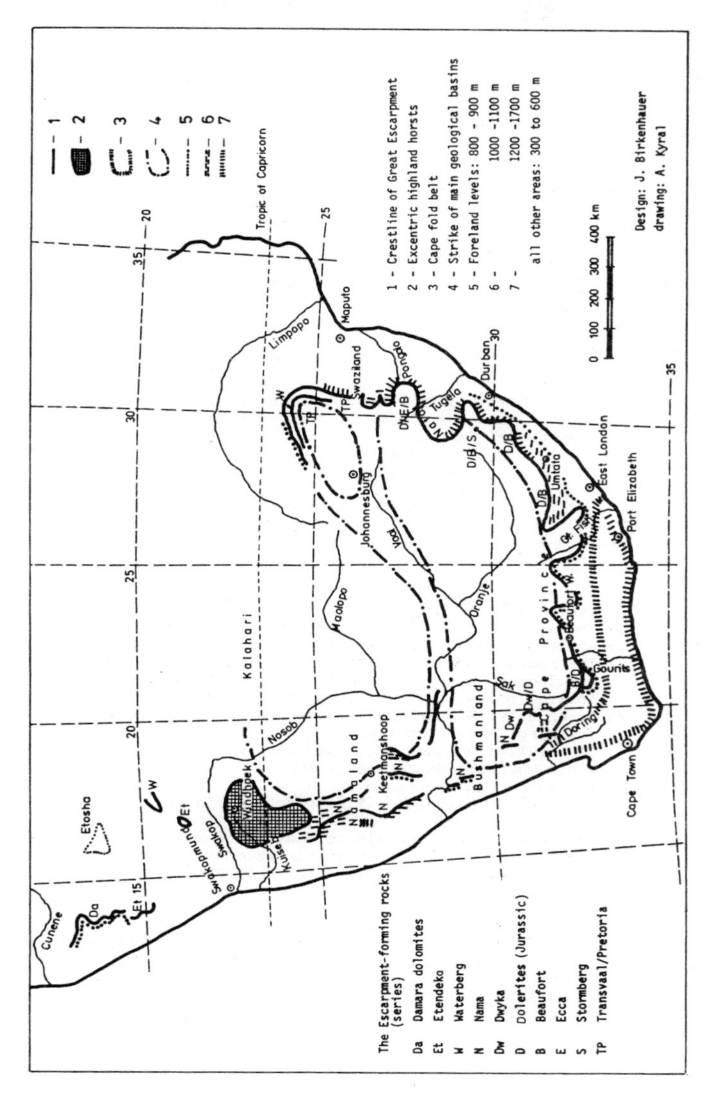

Table 1.1 Subareas of the Great Escarpment in southern Africa

Abschnitte der Großen Randstufe im südlichen Afrika

Subareas 1)	Heights (in m ab. s-l) 2)	Rocks	Geological Series and Systems
Waterberge	1500	sandstones, conglomerates	Waterberg
Strydpoortberge	1500/2100	) front: quartzites	Black Reef)Trans-)vaal
Transvaal Dra- kensberge	1500/2200	)) centre: dolomites)) back: quartzites)	Pretoria))
River basins in Transvaal and Natal	1600/2400	dolerites and quartzites over shales and crystal- line rocks 3)	Ecca (Karoo 2)
Natal Drakens- berg	3000/3400	dolerites, quartzites, sandstones over shales	Stormberg (Karoo 4)
Transkei, Witte- berge, Stormber- ge, Bamboesberg	2200/2800	dolerites, quartzites, sandstones	Stormberg and Beaufort (Karoo)
Cape Province: E and S 4)	1500/2500	dolerites, quartzites, sandstones	Beaufort (Karoo 3)
Cape Province: NW 5)	a) back: 1000/1600	dolerites, greywackes	Dwyka, Ecca (Karoo 1 a. 2)
	b) front: 1000/1700	quartzites	TMS (Cape 1) 6)
Orange river basin	800/1200	sandstones, greywackes (mudstones)	Nama (Late Pre- cambrian)
South West Africa: S 7)	1200/2000	sandstones, greywackes (mudstones)	Nama, Dwyka (Karoo 1)

- -

Annotations

1) The subareas are arranged from the NE over the S to the NW.
2) Above sea level; this abbreviation will be used throughout the text.
3) The rocks are arranged here according to their resistancies and their prominence. Wherever dolerites are mentioned, they are the main cuesta-forming rocks. The age of the dolerites is Stormberg (=Karoo 4).
4) Especially: Winterberge, Sneeuwberge, Koueveldberge, Nuweveldberge, Komsberg, Rogeveldberge.
5) Especially: Hantamsberg, Loerisfontein area.
6) Table Mountain Sandstone or Series; the abbreviation will be used throughout the text.
7) Especially: Huibplato, Tirasberge, Klein Karasberge, Schwarzrand, Tsarisberge.

General remark: For certain reasons dealt with in later chapters the subareas north of the mountain ranges mentioned are not listed up here. These subareas are especially: the Khomas Highlands, the escarpment gap, the Kaokoveld ranges.

NOTE REGARDING THE SPELLING OF RIVER NAMES: The names are spelt as on the Ordnance Survey maps. Formerly the Bantu prefix 'M' (meaning 'water', 'river') was written 'Um' as (e.g.) Umfulosi, Umgeni, Umkomaas, Umzimvubu, Umkuze etc. The older spellings are still retained in place-names like Umtata (etc.). The modern spelling of the rivers is (e.g.) Mfulosi, Mgeni, Mkomaas etc.

c) Stratigraphy along the Great Escarpment

Since, in the further course of this inquiry, some stratigraphic knowledge is necessary, it seems useful to give some basic information in advance. In doing so, we may well start from the fact that the forelands of the escarpment are shaped, to a rather large extent, by outliers, benchlands and mesas of resistant rocks which reappear in the very scarp itself. These mesa-forming rocks belong to the cover topping the usually crystalline basement. Cover and basement are made up of rocks of quite different geological ages - though, all over the run of the escarpment, from the north east to the north west, the fundamental relationship of cover ('Deckgebirge') and basement ('Grundgebirge') remains the same. Typical for the covering rocks are cuestas, mesas, tafelsbergs and spitskoppies. The latter appear - independent of the particular series and formation - wherever the cover has been strongly dissected and removed almost completely by an encompassing and circumvallating process of erosion, leaving only a last central spot of resistant rock.

Essentially, there are two types of resistant rocks:

(1) quartzites, sandstones, conglomerates (especially when they have become silicified);
(2) dolerites. (Fotos: 2, 36, 37.)

Where the climate is arid enough, i.e. in the west and in the northwest (Cape Province to the north of the Cape ranges and South West Africa) schists and mudstones (greywackes) are liable to form a third group. Under more humid conditions, however, schists, shales and mudstones are easily eroded away and therefore do not form cuestas, mesas and the like. (Foto: 43)

Table 1.2 shows the main rocks and their ages. (Cf. note 5 in Table 1.1.)

Some special remarks on the rocks as listed within the two groups above are necessary.

Ad 1:
The Table Mountain Series (TMS) is in Natal equivalent to the Natal Sandstone Series. This series loses its quartzitic nature from southwest to northeast. In this direction the whole series moreover thins out completely. Accordingly, the

Table 1.2 Stratigraphy of rocks along the Great Escarpment

Gesteine und deren stratigraphische Einordnung entlang der Großen Randstufe

The stratigraphical notes are arranged in the following way: In the first part (=I) the stratigraphy of the east and the south is dealt with, in the second part (=II) that of the northwestern Cape Province and of South West Africa. Two main groups of rocks are distinguished: (A) the covering rocks ("cover") and (B) the rocks of the basement ("basement"). In general, the covering rocks are sedimentary ones, the basement rocks are crystalline. These subdivisions are given for both parts.

Series and systems o South African terminology	International terminology Ages (in thousand million years)	Rocks and areas
I. East, south and west		
A. Cover		
Stormberg: Lebombo and Drakensberg lavas,	Jurassic	rhyolites, dacites, andesites, dolerites
Molteno		Cave sandstone
Beaufort	Triassic	shales and sandstones, sometimes quartzites and conglomerates
Ecca	Permian	shales - in the NE: shales and quartzites
Dwyka	Carboniferous	shales and tillites
Witteberg	Lower Carboniferous	quartzites
Table Mountain Series (TMS)	Ordovician, Silurian	quartzites and sandstones (the latter in central and northern Natal)
Nama	Cambrian, late Precambrian	shales, mudstones (greywackes), sandstones
B. Basement		
Namakwaland Series	1,1	gneisses, granite-gneisses, amphibolites (the latter in central and northern Natal)
Waterberg	1,4 to 1,6	sandstones, conglomerates, greywackes (northern Transvaal)
Transvaal System	2,0 to 2,3	quartzites, dolomites (north-eastern Transvaal)
Pongola plutons	2,3 to 2,6	granites etc. (northern Natal)
Pongola System (Mozaan Series, Insuzi Series)	2,9 to 3,1	gneisses, granites, quartzites, different lavas (northern Natal)
Ancient Gneiss Complex	3,1	gneisses (northern Natal)
Swaziland System	3,5 to 3,7	gneisses, schists, granites (northern Natal, Swaziland, eastern Transvaal)

Series and systems o South African and South West African terminology	o International terminology o Ages (in thousand million years)	Rocks and areas
II. Northwestern Cape Province, South West Africa		
A. Cover		
Ecca	Permian	shales, greywackes (Orange and Fish river basins)
Dwyka	Carboniferous	shales and tillites (Orange and Fish river basins)
Nama	Cambrian, late Precambrian	shales, greywackes, sandstones (Orange and Fish river basins)
B. Basement		
Namibian (Damara Sequence)	0,6 to 0,9	schists (central South West Africa)
Mokolian	1,1 to 1,8	granites etc., gneisses (southern South West Africa)
C. Intrusives		
Etendeka	Early Cretaceous	basalts (central and northern South West Africa)
Damara	Cambrian	granites (central South West Africa)

Sources of Table: The Table was compiled from the geological maps of South Africa and of South West Africa, scale 1 : 1 000 000.

Note on the usage of South West Africa and/or Namibia: In tables, figures and text this territory is usually referred to as South West Africa, according to the usage on the official maps.

TMS not longer forms tafelbergs, mesas and the like. In northern Natal, the place of the TMS is in this respect taken over by the quartzitic middle Ecca Sandstones. In the Transkei and in the adjacent areas of the Cape Province to the north, where the TMS is still covered by younger strata, the quartzitic sandstones of the Beaufort Series take the place of the TMS in order to form mesas and the like.

In the west and northwest of the Cape Province the TMS thins out again and the Karoo cover becomes pronounced again, especially so with the Ecca layers. Towards the Orange and beyond this river the far older sediments of the Nama Series immediately on top of the basement take over.

Wherever sediments have become folded or metamorphosed, as for instance in the Damara System, cuestas, mesas and tafelbergs disappear again.

Ad 2:
The dolerites are fundamentally of a basaltic nature. They appear as in- and extrusives within the Stormberg Series. Two sub-phases can be made out. The older one is the Drakensberg phase, the younger one the Lebombo phase. Though in both phases basaltic lavas are prominent, there may be changes to limburgites, rhyolites and so on.

Within the sedimentary rocks from the Ecca to the Stormberg Series the dolerites intrude as dykes (Gänge), cones and sheets or sills (Lavabänder). Dykes and sheets frequently are of a considerable thickness. The intruding dolerites expand the sedimentary rocks, adding to their heights - in some regions up to looo metres. The excessive height of the Drakensberge (as well as their bizarre rock formation) is particularly due to the dolerites.

The dolerite intrusions seem to have been dependent on the petrography of the covering sedimentary rocks. The intrusives penetrated into the rocks in between the layers, especially where these change from shales to sandstones or conglomerates. Such changes particularly take place in the middle Ecca, in the Beaufort and Stormberg beds. Wherever the beds consist of shales only, as in the lower Ecca and in the Dwyka Series, no or almost no dolerite intrusions can be observed.

The dolerites are very important for the geomorphology of the Great Escarpment, especially in its central part. Eventually we will turn back to this observation. The dolerite sheets are indeed important for the formation of the escarpment, though cuestas, however, will only occur where the sheets and sills are thick enough. This is the case farther inland only, especially to the north of the Cape ranges. The Cape Fold Belt seems to have acted as an effective barrier against dolerite intrusion. Dolerite sills gain their greatest thickness usually on top of the upper Stormberg layers - with thicknesses between 1200 and 1400 metres. Here the layers form the widest consecutive lava coverage on earth. It is this coverage that makes up most of the Natal Drakensberge.

In comparison with the particular groups of rocks dealt with above the rocks of the Cape System as well as those of the crystalline basement do not seem to have any special bearing on the forming of the Great Escarpment (as Fig. 1.1 will show).

2. GEOLOGICAL EVENTS OF GEOMORPHIC CONSEQUENCE

Fig. 1.1 may be used as a reference for this chapter.

It does not seem necessary to deal fully with the geological evolution of southern Africa; it rather seems sufficient to concentrate on those events or phases which have had some major bearing on the geomorphology of the Great Escarpment. Phases or events of such a major bearing seem to have been the following eight ones:

(1) the Cambrian orogeny,
(2) the peniplanation of Gondwanaland,
(3) the forming of the Cape and Karoo systems together with the warping of southern Africa,
(4) the Cape orogeny,
(5) the dolerite phase,
(6) the faulting phase,
(7) the rifting phase,
(8) the phase of erosion.

The following paragraphs will deal with these phases and the main events occurring within them.

1) The Cambrian orogeny

Southern Africa, within the Swaziland system, contains some of the oldest rocks ever found on earth so far. It is natural then that since then quite a lot of orogenies must have occurred, even still in the Precambrian. Yet with regard to geomorphic traces of some avail still today only the Cambrian orogeny has to be made mention of here. The reason is that it is this orogeny which formed the Khomas and Windhoek Highlands, extant today as an eccentric horst region. Most of the folded rocks in these highlands belong to the Damara and Sinclair sequences.Their folding is very intense. Far older rocks were folded together with them.

During the Cambrian orogeny prolific intrusions of granites and even serpentinites happened.

An earlier Pre-Cambrian 'trough' seems to have been 'respected' by the orogeny and became built into the geological structure. It makes itself felt today as the Swakop-Kuiseb Gap ('Randstufen-Lücke'). (For the old age of the gap and the trough: see TANKARD et al., 1983, 11, 276.)

Oddly enough, the sediments of the Nama group, with an age equivalent to that of the upper Damara sequence and situated immediately south to the area of the orogenetic events were not folded at all. At best, they became slightly tilted and thus form the vast sedimentary plateaus south of the Khomas Highlands and especially - together with the younger Dwyka and Ecca layers - the plateaus of the Fish river region. It is into these layers that the Fish river has cut its magnificent canyon, right through and down into the basement of Mokolian age (i.e. more than double the age of the Nama layers).

2) The peneplanation of Gondwanaland

Most of the mountain ranges that must have come into existence in the Cambrian orogeny were later bevelled away - more or less completely - by a thorough peneplanation. This event can be deduced from the fact that the younger sediments of the Cape and Karoo systems transgress smoothly over a surface that cuts all the older structures.

3) The Cape and Karoo systems

The Cape system (as has been mentioned above) thins out to the north and east, thereby showing that its main sedimentary trough lay in the south of present southern Africa. Without a marked unconformity the layers of the Karoo system follow on top of the Cape sequence. Their wide distribution to the north shows that gradually the northern parts of southern Africa became submerged so that the Karoo layers could evenly be spread over the peneplain mentioned in the previous paragraph. This happened already within the lowest series of this system, i.e. the Dwyka series with its famous Carboniferous tillites. As the tillite in-fill shows preexisting valleys were used by the ice-sheets and flattened out by the moving ice. MARTIN (1972) showed that this is the case in Damaraland (central South West Africa) and in the Swakop - Kuiseb Gap.

By later erosional processes the ancient valleys became excavated again and today are partly used by the Huab and Ugab rivers.

The sedimentary process continuously went on and without marked unconformities in the series following the Dwyka beds, i.e. the Ecca, Beaufort and Stormberg series - which means, internationally spoken, from the Permian right up into the Lower Jurassic; synchronously, southern Africa must have become warped into the shape of a great bowl or basin-like depression of a rather wide extension. This can be deduced from two facts:

a) the thickening of the beds towards the centre of the depression,
b) the circumferential strike of the series around this centre after the series became exposed by the onset of younger erosion cycles.

The depression actually consists of two intracratonic sub-basins (TANKARD et al., 1983, 276, 365, 366). They developed out of the older Kalahari craton and were separated from each other along the well-known Griqualand axis to the north of the Vaal river. The axis became formed during the sedimentation of the Karoo beds and thus is a further example for synchronous warping. It is important to point out that no younger landforms can have become effected by this warping, younger meaning: Lower and Upper Cretaceous as well as Tertiary (cf. ch. 5, 6 and 9). The southern sub-basin is today drained by the Orange river, the northern one by the Nosob and Molopo rivers. From geomorphic reasons the basins are therefore called after these main rivers draining them. Geologists would call them the Karoo (south) and Kalahari (north) basins.

The circumferential strike mentioned above is not prominent all around the latter basin; it here occurs only between the Orange river and the Khomas Highlands; around the Orange sub-basin however the strike can well be noticed all around the eastern, southern and western margins of this basin.

4) The Cape orogeny

The Cape orogeny began in the Upper Carboniferous and culminated in the middle Triassic (ca. 220 my BP). The folding included all the layers from the TMS up towards the Ecca and Beaufort, comprising both the quartzites (TMS, Witteberg) and the shales (Bokkeveld, Ecca, Beaufort) intermingling with them.

In some regions even the Nama shales became incorporated into the folded layers.

The folding as such concerns only the marginal parts of present southern Africa, though actually far wider regions of Gondwanaland became involved in the orogeny - today lying as far apart as Bolivia, Peru, Argentina, Antarctica, eastern Australia (TANKARD et al., 1983, 399). This configuration suggests that southern Africa was on the margin of the orogenetic activities of that age.

Because of the repeated change from quartzites to shales and the sifting out of the different rock resistencies by later erosion the typical Cape Fold Belt topography has resulted. Similar features can be found in the Swiss Jura or the Appalachians. It is from the latter region that MACHATSCHEK (1955) chose to call all regions with these traits as 'Appalachian'. Folding - and with it - the 'Appalachian relief' quickly lose their impact once the northernmost Cape range has been passed. In spite of the resistant quartzites of the TMS and Witteberg vaults planations were later cut across them; inter-montane basins were formed where, between the vaults, the shales were removed (especially Bokkeveld and Dwyka basins).

At the end of the Cape orogeny the main outlines of southern Africa as they exist today were completed.

5) The dolerite phase

As has been indicated above the dolerite intrusions happened within the Stormberg series. The intrusions began around 187 my BP (Lower Jurassic) and ended around 155 my BP (TRUSWELL 1977), i.e. in the middle Jurassic. The in- and extrusions are arranged circumferentially around the core area of Lesotho. In this core area the lava sheets are prominent over extensive tracts. In an inner belt around this core the masses of dykes occur, in an outer belt only single and often separate intrusions, formed cone-like today, show up.

6) The faulting phase

As the faulting phase preceded the rifting phase one may say that the final rifting and break-up of all Gondwanaland was prepared in the faulting phase. This phase began after the dolerite intrusions had come to an end, because the dolerites themselves were also subject to the faulting.

The faulting resulted in a host of major and minor faultlines, cutting through all older rocks right down into the crystalline basement as well. The faulting must have begun in the Upper Jurassic and it lasted into the Lower Cretaceous. It was the age of diastrophism all around southern Africa. Natal was affected by a fault system with complete horst and graben structures (BEATER and MAUD 1960, MAUD 1961). In the southern Cape, along the strike of the folds, basin-like troughs were formed, running from east to west. These troughs were synchronously filled up with very thick Lower Cretaceous sediments of a marine and lacustrine nature (MCLACHLEN et al. 1976, SHONE 1978, RUST and REDDERING 1983). According to DINGLE et al. (1983) faulting here also began in the Jurassic.

A northeastern branch of the faulting phase affected the Transvaal to the north and east of Pretoria. Here the faulting was also post-dolerite. As can be seen from the Stormberg series (which originally must have capped the older rocks of the Transvaal system in a horizontal way) this part of Transvaal must have synchronously been warped into a basin. Though the dimension of this basin is far less than that of the two other ones - mentioned in (3) - the circumferential strike of the rocks of the Transvaal system is - as a consequence of this warping - very pronounced indeed. The warping in this period may have reactivated an older one with the Bushveld Igneous Complex as its centre (cf. KAYSER 1986). As a consequence of the later warping and circumferential strike the Great Escarpment appears here, in the eastern Bankenland, as a giant treble cuesta, existing of huge tiers or echelons of quartzites, dolomites, conglomerates - one after another (cf. KAYSER, 1986, 14, 15). The sheet-like granites in the east of the Pietersburg highland were also formed into cuestas.

In South West Africa this age of diastrophism - as it is aptly called - was accompanied by the Etendeka basalts and granites which show up in the Erongo Mountains, the Groß and Klein Spitzkoppe, the Brandberg and the Messum Crater. The main Etendeka intrusives are earliest Lower Cretaceous, the Erongo ones are a little younger, but still pre-Valanginian. The Cape Cross intrusives are somewhat older, ie. Portlandian (late Jurassic). (KENNEDY and COOPER, 1975.) According to WINDLEY (1984) the so-called Kaoko lavas (including Brandberg and Erongo magmatites) are however Barremian (ca. 120 my BP).

The highly risen block of the south African interior, the highlands of South West Africa, the troughs and grabens of the Cape ranges, the fault systems of Natal and of the eastern Transvaal: all these features are a result of this age of faulting.

The monoclinal structure of some parts of Natal was another feature also formed in this period. The monoclinal nature can be deduced for instance from the down-warping of the Stormberg lavas in between the Stormberg layers. A. PENCK was the first to have pointed it out (1908).

To speak of all Natal as a monocline as KING continues to do (1972, 1982) is certainly not in accordance with the geological facts. A true monocline in the whole of Natal should show a simple down-warped fall of all layers. This is however not the case. Instead, in southern and central Natal, an S-like bending of the fault system is to be noticed, together with the horst and graben structures already mentioned and with different heights of the same rocks at different, though adjacent places. The TMS-blocks show this feature in an especially convincing way.

In northern Natal the descent from the highland towards the coast is also not of a true monoclinal nature, since major fault-lines bring the surface down in a broad step-like manner from one fault-line in the interior to the next towards the coast. This is especially the case in the neighbourhood of the Pongolo catchment area. The monoclinal structuring is here interspersed only between the major fault lines.

Since this age of diastrophism no later tectonic movements - either down or up, faulting, warpings and the like - can be proved. In northern and central Natal FRANKEL (1968) as well as MAUD and ORR (1975) showed that the Upper Cretaceous and Tertiary beds are no more subject to any tectonic movements; they rather spread evenly and unbrokenly across the older faults and all the low-lying horsts and grabens. (Fig. 3.1) The consequence of this fact is to be deduced easily: since at least the middle Cretaceous southern Africa must have always rested in the same heights ab. s-l as it stands today.

In the Algoa Bay and its hinterlands subsidence and faulting came to an end with the Valanginian (cf. SHONE 1978). The same seems to be true about South West Africa. MARTIN (1972, 297) states: "During the Tertiary the Atlantic coast of southern Africa seems to have been rather stable. There is not much evidence for uplift or warping. ... Tectonic adjustments seem to have been largely confined to the shelf."

TANKARD et al. (1983,440, 441) are however of the opinion (based on evidence brought forward by TANKARD 1975) that some movements did indeed happen through the later Tertiary. The authors refer to the so-called Transkei swell from which they deduce a continuous emergence based on the lack of well-developed marine strata and on a suggested elevation of the Neogene shoreline in the Bathurst region (southern Cape). According to these authors the elevation of that region amounts to 210 - 250 metres. The present author had a close look at this specific region (east of Port Elizabeth). He was however not able to confirm this view (cf. ch. 5). He feels justified therefore to rely on his own observations for the unfolding of his arguments.

Similarly, he thinks himself justified in neglecting RUDDOCK's (1968, 1973) and TANKARD's (1976) findings. The "repeated tiltings" in the hinterland of the Algoa Bay these authors speak of seem to ocur within the sediments and do not at all affect the main features of the topography as it prevails today. The present author feels rather sure of this finding from his own observations in this area (cf. ch. 5). In this opinion he feels justified moreover because it is corroborated by other research dealt with in sub-part 8.

TANKARD (1976) refers to tilting and warping since the Miocene in the area of the Ysterfontein and Elands Bays, north of Cape Town. These tiltings and warpings affect however only a very small coastal stretch and within that stretch only the low-lying Pleistocene coastal terraces.

Whether slight epirogenetic movements have occurred since the Cretaceous - as is suggested by MAUD - is open to question and with MAUD a matter of inferment only.

7) The phase of rifting

As has been shown, the intrusions of the dolerites and the Etendeka lavas as well as the diastrophism went ahead of the rifting phase, being preparatory to it and partly accompanying it. As a result of the rifting - and even before actual rifts came into appearance - the southern Atlantic and the Indian oceans began to open up around what is today southern Africa. Southern America - with the Falklands - was separated from around southern Africa - the Falklands originally lying off Durban (cf. MARTIN et al. 1980, 1982). Gradually the new African continent began to move into a generally northward direction.

The invasion of the sea into the at first very shallow trough between Africa and South America took place already in the very early Cretaceous (KENNEDY and COOPER, 1975;

HERZ, 1977). By Aptian times the spreading and rifting had become very pronounced as is suggested in the chart given by DINGLE et al. (1983, 191).

The marine Pomona Formation, 70km south of Lüderitz Bay, on the coast of South West Africa, with its marls, limestones and silicified conglomerates on top is a lower Cretaceous proof of the existence of the southern Atlantic ocean.

According to LARSON and LADD (1973) the initial rift must first have occurred during the Valanginian (130 - 125 my BP).

The general northward move of Africa since then can be demonstrated from the charts given by SIEVER (1983). 120 my ago the south coast was situated in the latitude of 57°, 60 my ago in 48° South, wheras today it lies under 35° South. These latitudinal positions show that Africa has moved northward by about 2500km. This naturally also means that southern Africa has gradually moved into tendentially warmer climatic conditions.

8) The phase of erosion

As has been indicated in the preceding paragraphs the age of diastrophism was connected with the last phase of uplift of southern Africa thereby shaping it - and especially its interior - into the high-lying block that has been in existence ever since then right up today. Up-lift must have come to an end in the Valanginian (Lower Cretaceous) at the latest. As has been stated above - but should be made very clear once more - this means that no further up-lifts, tiltings and (down-)warpings of any kind can have taken place since then.

Yet it is exactly such tectonic movements that still serve as a base for the morphotectonic assumptions of some authors - notably OBST and KAYSER (1949), KAYSER (1986), KING (1972, 1982). Moreover, in the view of these authors, these tectonic movements are supposed to have lasted right into the present.

Some other authors , however, doubt these assumptions, framing their doubts in rather a decided language. These are, for instance, MARTIN et al. (1981, 1982) and DE SWARDT and BENNET (1974). The latter expressly say: "The theory of coastal development advanced here implies that southern Africa stood relatively high ever since it formed the central and presumably most elevated part of Gondwanaland" (314; similarly: 321).

MARTIN et al. (1980, 192) state - rather categorically - : "The Great Escarpment did not evolve through a series of drastic up-lifts, but has been in a relatively elevated position since the Jurassic". DINGLE (1971, 183) writes: "...no tilting is evident for the Alexandria Formation". This formation came to be deposited on the southern Cape coast between the Eocene and the Pliocene. If, in spite of this rather long time, no tiltings are observable in the Alexandria deposits this must mean then that the whole coastal tract between Cape Town in the west and East London in the east was tectonically stable throughout this time; this has to be judged as another proof for the stability of southern Africa as a whole. And, moreover, it is in sharp contrast to the opinion of TANKARD (as mentioned above).

From all these statements it then emerges that southern Africa must have been a high lying mass of land since the middle Cretaceous at the latest. This naturally means that southern Africa since then has been subject - over all these 100 or even 120 my - to fierce erosion and denudation. Erosion and denudation must have started all along the new continental margins. Huge masses of eroded matter must have been carried away since then. This can be demonstrated for example from the Lower Cretaceous sediments. The original thickness in the Addo deep in the hinterland of the Algoa Bay amounted to 4000m; in the Kirkwood Formation an amount of only 1800 metres has been preserved however (SHONE 1978). Today the Lower Cretaceous sediments reach heights of about 600m ab. s-l. Originally the may have lain up to altitudes of 800m or even 1000m ab. s-l.

Summing up, the "youthfulness" of the coastal forelands of the Great Escarpment as well as parts of itself owes its origin to this last and very long cycle of erosion - in truth: a super-cycle.

3. PREVIOUS RESEARCH ON THE ORIGIN OF THE GREAT ESCARPMENT

a) Early German scholars

Research started from the most prominent features of the Great Escarpment, the Natal Drakensberge.

As early as 1871 GRIESBACH reached the conclusion that the Drakensberge were nothing but a huge cuesta. In 1885, E. SUESS, however, one of the most influential geologists of the time, considered the Basuto Highland scarp to be the result of a giant fault - one of the most gigantic ones on earth; he named it the Quathlamba Fault. For him it was one of the proofs for his theory of contraction.

The then leading geographer and geomorphologist in Germany, A. PENCK, felt himself challenged by SUESS's notion and tried to show (helped in this by the great South African geologist ROGERS) that the scarp was due to a monoclinal down-warping of all layers. Within this monocline A. PENCK did indeed notice some faults which he however considered to be of a minor nature. Therefore he did not come to notice the graded fault-lines of northern Natal or the horst and graben structure of central Natal. It is PENCK's notion of a simple monocline that later became one of the bases for KING's ideas.

In order to date the age of the monocline A. PENCK made use of two assumptions. The one is that the upper beds of the Karoo series had become eroded away by Upper Cretaceous times. By this erosion a true escarpment had been formed to the east and southeast of the Drakensberge. The erosion cycle that was held responsible for the forming of the escarpment thus had to be older than the Upper Cretaceous because the sediments of that age cap the previously cut Ecca layers.

The other observation starts from the fact that the Drakensberg lavas are down-warped as well. Therefore the down-warping of the monocline must have happened between the middle Jurassic and the Upper Cretaceous. A. PENCK fixed the date for the monocline to be Upper Jurassic because later on the surface of the monocline had become peneplained - a process which naturally needed some longish time to be formed.

The peneplain was deduced from the fact that the Upper Cretaceous sediments consecutively cap different layers of the Karoo series.

A. PENCK held the opinion that the old (i.e. Lower Cretaceous) peneplain was originally bevelled across the whole Drakensberg. The Drakensberg cuesta and the lower scarplands in front of it were cut out from this peneplain. In considering the Drakensberg a huge cuesta and not a fault-line A. PENCK returned to GRIESBACH's notions, thus confirming them. Later on, cuesta and peneplain were deeply incised by the valleys which still exist today and which give an appearance of complete 'youthfulness' to the whole region. According to PENCK incision - and rejuvenation thereby - was the result of a younger up-lift.

b) Rogers

ROGERS was the first scholar to notice that the Drakensberge are no singularfeatures in themselves but are part of a macro- or super-unit. ROGERS came to distinguish this super-unit by following the 'steep slope' and the 'step gradient' between the coastal belt and the interior plateau (or plateaus) all around South Africa. This continuous super-unit was first named by him as the Great Escarpment. (Later, it was proposed to call the escarpment the Rogers Escarpment in honour of the man who so successfully had laid the foundations of south African geology.) Yet, in his summaries of what was known about the Great Escarpment in 1920 or 1928 respectively, ROGERS does not give any definition of what should be understood by the term 'Great Escarpment' except the (merely) descriptive use of 'steepness'.

In the summary of 1928 ROGERS goes on to describe the rocks that make up the Great Escarpment. He noticed (expressly) that along its whole course the escarpment is successively built up of different rocks. Starting with the Black Reef quartzites in northeastern Transvaal he then mentions the dolerite sheets on the eastern front of the Basuto Highlands (today Lesotho), describes that then "the Beaufort beds, capped by thick gently inclined intrusive rocks of dolerite forms its face for some 230 miles" (320km) and that then, "near the headstream of the Dwyka river, the dolerites disappear, and the thick sandstones of the Lower Beaufort form the upper part ... in the Komsberg and the Roggeveld Berge". To the northwest the dolerite sheets appear again.

Farther north, the escarpment is divided into two branches; the lower one, in the west, is the Bokkeveld Escarpment, chiselled out of TMS quartzites, the higher one in the east is formed by the Klip Rug mountains. Nearing the South African border Lower Karoo beds follow, then the Nama sediments and the pre-Nama-basement. Outliers of Dwyka greywackes build up the Langeberg and the Hartslag Kop near Loerisfontein. From Spektakel onward the very hard quartzites of the base of the Nama system make the escarpment very prominent again for about 30 miles.

ROGERS noticed that wherever the escarpment passes on to crystalline rocks and thus "lacks a strong nearly horizontally bedded formation at its summit" the escarpment loses its prominence and wide valleys are "indented" on granites and gneisses. Not only is the prominence of the escarpment thus dependent on bedded rocks, but even the very heights of the escarpment result from such rocks. Everywhere it is highest where such rocks reach their greatest thickness. This is especially the case with the Stormberg lava sheets of the Natal Drakensberg.

Because of the dependence of both prominence and height on bedded rocks the escarpment can principally be defined - in the present author's opinion - as a series of cuestas along the great crescent of the continental margin. ROGERS did however not draw this conclusion.

Later on, following ROGERS' example, the term escarpment became applied to similar features in South West Africa and Angola by German geomorphologists (JAEGER, JESSEN).

c) German scholars between the wars

In 1930 BORN was the first to take up the term 'Great Escarpment', coined by ROGERS, as (in German) "Eskarpment". He tried to apply a new theory in order to explain its nature. This theory had been proposed by W. PENCK, the son of A. PENCK, in the twenties and published in 1925 ("The geomorphic analysis"). W. PENCK, being a geologist, tried to explain the occurrence of step-like planations by his 'Piedmonttreppen'-theory. According to W. PENCK, steps of successive 'piedmonts', climbing up as it were the slopes of mountain ranges, could be explained by the use of a single tectonic concept, viz. the concept of a gradual widening dome of up-lift. Everything else could be left to erosional forces. For these forces he thought the knick-points in the thalwegs of valleys to be the decisive elements; these knick-points would come into existence with the gradual spreading of the dome of up-lift. Successively, these knick-points would then become the base levels for the planatory processes bevelling into the mountain slopes, thus forming the 'piedmonts' there.

BORN, in 1930, tried to show that the escarpment owed its stepped nature to the processes outlined in W. PENCK's theory. BORN, in doing so, pointed out three earlier stages before an actual escarpment became formed. For these stages he went back to the findings of A. PENCK in Natal, partly making use of them. The three preceding stages should be: (1) the basin-like warping of the Karoo-beds in the south African interior; (2) a strong up-lift of this interior combined with a simultaneous down-warping towards the Indian ocean, and (3), before this event, a late Mesozoic peneplanation cutting across the Karoo beds.

The Great Escarpment proper should then have been formed by two younger stages: (1) a complete planation in the way of W. PENCK and planing for instance the monocline of Natal in the older Tertiary, forcing the escarpment to retreat and by this retreat first creating the erosive margin between the coast and the interior; (2) consecutive further up-lifts - with each up-lift resulting in a piedmont scarp and a planatory suface in front of it.

As will be noticed, BORN with this assumption of scarp retreat is the fore-runner of a similar notion that has been held by KING throughout the last decades.

As proof for his assumption BORN had tried to find out surfaces which for him had held similar or even identical positions within the piedmont forming process. He therefore compared the situations at places as far distant from each other as the Devil's Kantoor in eastern Transvaal, then near Beaufort West in the eastern Cape Province and finally on the crystalline rocks of South West Africa. Comparing the three situations BORN came to the conclusion that the escarpment everywhere descended towards the coast in the same distinct steps and in the same number of steps. He stated: "...there is no doubt that, in these places, we find the true up-lifted piedmont surfaces" (1930, 317).

What BORN however did not take into account was the different rock resistancies in all these places though he admitted (318) "that these surfaces rather often (sic!) occur in levels of soft clayey beds whereas hard resistant rocks are switched in below and above them". Ostensibly, BORN was not troubled by such an observation - nor did he trouble himself with connecting his proposed surfaces all over the vast distances between these places.

The present writer's observations at the self-same places have shown that the so-called surfaces owe their existence either to local excavating processes within the different strata of softer rocks or to surfaces that indeed can be connected through to all these places and between them. (Naturally they are not the same surfaces BORN talked about.)

JAEGER came to know the escarpment in southern Angola and explained it in his first paper (1914 resp. 1919) as a huge abrasional cliff; but in 1923 he revised this idea because he found that it was not in accordance with the fact that the valleys, descending from the interior plateaus, break through the coastal inselberg ranges. In his 1930 paper he had completely come under the influence of W. PENCK's ideas. He therefore looked at the whole southern African escarpment as one gigantic 'piedmont-treppe'. JAEGER, who was the first to translate 'escarpment' as 'Randstufe' into German, was opposed to the idea that the 'escarpment' should be regarded as a cuesta. He forwarded two reasons: (1) the scarp is developed across folded or metamorphosed rocks over long distances; (2) the scarp does not adhere to just one resistant rock bed but 'jumps' from one bed to another.

Within the 'piedmont-treppe' JAEGER described three steps: (1) the 'coastal' peneplain (2) the 'interior' peneplain cutting the crystalline rocks as well as the mesalike cover; (3) the oldest and highest peneplain on top of the Khomas and Naukluft Highlands of central South West Africa. The 'interior' plain consists of two sub-levels: the lower one indented into the higher one along the Fish river basin.

According to JAEGER, the 'coastal' and the 'interior' peneplains are supposed to be of Tertiary age (after some preliminary developments in the Cretaceous) and owe their existence to successive up-lifting.

JAEGER drew attention towards the difference in shape of the two peneplains: the 'interior' (and upper) plain was supposed to run along in more or less horizontally constant altitudes; the 'coastal' (lower) plain was supposed to rise gradually towards the scarp at an angle of 0.5 to 2 degrees, "an angle that even in mesa landscapes is not easily detected" (1930, 139). (According to the author's experience however even an angle of 0.5 degree is easily detected by the human eye - and especially in mesa landscapes.) JAEGER does not seem to have actually measured the angles, but reconstructed them obviously from maps, comparing the heights over rather longish distances.

By this reconstruction JAEGER became the inventor of what in German since then has always been referred to as the "Schiefe Ebene" (i.e. 'oblique plain'). Altogether, JAEGER does not seem to have been puzzled by the existence of such an oblique plain, explaining its existence simply by tilting. Instead, two other problems forced themselves upon him. The one is (1) that southern Africa must have suffered from an up-lift of 600m, then followed by a subsidence of more than 600m. JAEGER deduced this assumption from his observation that (a) the coastal plain was cut into the scarp at the present height of 600m ab. s-l by abrasion in the Upper Cretaceous and (b) that after this event it must have been remodelled by Tertiary planatary processes in the then neighbourhood of the sea, as testified by Tertiary marine deposits. The other problem (2) is posed to him from his observation that the escarpment is not as dissected by valleys and not as denuded as should be expected from the assumption that southern Africa must have lain rather high since the Upper Cretaceous. A solution for the two problems will later be given.

KORN and MARTIN (1937) came to a conclusion very much opposite to that of JAEGER's with regard to the age of the Great Escarpment. From observations in South West Africa they concluded that it was of a rather young age, viz. youngest Tertiary or early Pleistocene even, having come to be formed in connection with the most important and strongest up-lift, by which the old Cretaceous peneplain on top was finally cut into pieces. In view of later research however these findings were based on data that no longer can be held to be correct.

Though OBST and KAYSER could publish the results of their observations only as late as 1949 (the pre-war printing plates were destroyed in the bombing of Berlin), their research nevertheless had been done in the thirties, even before that of JESSEN. OBST and KAYSER were the writers who effectively propagated the term "Große Randstufe" in German; thus this term has become widely accepted since then. In using this German term they made use of KNETSCH's translation of the English term Great Escarpment as 'Große Randstufe' (meaning: great marginal scarp) in 1940. (It might be remembered that it was JAEGER who, in 1930 first used the term 'Randstufe', meaning 'marginal scarp').

On the whole, OBST and KAYSER followed along the lines having been drawn by the two PENCKs and by BORN. Consequently, they came to consider the escarpment as a manifold and multi-storeyed erosive scarp. This kind of scarp had come into existence since the Upper Cretaceous and Lower Tertiary as a consequence of several relatively young phases of general up-lift all along the margin of southern Africa. In connection with these phases, a wide piedmont surface came to be established in the forelands of the scarp, consisting of several sub-surfaces. The escarpment itself, all along its course, was considered by the two authors as to be divided into mainly two parts because of different rock resistancies. The one part is the escarpment formed over the crystalline basement with its nearly complete peneplation (before the Upper Cretaceous) and altogether a 'softer' nature of the sub-surfaces developed on these rocks. The other part consists of the marked cuesta-like scarps in the younger strata covering the basement.

Such surfaces and sub-surfaces which OBST and KAYSER believe to have shaped the escarpment and its forelands are shown in Table 3.1.

By the present author's observations neither the surfaces as such nor their heights nor their ages could be verified. It might nevertheless be pointed out that there are indeed some coincidences in the altitudes of particular levels. This will be shown later on.

Table 3.1: The surfaces and sub-surfaces ot the 'piedmonttreppe' along the 'Great Escarpment' according to OBST and KAYSER (1949)

Die Flächenniveaus an der Großen Randstufe nach OBST und KAYSER (1949)

Levels	Heights (ab.s-l in m)	Tectonic events	Ages
Hochland (highland)	3000	down-warping	Upper Cretaceous to Lower Tertiary
Hochlandrand (highland margin)	1800/2000	uplift, then standstill	Oligocene
Vorlandstufen (foreland scarps)	A: 1200/1500)	several phases of	Miocene
	B: 950/1200)	uplift and inter-	Mio-Pliocene
	C: 600/800)	mittent standstills	Pleistocene
Küstenebene (coastal plain)	below 600		Holocene

Source: Compiled from OBST and KAYSER (1949)

It should moreover born in mind that OBST and KAYSER apparently thought all their surfaces always to rise with a gradient from the lower heights towards the upper ones as given in Table 3.1. From the results of the present writer this is indeed the case with what OBST and KAYSER call the highland marginal level and the foreland level A. The other surfaces however do not possess any descernible gradient at all.

It is not quite clear from what OBST and KAYSER write whether they attribute these gradients to the original peneplanatory processes or to the tilting or to both reasons.

A particular question is whether there really exists a highest surface at an altitude of ca. 3000m ab. s-l in the Basuto Highlands. The next real surface below this altitude is the one in 1800 - 2000m, and this surface extends widely over the interior outside the dolerite capping. Therefore this surface can be considered to be older than the dolerite intrusions. If so, it is this surface that then must be regarded as the true Gondwana surface. OBST and KAYSER assigned this surface only to the highland margins. They supposed it to be of an Oligocene age. Yet, if it is older than the dolerite intrusions it really is of a Jurassic age.

JESSEN's book on the continental margins was equally published after the war (1948); it was equally based on fieldwork during the thirties, the fieldwork being executed in Angola (JESSEN 1936).

JESSEN was the first scholar indeed who occupied himself with the continental margins as macro-features (cf. BREMER, 1985). JESSEN based his findings on the ideas of A. PENCK and of OBST and KAYSER. Starting from Natal and the Transvaal JESSEN distinguished between four tectonic phases (1948, 50). These are:

(1) A first period of up-lift in the Upper Triassic and/or the Lower Jurassic respectively, followed by a long time of erosion by which the oldest surface was shaped, whose remnants are considered to have been preserved in parts of the Basuto Highlands;
(2) the phase of monoclinal tilting at the end of the Jurassic and in the Lower Cretaceous;
(3) a second period of sustained up-lift in the Cretaceous, followed by a new erosion cycle;

(4) several shorter periods of up-lift during the Upper Cretaceous, resulting in the shaping of several sub-surfaces.

For another part of the 'Große Randschwelle' (great marginal rise) called by him the 'Kap-Schwelle' (Cape rise or swell) he came to distinguish between the following six phases:

(1) First appearance of the 'Randschwelle' in the Lower Jurassic as a consequence of up-lift and tilting;
(2) the intrusion of the dolerites and a resulting doming;
(3) a phase of renewed up-lift at the end of the Jurassic;
(4) further up-lifts combined with faultings in the Cape Fold Belt in the middle Cretaceous and with an up-lift and tilting of the Cape peneplain;
(5) the development of the escarpment with some sub-scarps and a planation of its forelands at the end of the Cretaceous;
(6) a new and sustained up-lift accompanied by coastal down-warpings in the Lower Tertiary (1948, 51, 52).

Altogether, JESSEN seems to believe in four larger periods of up-lift and three separate periods of tilting and warping.

To the mind of the present author, the main objection against JESSEN's conclusions is not so much the fact that the ages fixed by JESSEN can only partly be verified, but the fact that there do not exist any clues for a differentiated development of the two 'Schwellen' (Natal - Transvaal region; Cape region) - at least, not as far as the Great Escarpment and its development is concerned.

d) Research after 1945

After several extensive journeys to South West Africa ABEL came to form his views with special reference to conditions in the Kaokoveld and the highlands and scarps south of the escarpmentgap (or Swakop-Kuiseb Gap). He reached the conclusion (in 1959) that the Great Escarpment came into existence during the Cretaceous (middle to upper). He based this conclusion on the study of different high-lying surfaces in front of the escarpment and on large inter-montane basins (which seemed to extend from the surfaces and to ingress widely into the escarpment). According to ABEL, the highest of these surfaces should be dated as Upper Cretaceous or Lower Tertiary respectively.

Moreover, ABEL tried to find conclusive evidence for the solution of another problem that had puzzled geomorphologists and geologists for a long time: the problem, namely, of the escarpment gap (Randstufenlücke) combined with the problem of the 'oblique plain' (Schiefe Ebene) in front of the gap and extending down towards the sea. The gap is situated north of the Khomas Highlands and has a width of 300km. (If however, one considers the respective ends of the cuestas in the north and in the south the gap extends over 500km.)

ABEL thought that the gap owes its origin to an old tectonic depression going back into the Carboniferous or even beyond it. ABEL compared the gap with another one farther down to the south, namely that of the Orange basin. The depression of this basin similarly seems to be of an old age (HAUGHTON and FROMMURZE, 1928; GEVERS, 1933, 1934, 1961).

The 'oblique plain' had formerly been explained by KAYSER (1949) as a true but down-warped peneplain ("echte Flächenflexur"). KAYSER had fixed the date for the warping as Oligo-Miocene. ABEL however came to the conclusion that the warping of the plain had taken place in the Upper Cretaceous.

MABBUTT (1955) occupied himself in some detail with high-lying surfaces in Namaqualand (northern Cape Province) and in South West Africa. These surfaces are similar in height to those mentioned by ABEL. MABBUTT is however not concerned with the origin of the Great Escarpment (which, by the way, is not very prominent in this region of the Orange depression). Therefore, his findings will not be dealt with here (though, certainly, in the following chapters).

Another author who however must here be dealt with at some length is KING. KING's notions about the development of the escarpment have shaped the views of scholars all over the world. His notions have however changed enormously in their details since KING's first publication in 1951. Even his summaries in 1972 and 1982 show some differences.

These changes in KING's opinions were mostly concerned with the following ones: phases of up-lift, heights, warpings and tiltings. DE SWARDT and BENNET (1974) aptly and painstakingly listed the changes. In spite of these changes KING's fundamental ideas have remained the same throughout these decades. These fundamental ideas may be summed up in the following four points:

(1) The monoclinal nature of Natal, expressly based on A. PENCK's early work;
(2) the successive retreat of the Great Escarpment as an erosive and denudative form from an original position rather near to the sea (an idea he shares with BORN 1930);
(3) the planations of the forelands, starting from the knickpoints in the thalwegs of the rivers and caused by several up-lifts (an idea he obviously borrowed from W. PENCK);
(4) the process of pediplanation, i.e. a planation due to the widening of pediments on the longer slopes of scarps and the merging of these pediments eventually forming wide plains (an idea being uniquely KING's own).

A. PENCK's notion of just one single and simple monoclinal and planatory event became adapted by KING in the way of several renewed phases of up-lifts, tiltings and warpings, each particular phase resulting in a new planatory onslaught on the scarps with the result that all individual scarps - and with them the escarpment as a whole - were gradually forced back into the interior. The scarp retreat then is a consequence of a series of erosion cycles rejuvenated with each successive phase of up-lift and tilting.

Based on the summaries of 1972 and 1982 Table 3.2 tries to sum up KING's ideas.

As has been mentioned above KING's ideas have become very influential - so much so that almost every scholar dealing with the morphotectonics of Africa south of the Congo basin felt liable to take KING's notions as a framework for his own research - whatever the distance of his own research from KING's Natal monocline might be. As a consequence, nobody so far seems to have been worried by the question whether there really are true connections between the levels and between the geological and/or geomorphological events.

Criticsm of KING's ideas may be based along the following lines.

Table 3.2 Landform development along the Great Escarpment Escarpment in Natal after KING (1972 and 1982)

Entwicklung des Randstufenreliefs in Natal nach KING (1972 und 1982)

Erosional periods, planations (1982)	Tectonic events	Age (1972; 1982)	Uplifts in metres 1972	1982
Gondwana-landscape[1] (=intermission I)	Break-up of Gondwana-continent, creation of the Natal monocline, heavy faultings (=active period A)	Late Jurassic		
Kretacic landscape (=intermission II) = Sani Pass surface proto-scarp, 500 metres high first valleys (formerly called: post-Gondwana surface)		Lower Cretaceous		
	uplift (active episode B)	Cretaceous	1200	1200
Mooreland landscape (=intermission III) plain, laterites; scarp: 1200 - 1500 metres high, scarp-retreat (formerly called: African surf.)		Late Cretaceous-Early Cenozoic		
	uplift (active episode C)	early Miocene	200-300	200-300
'Rolling' landscape (=intermission IV)				
	severe tiltings, uplifts, monoclinal steepening (=active episode D)		600	
'Widespread' landscape(=intermission V) pedimentation, scarp retreat, independant basins at various levels (formerly called: Congo cycle)		Pliocene		1700[2]
tilting, Plio-Pleistocene	mighty upheaval, strong monoclinal steepening (=active episode E)			
'Youngest' landscape (=intermission VI) rejuvenation		Quaternary		

1) formerly called 'surface'
2) no exact indications, computed by author from Fig. 34. in King 1982

(1) Tectonics

If one compares the periods and the datings given by KING and those given by OBST and KAYSER one immediately detects such great discrepancies that one is led to doubt the reliability not only of the datings, but moreover of the up-lifts and tiltings as well (in spite of the careful investigation and the wealth of details gathered over wide distances).

As the work of the authors referred to in the preceding paragraph shows it is apparently possible to arrive at two completely different models of how the Great Escarpment developed - models that (1) start from the same surface positions, (2) from the same underlying assumption of successive up-lifts and tiltings, models that in their results however can on nearly no point be harmonized.

If scientists (facing the same landscape) arrive at such different solutions it may be deduced that there must be a fundamental error somewhere.

(2) Morphogenesis

The process of pediplanation though fundamental to KING's model seems not to be very clear and is indeed not easily be justified from what one sees in the field. What one does see is the frequent occurrence of relatively long (up to 500m) and flattish slopes or slope sections rather. As it seems these sections have been termed 'pediments' by KING. The present writer had a close and very careful look all over Swaziland, Natal and the Transkei with a special reference to slope development and 'pediments'; but in spite of this he was not able to detect any pediment as a truly separate feature of its own. What however is seen everywhere in the area mentioned above is the occurrence of slopes with indeed rather extended flat lower sections. Yet these sections are no pediments because they do not merge into each other and therefore do not form any continuous line of pediments over a longer distance at more or less the same height, not even where the flattish slope sections are next to each other. Yet, such a merging of flat slope sections into a pediment-line is the only process on which the idea of pediplanation could be based. Consequently, if there are no true pediments there cannot have occurred any pediplanation. Consequently again, if there is no pediplanation no scarp retreat can have happened.

Moreover, a close inspection of what had been termed the 'African level' (i.e. the first surface plained into Africa after the rifting of Gondwanaland) showed that - even on this older surface - there could not be found any traces of pediments nor any remnants of possibly fossil ones.

Similarly, true laterites as proof for an older surface were also not found in these altitudes - the red soils that do occur are without doubt connected with either the weathering of the dolerites or the feldspathic sandstones of the TMS; i.e. they are litho-soils.

(It should be noted however that these findings apply to southern Natal only; there are indeed some planed surfaces with laterites on them - though on TMS rocks only - in central and northern Natal. These must however be treated within quite a different framework from that used by KING, as will be shown in a later chapter.)

According to the handbook on geomorphology by LOUIS and FISCHER (1983, 177-179) pediments occur as flat slopes with degrees up to 10 or 15 on the seams of mountainous areas in arid and semi-arid zones. They consist of single cones, joining each other. The cones are formed in the neighbourhood of valleys; streams, on the cone surfaces, run on thin covers of detritus transported down by the diverging streams. Not one of these criteria does pertain to any of the lower slope sections in Natal. The most prominent feature for the lower slope sections in Natal is their very marked flatness (degrees less than 5).

In spite of the criteria given by LOUIS and FISCHER these authors go on to concede to KING the notion of slope retreat (1983, 214) - which again shows how influential and apparently convincing KING's ideas have made themselves felt even on minds as cautious as the one of LOUIS. LOUIS, expressly referring to KING's 'Morphology of the earth', takes for granted (on KING's evidence) that slope retreat is "in the tropics and in the semi-humid tropics an obviously and widely conclusive circumstance", though LOUIS knows in fact that "this circumstance has been proved...in the humid middle latitudes...as widely not conclusive". Regarding the subtropical regions of Natal the conclusiveness of slope retreat is however also severely to be doubted - at least in the opinion of the present writer.

MOON and SELBY (1983) also challenge any simple view of parallel slope retreat and demonstrate a widespread occurrence rather of equilibrium slopes. They conclude that parallel retreat will only occur when rock mass strength is uniform. Even such an assumption has however to be doubted as will be shown in a later chapter.

Summing up, parallel slope retreat alone is not sufficient at all to account for all the aspects of escarpment formation.

From these comments we may pass on to other research which has some bearing on the evolution of the Great Escarpment.

At first MAUD is to be named, a pupil of KING, who - together with other workers - worked on the Natal fault systems (1960, 1965, 1975). MAUD's results are of important consequence for all models of escarpment formation and retreat that are based on views as held by BORN, OBST, KAYSER, JESSEN and KING. For MAUD concludes that there cannot have been any faultings, down-warpings and tiltings since at least the Upper Cretaceous - except some slight epirogenetic movements (deduced from the various heights of ancient laterites).

The observations of the present writer also show (Fig. 3.1) that whatever tectonic events may have happened they must have ended with the beginning of the Upper Cretaceous at the latest because the transgressive Upper Cretaceous marine beds have never become disturbed since then. The same can be shown from the younger Miocene and Pliocene layers (MAUD).

In 1986, KAYSER's more recent findings were published posthumously. As with the older book, the publication was again based on numerous and extensive travels in southern Africa during the last decades. The findings concern the eastern Transvaal. KAYSER arrived at the following model for escarpment formation, here summarized into six phases.

(1) Cretaceous to early Tertiary: Development of a major planation surface, probably by a multi-cyclic process; at the end of this period: the first large-scale tectonic doming, especially around the margins of the continent.

Fig. 3.1: a) The Upper Cretaceous in northern central Natal (near Hluhluwe)

Lageverhältnisse der oberen Kreide im nördlichen Mittelnatal (bei Hluhluwe)

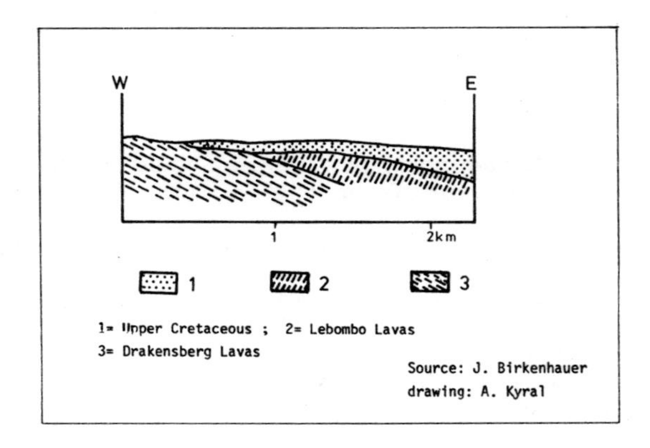

Fig. 3.1: b) The Upper Cretaceous in northern central Natal (near Mtubatuba) - schematic

Lageverhältnisse der oberen Kreide im nördlichen Mittelnatal (bei Mtubatuba) - schematisch

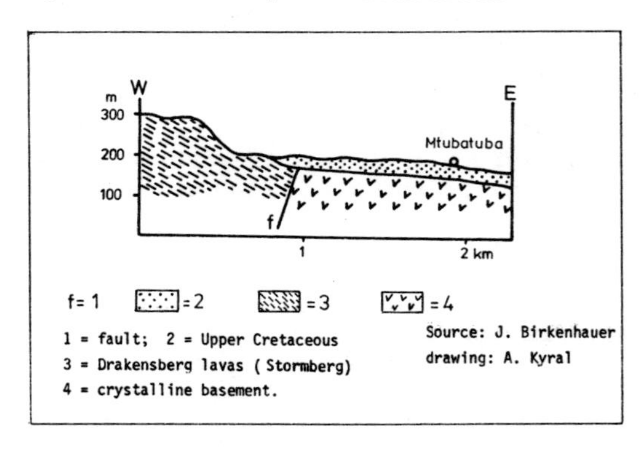

(2) Eocene: Local tectonic up-lift of the marginal swells, followed by the development of the 'highland marginal level' ('Hochlandrand-Niveau').
(3) Oligocene-early Miocene: Main period of asymmetrical marginal swell development in southern Africa. Development of the Great Escarpment on the outside of the swell. Scarp-steps ('Piedmont-treppen') in the sense of W. PENCK formed.
(4) Late Miocene-early Pliocene: Very little tectonic movement.Back-wearing of Great Escarpment, associated growth of the 'high escarpment level' (oberes Randstufenniveau).
(5) Pliocene: First up-lift of southern Africa as a whole. Development of the 'middle escarpment level' ('mittleres Randstufenniveau').
(6) Plio-Pleistocene: Second up-lift of southern Africa as a whole. Development of the 'lower escarpment level' ('unteres Randstufenniveau').

On the whole, the results summed up above are not much different from what KAYSER thought in 1949. Important seems to be that the five surfaces (or 'levels') are (1) younger than the Upper Cretaceous and (2) are all the consequences of different phases of up-lift, (3) that these again are younger than the Upper Cretaceous and go on into the Pleistocene. This altogether means that the Great Escarpment is considered to be of a rather young age.

If one now compares KAYSER's levels with KING's surfaces as the two most prominent reconstruction models of escarpment evolution it seems best to do such a comparison by way of a table. This is done in Table 3.3.

The juxtaposition of the two models of escarpment evolution shows some similar traits. One should however bear in mind - and this is the most important result of the comparison - that neither the datings nor the altitudes of the levels can really and easily be compared with one another.

In 1985 OLLIER and MARKER published a paper on the development of the Great Escarpment in which they distinguish between two main surfaces: (1) a high-lying paleoplain in an average altitude of 1500m (within which they seem to comprise all KING's surfaces prior to the break-up of Gondwanaland, i.e. Gondwana, post-Gondwana, African cycles); (2) the coastal plain of younger origin.

The old and the new plains are separated from each other by the Great Escarpment itself.

The escarpment - though in itself a major erosional form - was initiated by tectonic movements which the authors see related to the break-up of Gondwanaland and connected with an ensuing differentiated up-lift. The greatest amount of up-lift occurred in a zone roughly parallel to the Great Escarpment. The connection between these two facts is expressly stressed by the two authors. From this one gathers that they apparently regard this connection as the main thing in the escarpment evolution. The decisive phase of up-lift is supposed to have happened in the late Cenozoic. "Pronounced morphology (however) and actual location of the escarpment are a function of local favourable conditions, especially in areas of sub-horizontal hard rocks" (43) (Insertion of 'however': by present writer.) As a consequence of this function the "continuously sharply defined escarpment is missing" in the Swakop-Kuiseb Gap (43). This fact was already mentioned when dealing with ABEL's research and should here be duly accentuated, because ABEL himself did not think the rocks and their differences as causal to the formation of the escarpment, though ABEL had to admit that the scarp is most prominent when and where the dolomites and melaphyrs of the Damara system form large tafelberg and mesa areas in the Kaokoveld of South West Africa.

Table 3.3: A comparison of KAYSER's levels and KING's surfaces

Vergleich der Niveaus bei KAYSER und KING

KAYSER	KING
highland plain (=early Tertiary plain)	Gondwana landscape
highland marginal level	Kretacic landscape (formerly: post-Gondwana surface)
high escarpment level	moorland landscape (formerly: African landscape)
middle escarpment level	'rolling' landscape (formerly: post-African surface)
lower escarpment level	'widespread landscape' (formerly: Congo cycle)

Fig. 3.2: Models explaining the development of the Great Escarpment

Erklärungsmodelle für die Entwicklung der Großen Randstufe

	author, theory	conclusiveness
1) escarpment, forelands, fault, ocean	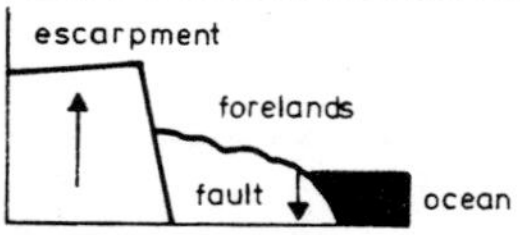 Sueß Theory of contraction uplift ↑ lowering ↓	no
2) erosional scarp, formerrelief, river, monocline	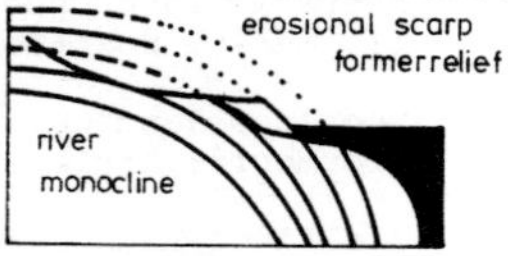 Penck, A. (1908) continental downsloping ("dip")	on the whole yes
3) scarp, several peneplains, old, young	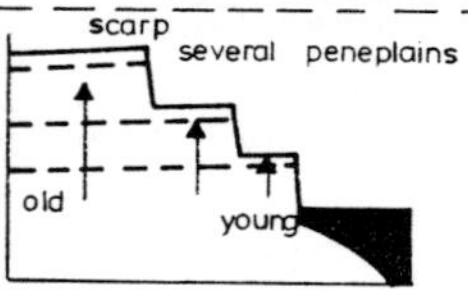 Obst/Kayser (1949) Born (1930) peneplanations, uplifts structure-assisted	no
4)	Jessen (1948) combination of 2) 3) 6)	no
5) formerrelief, situation today, Natal monocline	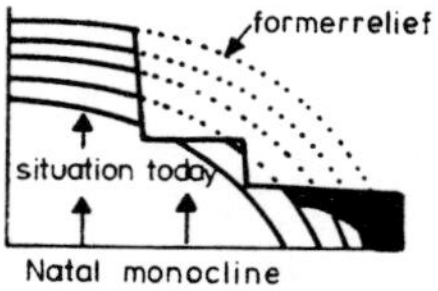 King, L.C. (since 1950) retreat of scarp because of pediplanation, uplifts and downwarping	no
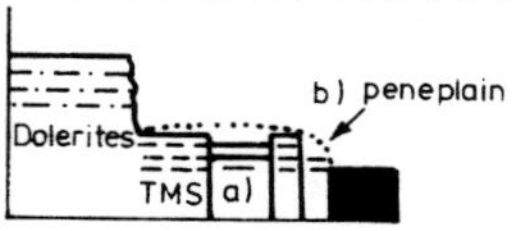 6) b) peneplain, Dolerites, TMS, a) TMS: Table Mountain Sandstone	Maud (since 1960) a) system of horsts and grabens b) cut by a peneplain, structure-assisted	yes no

Source: J. Birkenhauer
drawing: A. Kyral

4. DIRECTIVE NOTIONS

The theories and models that have been outlined in the previous paragraphs may be summed up by using schematized profiles serving as models for the theories. In Fig. 3.2 only the major ones of these theories are presented in this way.

The schematized profiles reveal the conflicting results and assumptions of these theories at one glance as it were and seem moreover to show that the riddle of escarpment evolution - that now has been posed to scholars for more than a century - has become more and more complicated within the research process. Looking at the wide range of conflicting notions a solution of the riddle seems to be hopeless. Nevertheless it seems possible at this stage of our inquiry to point out some insights that can be gathered from this century of research.

These insights, given below, may also serve as directive notions (or principles even) for the following parts and chapters. Naturally they again and again have to be tested against all the other facts that will be dealt with in the following chapters.

These directive notions seem to be the following ones:

(1) The southern African block must be considered to have lain at a rather high altitude since at least the middle of the Cretaceous.

(2) Since this time there have been no tiltings, warpings, up-lifts whatever. This block of land has been of a remarkable stability throughout since then. (MARTIN et al. write - 1982, 192 -: "The Great Escarpment did not evolve through a series of drastic up-lifts...", an opinion fully agreed upon by the present writer.)

(3) Since the middle of the Cretaceous southern Africa - and especially the area along the front of the escarpment - has continuously been subject to erosion (proved, for instance, by the huge masses of marine deposits on the shelf, especially in the Algoa basin and its vicinity - cf. MARTIN et al., 1982 -). There has not been just one late phase of erosion as is KING's notion.

(4) The escarpment is a result - amongst others - of this rather long period of erosion, yet more (orless) in connection with the larger drainage systems - which appear to be as old as the Great Escarpment itself - than with any retreat of slopes parallel to themselves.

(5) The locations of the main drainage systems have roughly been the same - at least throughout these 100 my (and perhaps even longer than that).

(6) The sinuosity of the escarpment is due to the denudatory and dissecting processes based on these old drainage systems.

(7) Any planifying processes (surfaces, terraces and the like) have had a rather ample time indeed to assert themselves.

(8) Any kind of planation (surfaces, terraces and the like) can suitably be used as a guiding level because of the tectonic stability of southern Africa within these 100 my or so.

(9) In spite of the stability of the block since Cenozoic times it should be born in mind that its very geological structure is subject to an older and complicated tectonic history. Important features of this history seem to be the following ones:
   a) the late Jurassic and early Cretaceous faultings and the fault systems created thereby in
      - Natal,
      - the Cape Fold Belt,
      - central and northern South West Africa,
      - northeastern Transvaal,
   b) all of them accompanied or even fore-run by intrusives,
   c) the Triassic folding of the Cape Belt,
   d) the very ancient nature of depressions as for instance the Swakop-Kuiseb-Gap or the Orange river basin.

(10) The importance of differentiated rock resistancies and, together with them, the cuestas, accentuating - and even dominating - the Great Escarpment along its whole course.

SECOND PART: INVESTIGATIONS AND ANALYSES

5. COASTAL PLANATIONS

a) Introductory remarks

In order to gain a better understanding of the processes leading to the formation of the Great Escarpment as well as the time-table involved in its formation it seems advisable to begin with the coastal forelands.

The reason for this is that it appears to be possible to find indications here for what happened and how and when particular formative events happened. They may help perhaps to form a more reliable base for all other observations. In doing so the present writer's theory (or theories even) can gradually be built up - together, naturally, with using the findings of other authors as well. All findings will be tested against each other and will tried to be fitted into a comprehensive model.

b) The phenomena

Several flat horizontal levels run parallel to the coast, one above the other (see fotos 3 - 13). As such they are very remarkable phenomena. They are the more remarkable since everywhere they appear at the same altitudes ab. s-l - and in this way over the course of hundreds or even thousands of kilometres. To be able to observe these levels over such vast distances was a unique experience in itself.

The altitudes of these persistent levels are the following ones:

- 100 m
- 170 - 200 m
- ca. 300 m
- 400 - 430 m
- ca. 500 m
- ca. 600 m.

The two lower levels, similar to river terraces, reach into the lower stretches of the valleys of all major rivers. The upper levels reach up into the middle thalwegs and are moreover prominent on the interfluves. Fig. 5.1 shows the distribution of the levels in Natal, Fig. 5.2 in the eastern Cape.

Table 5.1: Coastal levels in the Cape Province and the Transkei according to various authors

Küstenniveaus in der Kapprovinz und in der Transkei nach verschiedenen Autoren

Authors	SCHWARZ	DU TOIT	HAUGHTON	KRIGE	THOMPSON
Years	1906, 1916	1920, 1922	1925	1926	1942
Areas	Ciskei, Alexandria	Ciskei, southern Transkei, Alexandria	Algoa Bay and hinterland, Alexandria	Cape Province: E, S	Transkei
Altitudes of terraces (in m ab. s-l)	500/600	600	600/650	500/600 1)	-
		500	500/520	-	460/500
	450	-	460	-	430
	300/340	280/300	260/300	280/340 2)	300
	180	180	200/210	200 1)	-
	-	100	-	100/140 2)	150

1) Said to be 'very extensive and characteristical'.
2) Said to be 'very extensive and characteristical', with marine deposits said to be shaped by 'wave-cutting'.

Table 5.2: Coastal terraces in South West Africa and Angola according to various authors

„Küstenterrassen" in Südwestafrika und in Angola nach verschiedenen Autoren

Authors	JESSEN	BEETZ	BEETZ	ABEL	SPREITZER	GEVERS	BESLER	BIRKENHAUER
Years	1932	1933	1933 1935	1953 1954 1959	1966	1936	1980	1982-1985
Areas	Angola	Angola	Kunene river	Namib	Namib	Namib	Namib	Cape, Orange a. Fish rivers Namib
Altitudes (in m ab.s-l)				600	600			600
				500		500		500
		400		400	400			400
	300 1)	300	300 2)	300		300		300
			150/200	200	200		200	200
							150	
				50/150			100	100

1) JESSEN (1931) mentions FABER (1926) who called the plain in 300 m the 'Schiervlakte' (= 'sheer - or absolutely level - plain').
2) BEETZ expressly mentioned the absolutely level character.

It would be possible to draw similar charts for all the other coastal forelands of the escarpment. It suffices however to give only these two examples because of the very similarity of the phenomena everywhere. All other figures would really and essentially show the same features.

Moreover, the terraces or planations can be recognised in all poorts, wind-gaps and valley breachings, - whether in the Lebombo-Ubombo-chain, or in Swaziland, or in Zululand (Pongolo, Mkuze, Usutfu rivers, for instance), or in the Cape ranges or even in the Orange basin and its vicinities.

The benches (or terraces, or planations) do not show any signs of tilting or warping, neither up nor down, neither in the Bathurst region nor in the region of the so-called Transkei swell (cf. TANKARD et al., 1983, 440, 441). They cut across faults and rocks of different resistancies (Fig. 5.3., Fig. 5.4). As such, they are completely independent of any a priori structure. Therefore it may be safely stated that they are not at all structurally controlled. Recent and sub-recent morphogenetic processes are however and obviously structurally controlled.

Both the structural independency and the very height constancy of all these levels are - as has already been remarked above - indeed very extraordinary features. HELGREN and BUTZER (1972, 1977), for instance, similarly refer to the same heights of some levels and continuously over a distance of more than 500km, also on the Indian ocean shore of the southern Cape Province. They expressly mention the levels in 100 and 200m ab. s-l.

In northern Natal (Fig. 5.3) the highest level in 600m is equally completely independent of all faults wherever these may occur. Other profiles with similar results could easily been given as can be seen from Fig. 5.1. All higher levels do not only cut across the faults, but also across the TMS, the basement, the Karoo layers and the lower Cretaceous; the lower ones additionally cut across the Alexandria Formation (Eocene - Pliocene).

The benches are preserved best on quartzitic sandstones and quartzites (TMS, Beaufort, middle Ecca), especially so in the Transkei and in central Natal. DE SWARDT and BENNET (1974) also refer to these circumstances in these areas.

Fig. 5.1: Levels in Natal

Niveaus in Natal

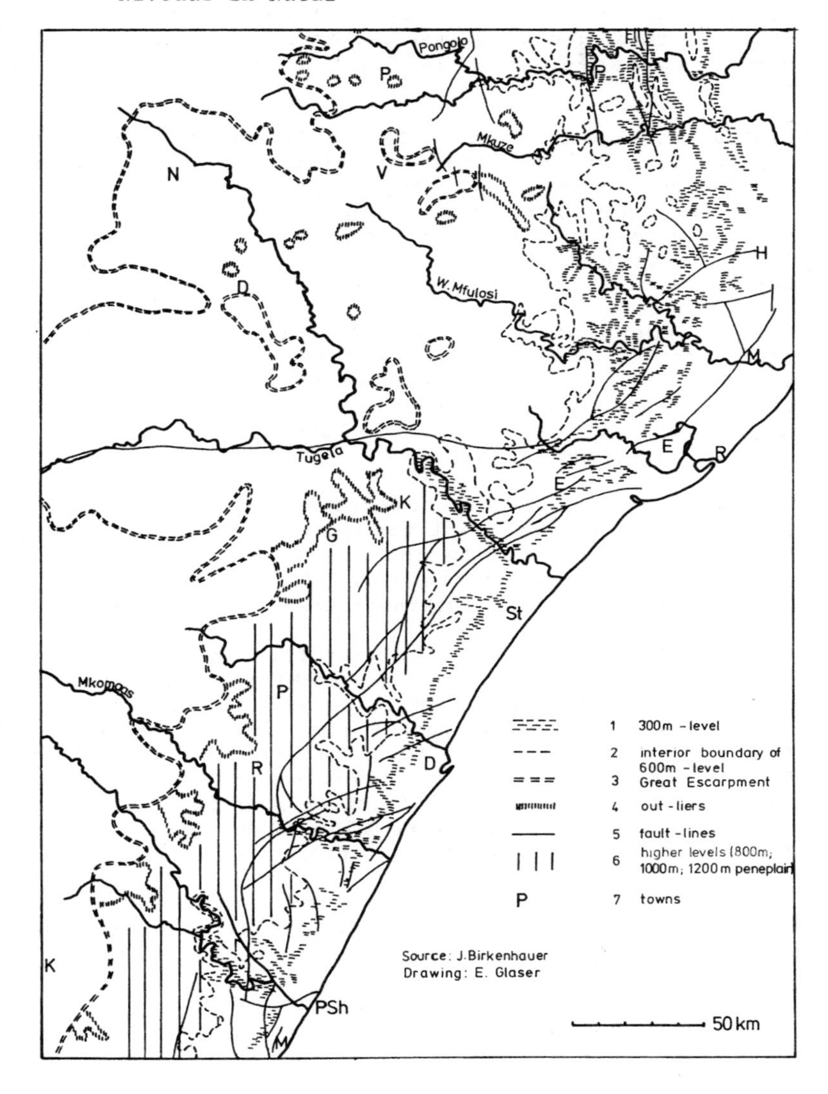

Terraces along the Natal coast were pointed out by various earlier writers, such as KRENKEL (1928), THOMPSON (1942), OBST and KAYSER (1949), MCCARTHY (1969). Similarly, the Cape Province coastal terraces are extensively dealt with by other writers. The sometimes rather lengthy descriptions are here compressed in Table 5.1 for the sake of a quick surview.

KING also, together with FAIR (1954), admits that there is "the well-known bench that follows the coast above approximately 2000 feet" (roughly 600m) and that "it can be traced into the upper Tugela and the upper Great Kei river basins virtually to the base of the Great Escarpment itself" (19). (At that time KING regarded this level as his 'African' surface.)

The agreement of so many different authors on the same altitudes of these benches and terraces is in itself a remarkable fact indeed. Following this agreement, it may be safely assumed that wherever planations occur in the selfsame altitudes they have to be identical with the ones in the same altitudes elsewhere.

It should be pointed out however that HAUGHTON (1925) believes the levels to slant towards the coast. According to him, this is especially the case with the highest one, sloping down from an altitude in 790m (or 2660 feet) to 460m (or 1500 feet) in the Winterhoekberge of the southern Cape.

DU TOIT (1920) described the very flat and wide plains of what he called the Upland-Ruggens-Plateau in front of the southernmost Cape ranges. He also was of the opinion that the levels were tilted. According to him, the level in 1000 feet (300m) gradually wins height along the plateau from the False Bay in the west (300m) towards the east in 380m (1250 feet), from Alexandria in 305m to 460m in the east, or again near Grahamstown even up to 670m (2200 feet) or more so (3000 feet = 915m) towards the interior (near Uitenhage).

Since all these heights are stated for areas around the Algoa Bay they suggest a warping and tilting around the bay and the interior basin.

The present writer however was not able to confirm the tiltings and warpings in this area. Rather, all terraces and benches do run along the same altitudes as everywhere else -

Fig. 5.2: Distribution of main levels in the eastern Cape Province and in the Transkei

Verbreitung der Hauptniveaus in der östlichen Kapprovinz und in der Transkei

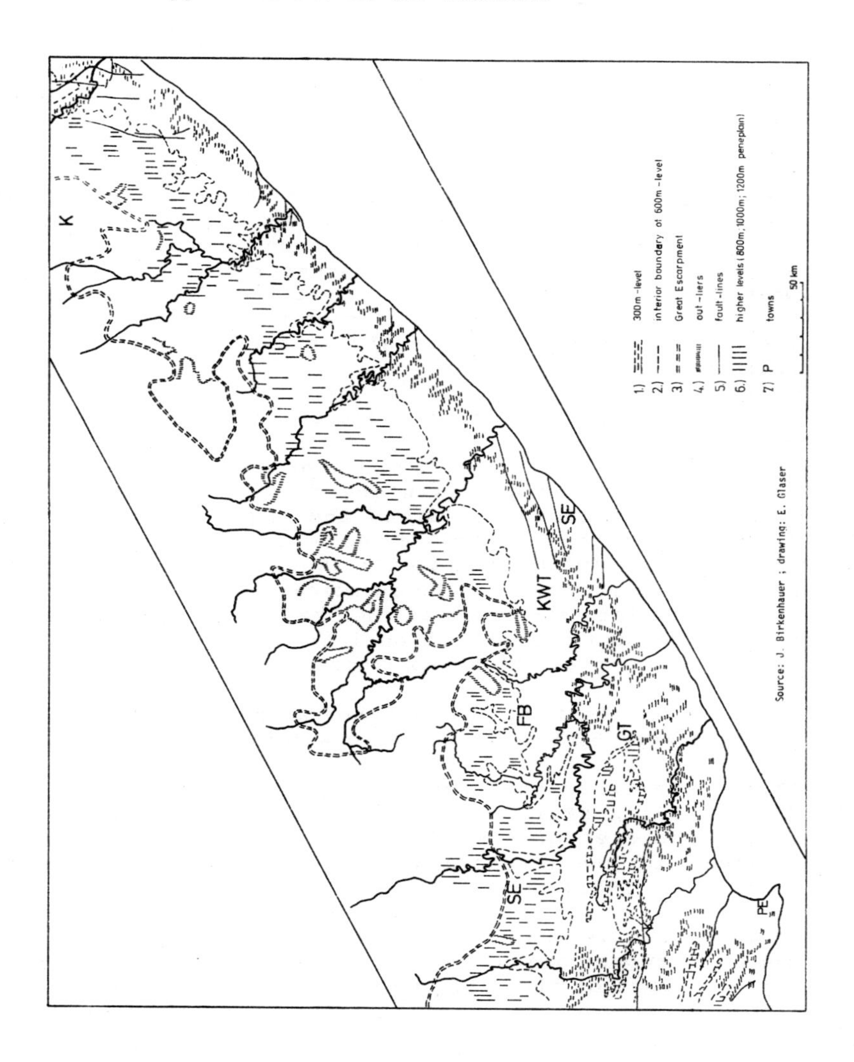

and right away at that from the False Bay in the west to East London in the east: in altitudes - everywhere - of 100 - 120m, 280 - 320m, 400 - 430m, 500m, 600 - 620m.

Though RUDDOCK (1968) described some younger tiltings in the immediate neighbourhood of the Algoa Bay they only seem to occur within the sediments. The present writer was not able to discern any deference at all in the heights of the terraces themselves, even not in the 100m terrace. (Cf. TANKARD et al., 1983, 441.)

Moreover - though DU TOIT described the benches as being not stepped and thus having no clear limits between them - there are indeed distinct scarps between the levels wherever one looks at them over the whole distance (if not scarps then slopes at least with a distinct steeper gradient). Thus, at whatever particular area one crosses over from one terrace to the next above one always crosses these slopes or scarps, the slopes (or scarps) also always occurring in the same altitudes.

It may be stated here that these observations do not only apply to the southern Cape Province but indeed everywhere throughout the research area of the present writer in southern Africa: i.e. right down the coast from the southern border of Swaziland all along the Indian ocean and then right around the Cape of Good Hope on to the coast-line of the Atlantic. The terraces will show up here in the northwestern Cape Province in the identical heights as elsewhere, they run on into the basin of the Orange river and its major tributaries and from there on into South West Africa, at least as far up as the Swakop-Kuiseb Gap.

In Table 5.2 the findings of other authors for the benches in South West Africa and southern Angola are summarized.

It is remarkable that some of the authors (FABER, 1926, JESSEN, 1932, BEETZ,1933) either mention the absolutely level character of the planations or at least give the altitudes of the levels always at the same heights (ABEL, 1953, 1954, 1959).

An objection against the existence of the planations as a chain of stepped benches one on top of the other seems to be the existence of the 'oblique plain' ('Schiefe Ebene') mentioned before (fotos 13, 15). As has been also mentioned

Fig. 5.3: Cross section in northern central Natal

Profil im nördlichen Mittelnatal

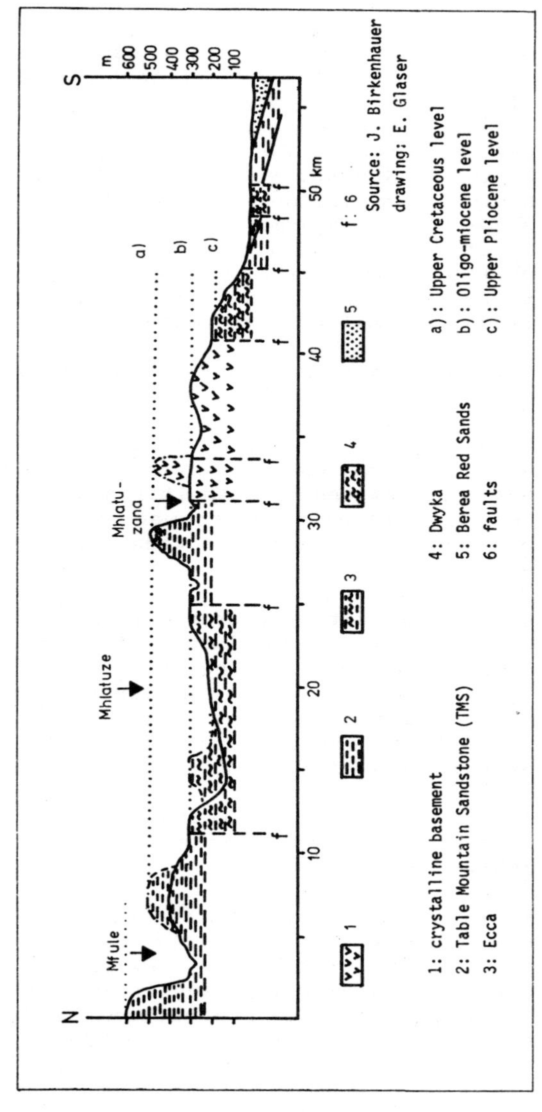

above, KAYSER (1970) referred to the 'oblique plain' as a truly tilted peneplain (cf. ch. 3c). Dealing with the same area, HÜSER (1977) dispatched with the levels in a rather off-handed way when, in referring to SPREITZER (1966; cf. Table 5.2) and the steps as pointed out by this author, he wrote that "such steps however were not described by any other author", a statement that cannot be held to be correct if one looks at Table 5.2.

Another witness for the stepped nature of the 'oblique plain' is GEVERS. He wrote (1936, 77):"...the slope of the Namib plain...is, however, not uniform, a number of distinct steps being discernable. One of the most pronounced terraces is that...at an elevation of some 300 metres" - a statement obviously and easily corroborated from Table 5.1. He went on to refer to the fact that "...there is a remnant of a terrace lying slightly above 500m."

The fact that some authors apparently were not able to discern the terraces may be explained by the observation that the terraces are slurred over. This effect occurs because a lot of débris is spread over the surface of the Namib desert. GEVERS for instance (1936, 73) mentioned a bore-hole driven through this débris that here exceeded 273 feet in thickness and showed that "level expanses of sand and débris are formed between the elevations".

Despite the 'expanses of sand and débris' the levels make themselves felt by characteristically very flat areas - all of these areas lying in the altitudes between one slurred-over slope above and the other one below, thus indicating a terrace between them. An outstanding and characteristic example of the described circumstances is the Welwitschia Plain in an altitude of 600m on the northern bank of the Swakop river (foto 13).

In Angola, the flat altitudes in 400m and in 600m are expressly called the 'Namib peneplain' by BEETZ (1933, 144, 145).

As has been pointed out above, the slurring-over of slopes and indeed the often nearly complete missing of scarps and steps between the levels - except however in the neighbourhood of active valleys and inselbergs - is due to the filling-up of all relief unconformities by the desert sand and the fine gravel (foto 50). BEETZ (1933, 146), for instance, drew attention to the fact that, in Angola, the valleys of an older landscape were completely filled up by sands, reaching a thickness of over 150m. A thickness even of 200 - 300m is in some places thought to be possible by him. RUST, in a recent talk given on the desertification in the neighbourhood of the Namib, showed that huge masses of sand and very fine gravel are driven into and shed over the deeply dissected valleys, a recent and still on-going process by which the valleys gradually become fully flattened.

The process seems to advance rather quickly. From this single example one is led to imagine what surely must have happened to all steeper landforms through all those millions of years during which the Namib desert has been in existence.

Therefore 'oblique plains' do not at all seem to be so very remarkable under desert conditions. Therefore again one should not - perhaps prematurely - jump to conclusions from an observed 'obliqueness' to the tilting of a peneplain. 'Obliqueness' should rather be considered as an indication of arid morphodynamics.

Smaller examples of the same phenomenon can be found at other places. Though far distant from one another all of them show the same characteristics. All the examples occur within a similar topography, i.e. valley embayments. Within these embayments the surface today leads up from the base level to the surrounding heights in a very smooth and flat way. These embayments may be termed 'low gradient embayments' ('Flachböschungsbuchten').

Three examples may be expressly referred to (there are more of them). The Kneersvlakte (smoothly grading over into Lepel se Vlakte) is a first very extensive example. (The Boer term 'vlakte', by the way, means nothing but such an 'oblique plain').

The example is to be seen south of Kliprand, 50km due east from Gariep (on the N 7 - road in the northwestern Cape). The two other prominent examples can be found on the southern valley banks of the Orange river, the one south east of Vioolsdrif, the other more to the east, i.e. near Pella and Pofadder. (Fotos 26 - 29; Fig. 8.1).

All the 'low gradient embayments' are only found north of the latitude of 32 S, though never on the semi-arid eastern sides of the mountains,because it is only on the western sides that fully arid conditions prevail.

The embayments mentioned can easily be detected on Fig. 12.3.

In the 'low gradient embayment' of Kneersvlakte and Lepel se Vlakte the gradient moves slowly up from an altitude of 200m (here crossed by the N 7 - road) towards a level of 700m at a degree of 2 points or even less. This gradient is indeed very similar to the one of the 'oblique plain' west of the Swakop - Kuiseb Gap. (The 700m level is one of the intermediary levels: cf.ch.8).

In spite of the 'oblique plain' between the Gap and the coast it is remarkable that along the thalwegs of the Swakop and Khan rivers each lower level is distinctly entrenched into each higher level. The entrenchments form trumpet-like embayments towards the coast and narrow hose-like ones upwards into the canyons. The entrenchments (or embayments) in 400 - 600m possess widths of 20 - 25km; the hose-like funnels have similar lengths. These well observable circumstances are taken by the author as sure indications for the existence of the levels even in the neighbourhood of the 'oblique plain'.

Now, similar phenomena - i.e. the hose-like 'funnelling' of each lower terrace upwards into the valley bottoms of the river - can be found with all other rivers. They are accompanied by a very characteristic valley bottom widening. It is surprising however that these features seem to have never before been noticed in spite of the very detailed references so many of the older authors have presented (cf. Tables 5.1 and 5.2). Even TURNER (1967) in his very detailed description of the geomorphology of Natal did not mention them at all and neither did KING in his monograph on the Natal monocline (1972 or 1982 resp.).

It is surprising moreover that the planations themselves have never been described and surveyed so far as just one common and general feature. Again one looks in vain to the treatises by TURNER and KING.

Yet the planations do not only exist within the resarch area of the present writer, but in other areas outside it as well. MEYER (1967), for instance, in his treatment of the morphology of the Pietersburg area in northeastern Transvaal, described planations widely occurring in the altitudes of 300 and 600m. He also noticed the distinct scarps between the levels (judging them however as escarpments sui generis). He also pointed out that the 300m level deeply ingresses into the higher Veld along the valley bottoms, dissecting the scarp there (cf. the present writer's description of the valley bottom widenings).Because of this the escarpment here becomes very sinuous and highland spurs digitate into the lowlands (which, by the way, is the same in Natal and the Transkei).

KAYSER (1986, 74) also described planations in these heights, i.e. the Low Veld in 300m and surfaces between 400 and 600m between Pietersburg in the north and Swaziland in the south. Again, the levels also occur as terraces along the valleys breaching the Lebombo range and along the river poorts through the escarpment proper, notably the Olifants river.

The top of the Lebombo range is cut moreover by the planations from 400 - 600m. According to KAYSER they are here accompanied by gravels stemming from the Great Escarpment. Apparently KAYSER thinks that, at one particular period, the Low Veld became filled up with these gravels, up to these heights of 400 - 600m. They were - according to KAYSER - later on excavated again when the Low Veld became transformed into another 'Schiefe Ebene' ('oblique plain'). Here, at least, the term is however nothing but a figure of speech since KAYSER himself made it very clear indeed that this so-called 'schiefe Ebene' possesses a step-like topography between the altitudes of 300 and 800m. The Low Veld proper is considered by KAYSER as a landform that owes its existence in the altitudes between 300 and 350m to a young removal of rocks (and perhaps the gravels) (1986, 55). (To the mind of the present author however the Low Veld, between the altitudes given by KAYSER, had never been filled up by fluvial deposits, but was excavated rather by processes in which the 300m planation itself came to be formed.)

The 400 - 600m planations, aaccording to KAYSER, can also be found in the inter-montane basins around Barberton and on the flats around Manzine in Swaziland. In this connection it may be pointed out as well that KAYSER coined the term 'unteres Randstufenniveau' for all the planations between 400 and 600m (meaning in English: 'lower escarpment level'). Here (as everywhere) however it is not clear whether KAYSER at all noticed the distinct scarps within his 'lower escarpment level' - thus forming distinct sub-levels - or whether he thought the level to be just one 'super-level' or another 'oblique plain'.

For the sake of completeness and for clarifying reasons some additional remarks have to be made.

The first is to point out that naturally there are still more levels below the ones given in the tables. These lower levels have for instance been dealt with by KRIGE (1926) and MAUD (1975), the former for the southern Cape, the latter for Natal. For the purposes of the present inquiry it is however not necessary to deal with these Pleistocene terraces.

The second remark refers to the detailed study by MARKER and SWEETING (1983) of the karst landforms in the Alexandria district between Port Elizabeth and East London. As a side-issue of their work they mention the altitudes of several coastal platforms and higher benches of more resistant rocks. The present writer is however not able to confirm the heights given by MARKER and SWEETING, especially those given in their chart. To him, all planations and benches do fall in either with the 200m level or the 300m level. The typical karst hills appearing here on the beds of the Alexandria Formation have summits, that according to this author's observations, do also fall in with the 300m level. Therefore it may well be assumed that the hills became eroded into the limestone beds after the bench had been cut across the limestones. If there might be any benches owing their existence to particular resistancies of the limestones they are not at all prominent. Similarly, the lower benches fall in with the terraces in the heights of 90 - 100m or 200m.

Summing up, the present writer feels justified that there are several distinct levels that horizontally stick to their very heights wherever they occur, levels running around the whole of southern Africa from the Limpopo basin in the northeast towards the Swakop - Kuiseb Gap in the northwest. These levels are those in roughly 600m, 500m, 400m, 300m,

200m and 100m ab. s-l. (These rounded numbers are of course chosen for the sake of convenience.)

For the purposes of this inquiry it is however not necessary to deal with all of these levels. It suffices - in the following chapters - to deal only with the upper levels, starting with the 300m level.

c) Dating the levels

Dating best starts with the 200m level. In Natal the level is characterized by the Berea Red Sands which reach up to it. According to MCCARTHY (1969) the sands were deposited in connection with a marine transgression in the Upper Pliocene/Earliest Pleistocene.

The terrace in 280 - 320m can also be linked with a transgressional period. Equivalent deposits occur in the Cape Province and in Zululand. Since the deposits gain their greatest prominence in the Alexandria district (between Port Elizabeth and East London) they have been named the Alexandria Formation. Here the formation was first studied in a more detailed way by DU TOIT (1920) and HAUGHTON (1925). According to TANKARD et al. (1983, 440) the marine sediments reach an inland margin of 305m ab. s-l.

The difficulty with the Alexandria Formation is that it contains fossils which do not point to just one period during the middle Tertiary, but to three or four of them. Eocene, Miocene and Pliocene fossils are all represented in the formation, though in different heights. Since all fossils are of marine origin this leads to the conclusion that the formation owes its existence to several transgressive periods in the Tertiary (cf. RUDDOCK, 1968).

Each successive transgression seems to have partly reworked the older beds. It is by this assumption only that the differences in opinion between geologists about the age of the deposits can best be overcome. Geologists who concerned themselves with the formation were BIESIOT (1957), FRANKEL (1960), BOURDON and MAGNIER (1969), KING (1970, 1972), LOCK (1973), RUDDOCK (1968, 1973), MAUD and ORR (1975), SIESSER and MILES (1979).

It is however not necessary to go into the details of the sometimes rather vivacious controversies. Some of the authors estimate the deposits to belong to the earlier Miocene or the Upper Oligocene (at least the main bulk of them), only TANKARD et al. (1983, 440) suggest an Upper Miocene to Lower Pliocene age.

From southern Angola, BEETZ (1933) reported fossil-bearing beds in the same altitude (300m ab. s-l) which he also dated as Lower Miocene. In this inquiry the deposits are regarded as Oligo-Miocene in age.

For the next higher level, the one in 400m, there is only one place along the whole coast of southern Africa to date it. This place is to be found at Needs Camp, near East London, in the eastern Cape. Here an indented flattish interfluve is covered by limestone beds with big shells in them. The limestone covers the interfluve between 320m and 390m, thus almost leading up to the level. The fossils were dated as Maastrichtian, though they also seem to have been reworked by an Eocene transgression (reaching up to about 340m ab. s-l). (Cf. LOCK 1973; DINGLE 1971; SIESSER and DINGLE 1982). TANKARD et al. (1983, 441) believe that the interfluve was subject to even three transgressions, one in the Upper Cretaceous, the other two in the Eocene.

The levels above the one in 400m must then be of a similar age. FRANKEL (1960) thinks the 600m level that cuts the Hlabisa hills in Zululand also to be of an Upper Cretaceous age, because some other sediments below the level, on the slopes of the Mfulosi valley, are of an Oligo-Miocene age.

Apart from the spot at Needs Camp there is no other direct proof for the Upper Cretaceous age of the 400 - 600m levels. JAEGER (1930), as far north as southern Angola, also thought the 600m level to be of this age, but by inference only.

Another more direct proof can be won from the fact that these levels cut, as planations, across the Lower Cretaceous deposits in the hinterland of the Algoa Bay. Therefore the levels must certainly be younger than these sediments. Today no Lower Cretaceous beds appear above the height of the highest of these terraces (the one in 600m) though the sediments must have reached far higher altitudes once, at least 800m or even 1000m (as will become clear in the following chapters).

WELLINGTON (1955, 112, 113) however believed that all levels between 200m and 600m are of a Miocene age, with reference to the Alexandria Formation. In two cross-sections of the Cape Fold Belt WELLINGTON (in his book Fig.15 on p.113, here presented as Fig. 5.4) connected the levels in such a way to suggest that they form just one Tertiary plain, only tilted towards the sea. Apparently, once the notion of tilting has been crept in then all other evidence is subjected so that it fits in with this super-notion. (It should also be pointed out that this so-called plain is capped with silcretes, called 'surface quartzites' by WELLINGTON. Apparently he did however not view the 'quartzites' as really being separate ones, each of them belonging to a separate level and each having differing proprieties - cf. ch. 7 b.)

The missing of Upper Cretaceous sediments all over the higher levels and all along the coast of southern Africa is certainly a very unsatisfactory fact. The fact may however be explained in the following way. Obviously, deposits have only been preserved under favourable conditions, i.e. where they were sheltered from erosion such as in bays and embayments or small pockets as near Needs Camp. In comparison, the case of the Lower Cretaceous sediments is somewhat similar. These also must once have lain all over the Cape Fold Belt. Yet today they have only been preserved within the larger and deeper sedimentary traps that are formed by the grabens and half-grabens stemming from the faulting-time. Since all faulting was over by Upper Cretaceous times no efficient traps could be formed for the following sediments. It is no wonder then that the Upper Cretaceous crops up so sparsely in the higher altitudes.

The case is however quite reversed when one looks at the situation lower down, i.e. in the flat basins of Zululand or on the very shelf. Here, because of low gradients and the basined-in structure, the sediments were not eroded away. On the higher levels however it is as DE SWARDT and BENNET (1974) stated: "...the bulk of the earlier sediments were eroded away, which accounts for the paucity of sediments on land".

Even the Lower Cretaceous sediments - in spite of their more favourable position - were eroded away, as we have seen, and, moreover, nearly completely even in some of the basins. The Worcester basin in the Cape Fold Belt is one example. The Lower Cretaceous sediments here also must certainly have reached as far upwards as they today do in the Sundays river basin or in that of Oudtshoorn. Yet today only some patches (for example the patch, where, in 300m ab. s-l, the water tank was built in the north of Worcester) remain, and at very low altitudes at that.

The eroded-away nature of all sediments older than the Tertiary is to be seen in relation with the so remarkably youthful landforms. This nature is one of the most striking features all along the coast of the Indian ocean, mentioned therefore by many authors (cf. TURNER 1967; KING, 1972; DE SWARDT and BENNET, 1974; DURY, 1981). (Cf. Fig. 5.6.)

DURY (1981, 129) expressly refers to the steady rejuvenation of all forms by an erosional process that was renewed again and again as the base level was continuously laid deeper and deeper as a consequence of the vast recession of the sea during the Tertiary. (This recession is a phenomenon we shall have to return to later on.)

Within the present inquiry the levels between 400m and 600m are taken to be of an Upper Cretaceous age. Some later chapters will furnish some more inferential proof for this dating.

d) Explaining the levels

The very characteristics of the levels described in the previous sub-chapters are a direct indication as to the conditions under which they were formed. These characteristics are now dealt with. The first is that they were formed in connection with marine transgressions. The second one (and a very striking one) is the very horizontality of the levels. Such a characteristic (as in all similar cases all over the world) can only be explained by the forming of the levels in the immediate neighbourhood of the sea. DE SWARDT and BENNET (1974, 321) even stated: "that the stages of greatest perfection of erosion surfaces in South Africa correspond broadly with maximum transgressions of the sea". Their so-called Dalton-Richmond level in Natal (lying in the altitude of 600m ab. s-l) "attained its greatest smoothness at a time of maximum transgression" (1974, 320). By this they seem to imply that the whole surface was formed under direct influence of the sea, i.e. as wave-cut platforms. Similarly, DU TOIT and KRIGE refer to the lower benches as having been 'wave-cut'.

Fig. 5.4: Cross sections from the Cape Fold Belt towards the Great Escarpment by WELLINGTON (1955)

Profile von den Kap-Ketten zur Großen Randstufe nach WELLINGTON (1955)

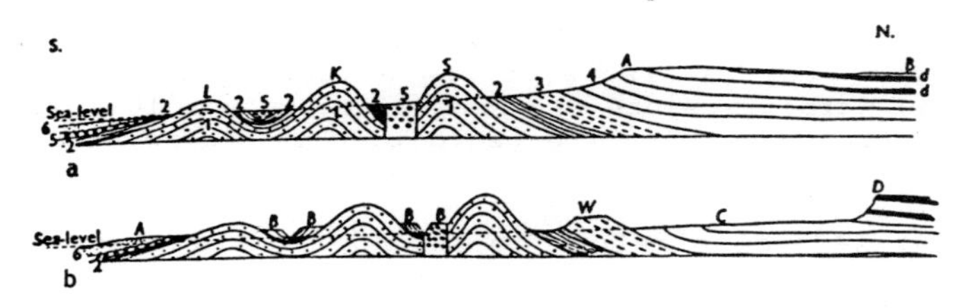

Schematic representation of the probable pre-Miocene (a) and present (b) surface of the eastern Cape Fold Belt.

(a) 1, T.M.S.; 2, Bokkeveld; 3, Witteberg; 4, Karoo beds; (d, dolerite sheets); 5, Enon; 6, Tertiary. L,Langeberge; K, Kammanassie Range; S, Swartberge; A,proto-Great Escarpment; AB, interior plateau.
(b) A, Tertiary marine beds of coastal plateau; B, Miocene erosion surface capped with surface quartzite; W, planed surface of Witteberge; C, Great Karoo basin; D,Great Escarpment. The present curve of river erosion is shown by the broken line.

Annotations: a) Enon: Lower Cretaceous
b) surface quartzite = silcretes (cf. Fig. 7.2)

Source: Wellington 1955, Fig. 15

Whether such wave-cutting can indeed be held responsible for the very broad planations and their very smoothness over rather large distances into the hinterland is however open to question. This question will therefore be dealt with later on.

The main feature to be stressed here is the obvious evolution of the planations in the neighbourhood of a then sea-level in connection with a marine transgression.
The remarkable horizontality is proof enough for it - and this horizontality is also a strong argument against any tiltings and warpings since the Upper Cretaceous. This fact is in good accordance with the geological events related in ch. 2.

BEETZ (1933, 147) also refers to the horizontality of the marine layers in Angola, both, the Tertiary and the Cretaceous ones. For BEETZ, the table-like mountains and mesas of southern Angola also stem from the horizontality of the layers.

DURY (1981, 129) even goes as far as to state that the observed fact of the steady rejuvenation is not at all a consequence of repeated up-lifts, but rather the consequence of a periodical recession of the sea (as was mentioned above), at least for the Tertiary. He does not at all believe in a series of up-lifts.

Yet, up to recent times, as the works of KING and KAYSER show, all the levels along the coast were taken to be proofs for up-lifting, each level as a proof for a standstill in this process, the height-differences being taken as measures for the amount of the respective up-lifts.

If however no tectonic events can be tied to the levels (or planations) - at least not since the Lower Cretaceous - the conclusion has to follow that the platforms indeed owe their existence in some way or other to active transgressions of the sea - with one of these transgressions having gone as far up as 600m ab. s-l. The belief in successive up-lifts as an explanation of highlying terraces is a belief strongly honoured by time - and it is indeed very difficult to oppose such a long-cherished idea. And yet, as has been shown from the compared models by KAYSER and KING, for instance, it is this one particular notion that has led to a host of the-

Fig. 5.5: Land Surfaces in Natal according to FAIR and KING (1954, Fig. 1, p. 21)

Niveaus nach FAIR und KING in Natal

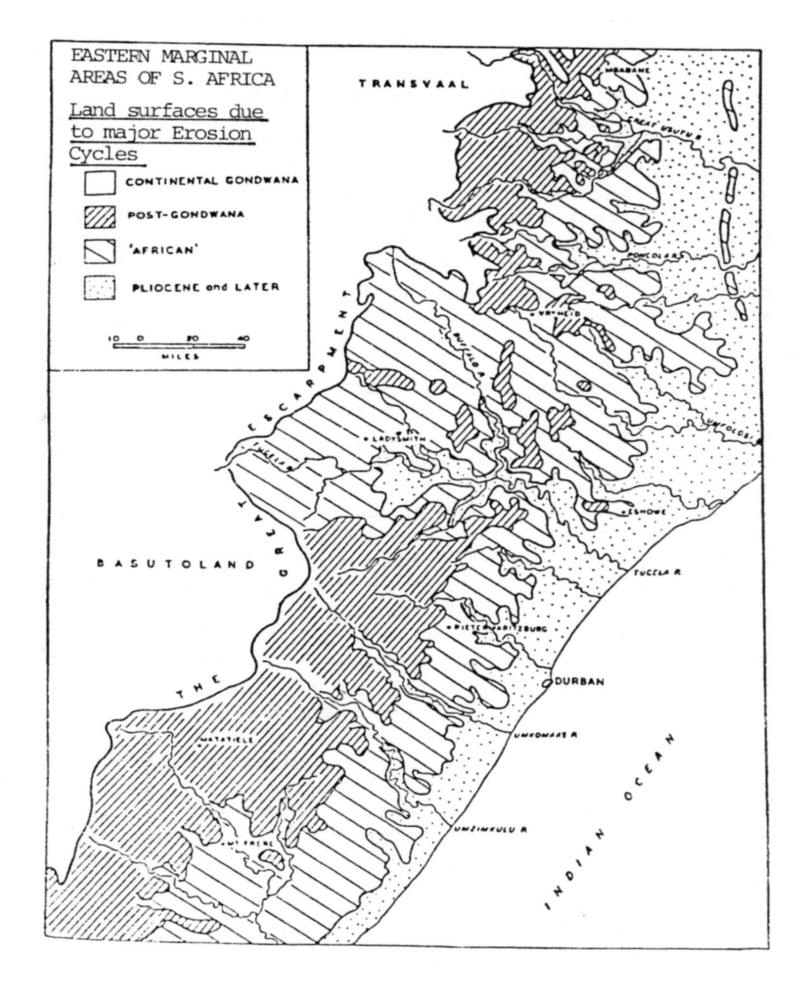

Source: Fair and King 1954, Fig. 1

ories about the evolution of the Great Escarpment (and of the landscape in southern Africa in general).

To the mind of the present author, these conflicting models must however be discarded (more or less completely) in view of the fact of active sea-level changes.

It remains to be seen whether this conclusion will enable us to reach solutions that will be better suited to solve the old riddles. They are certainly not solved by taking for granted up-lifts and tiltings and warpings, at least not since the middle of the Cretaceous. For if the working idea is correct that all these terraces were formed in the neighbourhood of a then sea level and if the other observation is also correct i.e. that all these planatory levels are completely undisturbed for thousands of kilometres then tectonics must definitely be ruled out for the past 100 my or so.

Sediments moreover "in off-shore bore-holes suggest that there has been a repeated alternation of major and minor transgressions since the late Jurassic" (sic!) and "early Tertiary and late Mesozoic sediments extended well inland from the present coastline prior to the Oligocene-Miocene transgression" as DE SWARDT and BENNET (1974, 320) expressly say. According to them two main periods of transgression emerge from the bore-holes: one in the Upper Cretaceous, another in the Oligo-Miocene (1974, 311, 317, 320). (It may be remembered that TANKARD et al. (1983) judge the transgression to be of Mio-Pliocene age.) DE SWARDT's and BENNET's datings confirm our datings. Moreover they corroborate our conviction that the two terraces - everywhere along the coast - are equivalent to these two main transgressions. These two equivalent terraces are also the most prominent ones (i.e. the one in 300m; the ones in 400 - 600m).

Both from DE SWARDT and BENNET and from our own observations it follows that the Upper Cretaceous transgressions went up highest. Its late Cretaceous age is also confirmed by DINGLE (1971) and by DINGLE et al.(1983).

The situation to be observed on the eastern flank of the Lebombo range can be used as another proof of what went on during the transgression. Here, in the neighbourhood of Jozini, for example, conglomerates (dated as Upper Cretaceous) cover the flank of the range from an altitude of 50m ab. s-l to that of 260m. The rather coarse conglomerates do not show any sign of later tilting, but move up over the pre-existing slope. The same may be said of the interfluve near Needs Camp. The sea obviously reached its highest surface in the Maastrichtian. After a regression, setting in at the end of the age, the sea began with a new advance in the middle Tertiary, confirmed by the Alexandria Formation. (From elsewhere, SCHOFIELD (1968) and ROBERTS (1973) report a Miocene transgression gaining equivalent heights of 300m.)

DURY (1981, 129) came to the conclusion that the total regression of the sea since the early Tertiary amounted to about 500m, in comparison with the present sea-level. Thus the platform in 500m can inferentially be also connected with the Upper Cretaceous transgression.

All this is evidence from southern Africa only. Yet the transgressions can be seen within a worldwide context. For exactly doing this VAIL et al. (1977, 1979) used material and data from various off-shore bore-holes and seismic experiments gathered from numerous places. On the base of these data they constructed a graph (Fig. 5.6) that shows the world wide sea-level fluctuations. Because of these world wide data they termed the sea-level changes 'eustatic', meaning 'sea-level changes of world wide occurrence' (VAIL et al., 1979, 71).

HANCOCK and KAUFFMAN (1979) directly dealt with 'the great Cretaceous transgression' (the title of their paper). They tried to compute the heights of the transgressions (of the Upper Cretaceous) by using contemporanean sediments both in Europe and in North America. They came to the result that the sea advanced against the land in three distinct phases, first to a level of about 400m, second to one in about 500m, and third one in about 600m. (Fig. 5.7). This third and highest transgression is of Maastrichtian age. Since then a strong regression occured and it must have set in very quickly. The altitudes given by HANCOCK and KAUFFMAN correspond surprisingly well with the levels of the same altitudes in Africa.

Once more the conclusion is forced on us that the terraces between 400m and 600m are indeed of an Upper Cretaceous age.

The coastal platforms or plains here dealt with so far possess different widths. The Pliocene one is the narrowest with a width of some hundred to about a thousand metres, the Oligo-Miocene one is broader with several thousand metres and the Upper Cretaceous ones are the widest (sometimes tens of kilometres, especially the one in 600m). These different widths or depths of the planations correspond very well with the lengths of time during which the transgressions lasted. For the Tertiary ones one may compare the graph from VAIL et al. (= Fig. 5.6), for the Upper Cretaceous one that by HANCOCK and KAUFFMAN (= Fig. 5.7). The whole transgressional period of the Upper Cretaceous lasted for about 20 my.

Because of the enormous heights the Cretaceous transgressions reached quite a lot of the continental surface over the whole world must have become drowned. Responding to the extension of the transgressions and the long period during which they lasted the Upper Cretaceous has aptly been called the 'age of thalassocracy ' by older geologists (i.e. BUBNOFF, 1948; WAGNER, 1960).

The Tertiary transgressions did on the whole not invade the land as far as that, but for Mocambique DU TOIT (1920) mentioned an invasion inland as far as 160km. In 1971, DINGLE, dealing with sediments on the southern coast of Africa, came to the conclusion that during the Tertiary two major transgressions must have happened, by him put into the Eocene and the Miocene.

e) Some consequences

The consequences to be arrived at on the strength of the previous paragraphs can be summed up as follows.

(1) There cannot be just one and as such decisive phase of erosional rejuvenation in the Tertiary as KING (1971) believed - and that at its very end only -, but there are successive ones, beginning with the regression from the highest rise in the Maastrichtian.
(2) Since all the planatory levels swing back into the valley-basins of the larger rivers these basins must also have been invaded by the sea. This again means that all these valley-basins (all around southern Africa) must have been in existence before the Upper Cretaceous.
(3) The sinuous form of the Great Escarpment (its 'sinuosity') must then be of a similar old age. This moreover means that the Great Escarpment must at that time have

Fig.5.6: Global sea-level changes according to VAIL et al. (1977)

Globale Meeresspiegelschwankungen nach VAIL et al. (1977)

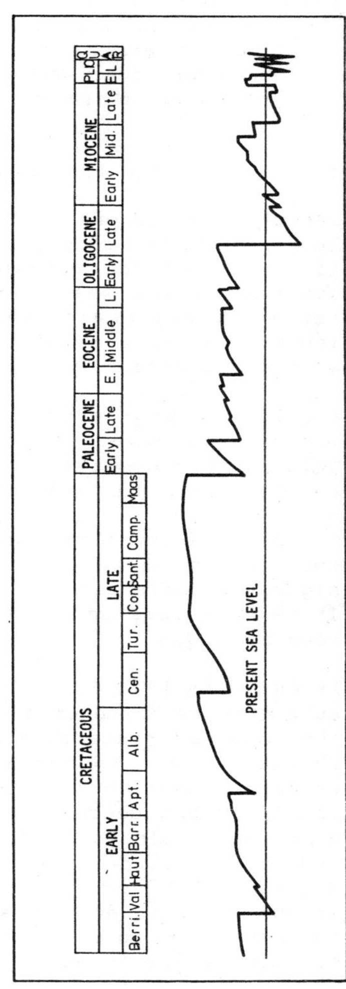

already lain much farther back into the interior than either KING (1972, 1982) or KAYSER (1986) assumed.

(4) The Great Escarpment must be of a far older age than the two authors mentioned believed.

(5) The completely level and horizontal nature of all the coastal planations over such vast distances all around southern Africa is understandable only when they are connected everywhere with contemporanean active transgressions of the sea against the land.

f) Origin of the transgressions

For the Pleistocene or 'the Great Ice Age' sea-level changes have long been accepted. They are explained by the assumption that a large amount of water was bound up in the huge ice masses and was then released again when the climate became warmer again. When the waters were bound into the ice-sheets the sea-level sank by about 100m, for instance, when the Würmian Ice Age reached its greatest extent.

Equally the sea-level would rise for about 70 - 100m above its present level if all ice around the poles would melt. This is more or less exactly the level of the 100m coastal terrace given in Tables 5.1 and 5.2.

Yet all sea-level changes exceeding these heights cannot be explained by mere glacio-eustatic alterations. Therefore higher eustatic changes are attributed to a tectonic origin. It is from this reason that the term 'tectono-eustasy' has been coined.

As early as 1973 FLEMMING and ROBERTS devised a theory for such changes based on plate tectonics. They tried to compute the changes in starting from the global hypsographic curve. They studied a systematic series of deviations from the present curve by varying the continental area and the maximum ocean depths independently of each other. They were eventually able to show that even small departures from the present distribution would produce enormous changes in the degree of flooded continental margins. An increase for instance in continental area by 10% accompanied by a 6% decrease in mean ocean depth would lead to the submergence of all continents (1973, 20).

Fig. 5.7: Sea-level rises after HANCOCK and KAUFFMAN (1979, Fig. 5, p.184) in the Upper Cretaceous

Meeresspiegelanstiege in der Oberkreide nach HANCOCK und KAUFMAN (1979)

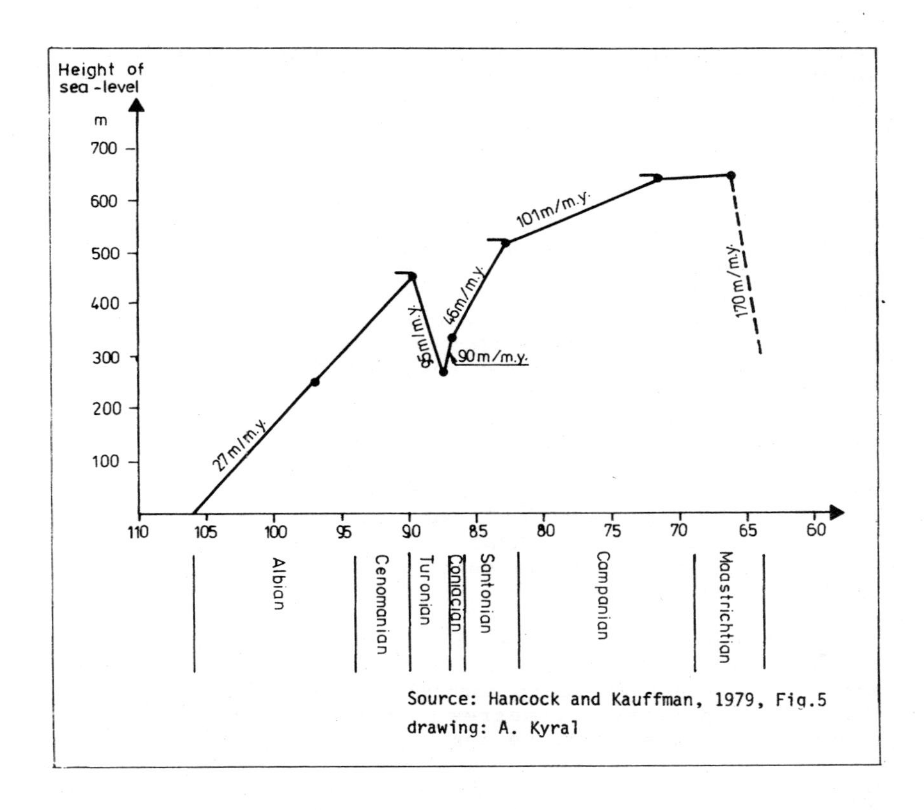

HAYS and PITMAN III (1973) demonstrated quantitatively how the world-wide Cretaceous transgression could be explained by plate tectonics. They concluded that in the Upper Cretaceous more than 40% of the continental area was covered by several hundred meters of water, up to more than 500 metres (1973, 20).

Regarding the high-lying planations around southern Africa in the light of these findings one is led to call them tecto-eustatic terraces or planations.

g) On the origin of the tecto-eustatic planations

Some of the so-called coastal 'platforms' are so wide and penetrate so deeply into the interior that they may well be termed plains, as was really done by many of the scholars mentioned in the first parts. Now there is no doubt now that all these plains - both the wider and the narrower ones - possess all the characteristics of etch-plains. Yet it must wellbe open to doubt whether they are peneplains in the Davisian sense or in the sense of LOUIS (1964) or BÜDEL (1958) - or whether they are something else. (The question will be dealt with in the next chapters.)

Another open question is whether the coastal plains were etched out by the sea itself. Some authors, for example DU TOIT (1920) or even DE SWARDT and BENNET (1974), expressly referred to them as 'wave-cut'. KRIGE (1926) called them 'marine-planed'.

Such ideas pose however serious problems as for instance DONGUS (1972) pointed out in referring to the planation in front of the marine scarp (or 'cliff-line') running along for several hundred kilometres on the southern side of the Swabian Jura in southern Germany.

What in spite of these problems (and others to be dealt with further below) cannot be denied (and should constantly been born in mind therefore) is the fact that all these planations or etch-plains must have originated in a position very close to the then sea-level. Therefore they can perfectly reasonably be called 'thalassoplains'.

With regard to such thalassoplains elsewhere GIERLOFF-EMDEN (1979, 909) stated: "The occurrence of horizontally flat rocky platforms of a width of several kilometres is still unsolved", but their existence cannot be denied. GIERLOFF-EMDEN did not indicate however into which direction one should look for a solution of the problem.

BIRD (1984, 95 ff.) tried to be more precise. According to him the rock structure is liable to be broken up by way of continuous wetting and impregnation with salt that crystallises in the joints after drying up, with the crystals gradually widening the joints. Within such a level of continued and renewed soaking a platform may be gradually formed. Beneath the soaking level the rock is not attacked whereas above it it is peeled away so to say. Since the soaking level penetrates through all structures (faults, folds etc.) this mechanism is able to explain why such platforms can cut across all structural differences. As we have seen it is exactly this feature that is so striking about the thalassoplains of southern Africa. It is indeed one of their two fundamental characteristics - (the other being their horizontality). The mechanism as outlined by BIRD is moreover able to explain why the terraces can remain horizontally flat even when they are farther inland. Yet again, as BIRD goes on to explicate (97), the horizontal flatness cannot fully be explained by this mechanism alone. Some other circumstances have to be considered. The circumstances mentioned then by BIRD do however not meet the requirements for the platforms in southern Africa, especially if they cut - as they do - across sandstones and quartzites.

For, with these rocks and with the crystalline ones one has to consider their tendency to be decomposed into grit when subject to intense weathering. It is such a decomposition that makes plausible a quick peeling-off by the transgressing sea, if the weathering (and a deep one at that) took place well ahead of the transgression.

Since the decomposition of shaly beds (i.e. the Bokkeveld or Dwyka series or the lower Ecca beds) and of crystalline rocks (including the intrusives, as the dolerites) is even easier the deeper ingressions of the plains against the interior in the form of large embayments or inter-montane basins wherever these rocks occur can then be explained in a far better way. Examples are the Dwyka and Bokkeveld basins within the Cape Fold Belt or the wide crystalline embayments on both sides of the Orange river.

In several places observations could be made that show that even today processes go on - and even under arid and semiarid terrestrial conditions - that lead to a gradual widening of the flat surface forms by forcing slopes backwards and levelling the parts in front of them, i.e. within the original near-marine level. Such cases could be observed in different altitudes and on different rocks, i.e. sandstones within the 100m-terrace, crystalline rocks and dolerites within the 600m-level. The first example occurs in the neighbourhood of the N 7 road near Gariep in the northwestern Cape, the other one in the low gradient embayment of Vioolsdrif on the Orange (already mentioned above).

In the cases mentioned streams and rivulets (which are naturally only periodical here) do not possess any well-defined channels.

Similar instances were described by GEVERS, BEETZ and SPREITZER in South West Africa and Angola for former times. BEETZ (1933) drew attention to the fact that the fossil bearing beds (Miocene) of the 300m-level are connected with horizontal deposits of extensive river systems with no well-defined channels. GEVERS (1936, 79) remarked: "The uniform and level fluviatile plateau-gravels around the Khan and Swakop rivers and their very wide distribution away from the present canyons prove that the Khan and the Swakop, and no doubt also the other rivers, originally did not flow in well-established channels, but meandered over a very wide level area, or, more probably, changed their shallow and wide channels very frequently with each 'coming down' flood, and in this way spread a thick sheet of river gravels and grit over a very wide area". It is not quite clear to what level the phenomena belong that GEVERS described but from the present writer's knowledge it is either the one in 300m or the one in 600m. SPREITZER (1963, 343) mentioned gravels from rivers that were certainly not incised in canyons at that time, belonging to the plains in 400 to 600m. GEVERS moreover observed that, above the Swakop canyon, "...definite terraces of fluviatile gravels and grits...form an almost level platform which today is 200m deep" (1936, 77). "Interbedded with these sediments occur several layers of rock salt and gypsum, pointing to lagoon conditions near sea-level" (77). GEVERS drew the conclusion (78) that there was in existence "a low, wide plain below the escarpment". The gravels and the grit on the level were also - in the neighbourhood of the sea - silicified. Silicification went together with intensive kaolinisation beneath the silicifying surface.

The observations by the present author and the reconstruction from the past by GEVERS show that broad plains were being formed near the sea-level under the conditions mentioned. These conditions may explain why a plain was formed farther away from any actual 'wave-cutting' (which, anyway, can extend only to a few hundred metres at the utmost).

In preferring the name of 'thalassoplain' for all these coastal etch-plains the author wants to show that these plains comprise a combination of certain features. These are:

(a) The planations were formed near sea-level.
(b) Their width (or depth) is not due to wave-cutting.
(c) Wave-cutting will have affected only a very small seaward margin along the course of the planations.
(d) They were certainly not formed by terrestrial processes in the way suggested by DAVIS, or BÜDEL and LOUIS, or KING and W. PENCK.
(e) They rather owe their origin to a combined effort of all the processes mentioned by BIRD and GEVERS. (For further summary: see ch. 17, 2 b)

h) Evidence elsewhere

From his field-work in the Rhenish Massif of Central Europe the author formerly described very similar plains, which certainly were formed in connection with active rises of sea-level. (For further information: see Appendix C).

i) The peneplain of Grahamstown

HAUGHTON (1925) and FRANKEL and KENT (1937) were the first to speak of the 'peneplain of Grahamstown' in the eastern Cape Province. They mentioned it in connection with the silcretes found on this peneplain. The term (i.e.'p. of G.') was taken up by HAGEDORN and BRUNOTTE (1983).

These two authors dealt with the peneplain in connection with some other coastal platforms. At first they describe a 'coastal plain' with a width of 30km, parrallel to the coast, with a dip from 450m in the north to 170m in the south. This plain seems to merge with another plain, the so-called Fish river plain that again shows a dip from 450m in the west to 300m in the east. Above these two plains, in a broad valley-like basin, in 630m ab. s-l, there lies what the two authors call the 'peneplain of Grahamstown'. At its lower end, this peneplain ends in so-called "Pedimentstümpfe" (pediment stumps), with which the upper plain breaks away to the lower one. For the two authors the upper plain is a true peneplain in the sense of DAVIS, BÜDEL or LOUIS.

To the present author's mind the peneplain is however nothing but a part of the highest Upper Cretaceous thalassoplain that here forms a larger embayment within the Cape Fold Belt because of the removal of the less resistant shales. Moreover, the two authors do not seem to have noticed the obvious steps or scarps that do occur within their plains - a very surprising fact because they do mention the very heights of the terraces in the same altitudes as they were observed by the present writer, i.e. in 170m, 300m, 400 to 450m and also refer to the very marked scarp in 550 m to 600m (where their 'pediment stumps' occur). Yet the scarps between the other levels can also be clearly observed (Fig. 5.2) - and with no dipping of the levels either.

The two authors believe to have evidence that the 'p. of G.' is a true peneplain in the sense of BÜDEL or LOUIS because, under the silcretes of the peneplain a kaolinite dominance makes itself felt, suggesting a savanna climate with sheet floods and deep weathering as steering elements for the formation of such a peneplain. Yet such a kaolinite dominance is not proved from the samples the present author took from under the silcretes of the 'p. of G.'. And if kaolinisation does appear below the duricrusts - as in some samples it does - then such a kaolinite content may equally well be explained from the observation of GEVERS (1936) mentioned above that kaolinisation is a process accompanying silicification under arid or semi-arid conditions. Therefore, a kaolinite content under silcretes cannot be taken as proof for a savanna climate. (Duricrusts and the conditions under which they were formed in southern Africa will be dealt with systematically in ch. 7).

The datings of the plains as given by the two authors rest on KING (1972). KING's datings did however change during the decades, as we have seen above, and they were moreover doubted again and again (FRANKEL 1960, 1968, 1972; LOCK, 1973; MAUD and ORR, 1975). Apart from that, KING's datings do not at all seem to be suited for comparisons at such a distance from Natal without the help of connecting levels. Of such, the two authors do not make any use.

The present writer's observations are (apart from establishing connecting levels and verifying them) corroborated by the two authors in as far as they point out that the coastal plain is indeed very level where it cuts across the Witteberg quartzites, in spite of their foldedness. The same is the case, by the way, with the TMS quartzites. On them also a typically very wide extension of a very level surface can be observed, be it on the folds near Uitenhage or near Port Elizabeth or next to the Gamtoes river.

In the east of their Fish river plain, on the sides of the tributary valleys, the two authors see the flats there 'merge' into each other (which certainly is the case) but go on to say that by this 'merging' process the coastal plain was here 'switched down' towards a lower level. Yet they do not make it clear what 'merging' really means and in which way it was brought about. Nor it is clear which particular area is referred to by them. If they refer however to the wider area around Committee Drift on the Great Fish river then again this area is nothing but the huge ingressional embayment in the altitude of 170m - 210m.

According to the two authors the Fish river plain shows no traces of kaolinisation . The present writer can fully agree with this. Yet if this is the case then this is another proof (obviously not seen by the two authors) that no savanna climate can have prevailed in the time when this peneplain came to be formed. A further proof of quite another climate are the pediments the authors talk about. Though the present author is rather certain of the fact that in this whole region there are no true pediments at all (but only the slopes and scarps between the different levels) he should like to point out that 'pediments' if and where they occur exclude a savanna type climate. Why the author feels so sure about this argument will be dealt with in ch. 10.

6. HIGHER COASTAL PLANATIONS
(800m to 1000m ab. s-l)

a) Describing the levels

Inland from the coastal planations dealt with so far some higher plains can easily be observed. They lie at distinct altitudes, the one in 800m, the other in 1000m. In comparison with the lower terraces the following characteristics can be made out about them:

(1) They extend farther inland, sometimes very far indeed.
(2) They often form the forelands of the Great Escarpment immediately in front of it. (Cf. Fig. 13.1.)
(3) On several occasions they run around the escarpment and extend in the interior behind the escarpment.
(4) They are far wider than the coastal plains below - especially the one in 1000m. They stretch over widths of at least tens of kilometres.

Since, as has been said, these higher planations often form the immediate forelands of the Great Escarpment the may be regarded as to be in a close temporal connection with the evolution of the escarpment. Therefore their ages and their origins will be of particular interest and will probably throw some light on the origin of the escarpment itself. At least they will come in useful for the fixing of the age of the escarpment, i.e. where they (together with the embayments extending from them) join the foot of the escarpment.

In describing these planations in somewhat more detail it seems best to start with the larger inter-montane basins within the Cape Fold Belt to the west of Port Elizabeth. Entering these basins from lower altitudes near the coast, the different surfaces within the basins step up from one clear level to the next, each level similarly and clearly separated from the one below or up by a distinct scarp. Going forward in this way one eventually leaves the levels of the coastal plains and finds oneself presently within higher levels which are also separated from each other and the coastal plains below by distinct scarps.

The first of these levels one comes to is that in 800m. This level soon becomes very prominent in the inter-montane basins of the eastern Cape Fold Belt. Within the basins, it usually forms the highest planatory 'storey' out of which

Fig. 6.1: The 800m and 1000m levels in the southern Cape Province

Die 800 und 1000m - Niveaus in der südlichen Kapprovinz

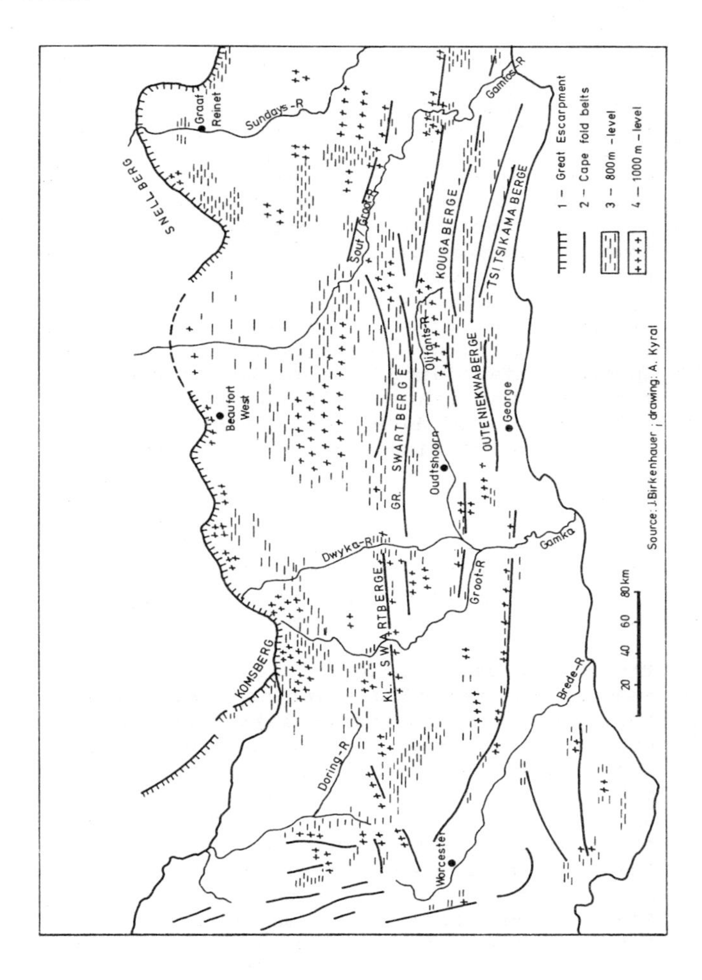

the slopes of the Cape ranges suddenly and steeply rise up. The steepness of the slopes is enhanced by the rocks the ranges usually consist of: TMS quartzites in the southern ranges, Witteberg quartzites in the northern ones.

Within these ranges (formed of very resistant rocks) one finds valleys that breach the ranges, forming gorges and poorts through them. The rivers are usually very deeply incised in narrow beds, whereas, on top of the valley flanks, one finds the 800m planation following the valleys and cutting over the very hard and intensely folded rocks. (The same is true, by the way, for the lower planations, i.e. if they reach as far up into the interior.) Within the poorts and gorges the 800m level does not at all look like a plain, but more like a terrace. (The same is true for the lower planations.) In spite of that the level cannot be regarded as a river terrace proper, since any terrace formed by fluvial processes should possess a gradient, however slight, down along the thalweg. These levels however do not show any gradient. These landforms must therefore be looked upon as being quite unique. It is moreover not possible to compare them with peneplains either in the sense of DAVIS or in the sense of BÜDEL and LOUIS. For, according to these authors, peneplains proper cannot be terraces accompanying valleys.

Yet these terraces can be seen everywhere along the larger rivers such as the eastern Groot, the Tanka, the middle Groot, the Gamka, the western Groot, or the Hex, the Tankwa or the Doring rivers. (The examples of these rivers are taken from all over the Cape Fold Belt.)

The 800m level reaches right back into the interior along and through the gorges and poorts, and, once it is through them, it begins to spread like a true plain, sometimes for tens of kilometres, until it reaches the foot of the escarpment itself (Fig. 6.1). Between the northernmost Cape range on its south and the escarpment on its north the level spreads in the form of vast inter-montane basins, all of them formed along the catchment areas of the present rivers. These basins can well be observed around Somerset East (to begin in the east), Pearston (Coetzeberge), Graaf-Reinet, Aberdeen, Beaufort West, Laingsburg, Touws, and, after the turning round of the escarpment to the northwest, in the basins of the upper Tankwa and Doring and in those of the tributaries of these or else in the river basins near Springbok and Steinkopf in Namakwaland.

In the southern Cape ranges, where they strike from east to west, some more higher planations can be observed, all of them cutting across the quartzites. These levels are: one in an altitude of 1000m, another one (with a gradient) between 1200 and 1500m, and an uppermost one in altitudes between 1800 and 2000m. The two higher ones only occur in traces. The one in 1000m is more frequent, but rather scarce in comparison with the one in 800m. On leaving the Cape Fold Belt the 1000m level becomes wider and wider the farther north or northeast or northwest one travels until it becomes the really dominating level of the interior plateau, over tens of kilometres or even over hundreds of them. In these areas it is this level with reaches back to the very foot of the escarpment itself.

In great similarity with the 800m level the one in 1000m also forms large embayments and inter-montane basins along the catchment areas of the larger rivers (Fig. 6.1). These embayments and inter-montane basins can be observed from as far east as the Great Fish river (i.e. the area around Craddock), along the Nuweveldberge towards Beaufort West, in the neighbourhood of the upper reaches of the Gamka drainage system, and again in the area of Calvinia, after the turning of the escarpment to the north.

The Calvinia area is also a good example for the fact that here the 1000m level runs right around the escarpment over to its backside, forming another escarpment (or cuesta) on the back of the Hantamsberge. (Such a cuesta on the back side is called 'Achterstufe' in German.) From this 'Achter-stufe' of the Hantamsberge the level spreads over hundreds of kilometres along the Sak valley right on to the Orange river near Upington. Again it is very prominent all over Bushmanland between Upington in the east and Springbok in the west.

North of the Orange, the 1000m level (as well as the one in 800m) are very prominent in the upper reaches of the Fish and Konkip river basins, and again in the basins and upper reaches of the rivers Tsondab, Kuiseb, Swakop and Khan. Along the latter the 1000m level reaches right up to the steeply rising flanks of the Erongo Mountains.

Fig. 6.2 a - d. Profiles along some interfluves from the Drakensberge to the Indian ocean in the Transkei and in Natal

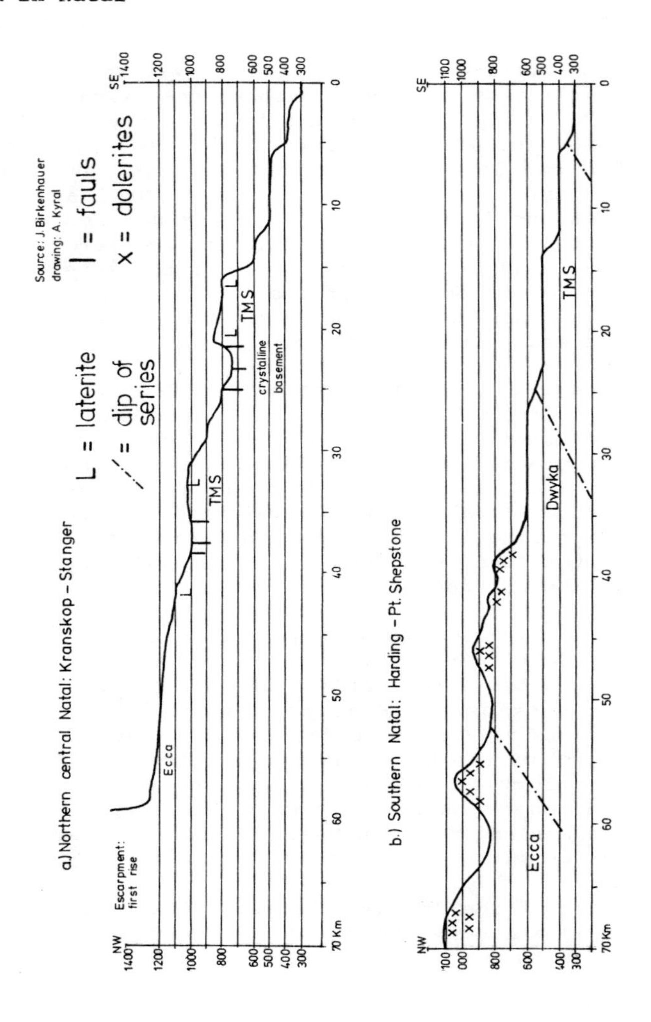

Profile zwischen den Drakensbergen und dem Indischen Ozean in Natal und in der Transkei

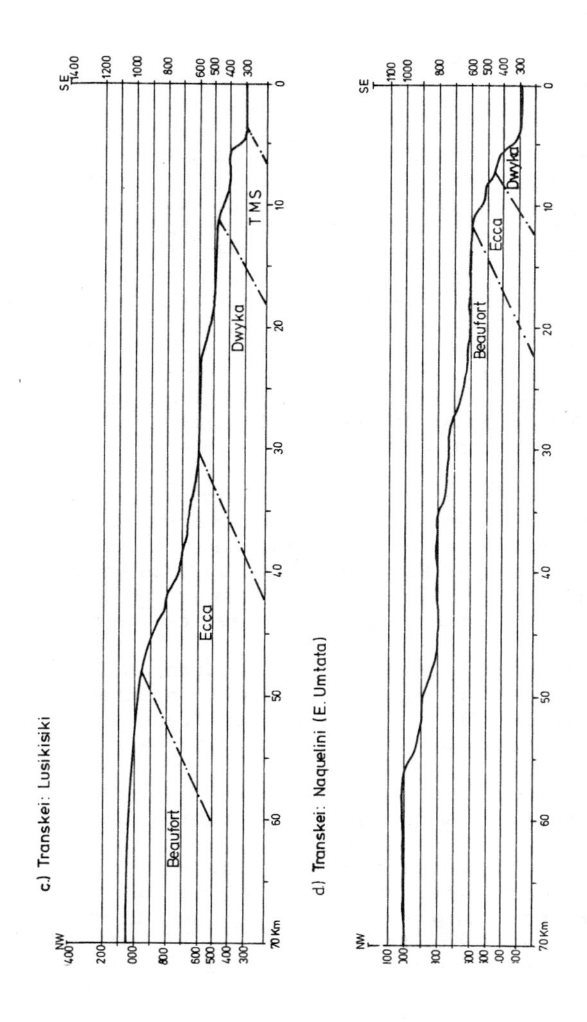

In spite of the 'Schiefe Ebene' ('oblique plain') the rivers (i.e. Khan, Swakop, Kuiseb) are entrenched into each higher level in the same way as has been described about the lower levels in ch. 5 b.

The description of the higher levels was begun in the Cape Province and extended into South West Africa because they can be observed there best. The eastern Cape, the Transkei and Natal are not so well suited for the recognition of the higher levels. The reason is the rugged nature of the landscape there. The rugged nature is due to two facts already known: (1) the prominence of the dolerite dykes and (2) the extreme youthfulness. With regard to the first fact KING and FAIR (1954, 23) noticed that "Dolerite capped koppies and hills reach to all altitudes between 4500 and nearly 6000 feet above sea-level" (i.e. 1400 and 1800 metres). For this reason they spoke of the "futility of attempting to correlate summit levels in these parts". Moreover, the different cycles "interdigitate with mountainous spurs projecting eastward from the Great Escarpment" making the task even more difficult. Everybody acquainted with this landscape well enough will heartily agree with this description.

Yet it is nevertheless possible to find threads for guiding one on in even such a landscape, i.e. the long interfluves projecting from the escarpment ('interdigitating') towards the lower coastal plains and forming the divides between the major river systems may be used as such threads. Examples of such interfluves are the ones from

- Kranskoop to Stanger in central Natal,
- Harding to Port Shepstone in southern Natal,
- Lusikisiki in the northern Transkei, and
- the one east of Umtata in the central Transkei.

Profiles drawn along these interfluves (Fig. 6.2 a - d) clearly show the same characteristics:

(1) longish stretches of flattened-out parts, separated from each other by steeper slopes or even scarps in between,
(2) the complete independance of these flat levels from rocks and fault-lines,
(3) the occurrences of fossil laterites on them which prove that ancient landforms have been preserved along the flattish parts - and the laterites with them.

(4) All flattish sections (as will be easily noticed from the profiles), without any exception, occur in the same altitudes known to us already, both of the coastal plains and of the ones in 800m and 1000m. Thus these two higher levels are landforms of their own also outside the areas where they are easier to be detected.

A profile for northern Natal (Fig. 12. 1 a), by the way, shows the same configuration.

b) Findings by other geomorphologists

In mentioning the Bushmanland plateau and in referring to examples from South West Africa we have entered areas where other geomorphologists have done valuable field-work.

The first to be turned to is MABBUTT (1955) who began his studies in Namakwaland and Bushmanland in the northwestern Cape and then extended them into large areas of South West Africa. MABBUT, from these studies, described levels in 500m, 600m, 800m and 1000m (apart from others still higher up). Since the one in 1000m spreads over wide parts of the Bushmanland plateau MABBUTT called it the Bushmanland surface. It is here developed over the older granites and the Karoo rocks. North of the Orange it bevels the older granites and the Dwyka series in the Warmbad district. Eastwards it is well preserved on the lower Nama beds of the Blydeverwacht Plateau. The 600 and 800m levels are here treated as sub-stages of MABBUT's Bushmanland surface.

JAEGER (1932), for the drainage systems of the Fish and Konkip rivers, pointed out the extraordinary flatness and width of the basins, each of them as wide as the Rhine graben; he doubted that they had been solely formed by lateral erosion.

OLLIER (1978) concerned himself with the history of some inselbergs in the Namib desert and, in trying to fix their age, referred to two erosion surfaces, the one being the Namib Unconformity Surface (as called by OLLIER), the other being the Tsondab Planation Surface. Both surfaces appear in the upper reaches of the Kuiseb and Tsondab valleys. In a cross section (on page 167) the Tsondab Surface is depicted as being above the Namib Surface - though it is younger than the Namib surface and should therefore (as a rule) appear rather below it. Deplorably, in this section heights are not given, but from the position of the Basement Conglomerate in this section and its comparison (in the text) to the Kuiseb Conglomerates in the height at 1000m (resp. a little below) one is able to decide on the height of the Namib Surface as roughly 1000m. This is the same height as the one in which the present writer came across the conglomerates.

For the Kaokoveld in northern South West Africa ABEL (1954) made it clear that the surfaces there appear in similar altitudes over wide tracts wherever one meets them. They can ostensibly be identified by these altitudes - the altitudes of the surfaces being given by him as 200, 400, 600, 800 and 1000 metres. The lower ones are already known to us as the coastal terraces. The surface in 1000m is by ABEL called the 'pre-front level', because it forms the foot-zone next to the escarpment. The level cuts away also across the tafelbergs and mesas nearer to the coast. Sometimes it also occurs in the form of a very flat inter-montane basin, for instance, between the Etorocha and Omuhonga mountains. The 800m level is best developed in another inter-montane basin, that of Orupembe.

(Above the 1000m surface another one is mentioned by ABEL as well as one between the 800 and 1000m levels. The latter, in 900m, was called the Etendeka surface by ABEL. We will deal with these intermediary levels together with some higher levels in a more systematic way in ch. 8 - and, indeed, there are some more of these levels than mentioned by ABEL.)

In 1959, ABEL published his findings from his later visits to central and southern South West Africa. Excellent fotos demonstrate the very flatness of the levels mentioned above. They also reveal that the levels are clearly separated from each other by distinct scarps between them. By this, the present author's observations - as stated above in part a) - especially for the Cape Fold Belt are confirmed for another region. In the south (i.e. Namaland, Orange catchment area) ABEL distinguished two main levels in 1000m and 800m (with sub-levels in 700m and 900m). The one in 1000m was called by him the Keetmanshoop level, the one in 800m the Seeheim level. These levels were here regarded by him as the upper and the lower high terraces ('obere und untere Hauptterrasse') of the Fish river - thus inferring that these terraces owe their existence to fluvial processes. This however cannot be the case as can be seen from the very height constancies of these levels. (Another level, higher than the one in 1000m, was mentioned by ABEL also; it will be dealt with in a more systematic way in another context.)

BEETZ (1950) - in trying to solve the riddles the course of the Kumene river poses - mentioned that the river flows in an altitude of 1000m near Humbe, still well to the east of the escarpment. Coming out of this (characteristically) very flat area the river goes on to break through the escarpment in an impassable canyon.

MARTIN (1968), in proving the existence of wide paleo-valleys covered by the Dwyka tillites, represented some profiles (p. 125) for the areas of the Huab and Hoarusib valleys in the Kaokoveld of northwestern South West Africa. In these cross sections the rather broad tracts of land in the altitudes of 1000m and 800m are again very prominent.

In the eastern Transvaal KAYSER (1986, 53 and 55) observed planations in exactly the same altitudes (i.e. 800m, 1000m) which by him were termed 'oberes Randstufenniveau' ('high escarpment level'). The inter-montane basins of Badplaas and Amsterdam in the Barberton highlands are part of this level (or rather: levels!) as well as similar basins in Swaziland (for instance south of Mbabane). KAYSER also pointed out the gigantic amphitheatre-like valley closes in the same altitudes, as for instance that of the Sabie river. The same phenomenon was observed by the present writer when he first came to know this area. He thinks it possible to say that these valley closes (as indeed all similar ones in all other river catchment areas around southern Africa and at whatever heights they appear) are sure signs of how far diverse transgressional events reached up along the pre-existing valley-floors. (For a systematic treatment: see ch. 12.)

Summing up all these findings of other authors one feels justified to say that the 800 m and 1000 m levels are phenomena that are well preserved around the whole of southern Africa. It must however be pointed out here that the authors mentioned above (i.e. MABBUT, JAEGER, OLLIER, ABEL, KAYSER) believe that all these surfaces gently slope down into lower depressions or up into higher swells. According to the present author's findings this is nowhere the case however. MABBUT moreover used the very heights themselves in order to identify the surfaces over larger tracts of land. This becomes clear when he spoke, for instance, of the '2600 feet plane' or the '2000 feet sub-cycle' and so on.

c) The 'rule of series relation' and its consequences

In order to give some more inferential proof both for the very flatness and horizontality of all levels dealt with so far as well as for their occurrence at the same altitudes over such vast distances it seems advisable to draw attention to an Austrian geologist, VON KLEBELSBERG. VON KLEBELSBERG (who dedicated himself to the exploration of the Alps) came to observe certain high-lying levels (or 'storeys') which in different areas appeared however in different altitudes. This posed the problem of connecting levels of presumably the same age at differing heights. First, VON KLEBELSBERG noticed that a particular 'storey' showed the same morphological features in whatever altitude it appeared. Second, and for our problem more important, the heights between the 'storeys' remained the same. 'Storey A', for instance, was always 300 m higher than 'storey B', and that again was always 500 m above 'storey C' - in complete indifference from what the present altitude of 'storey C' would be: in one area of the Alps it might occur at 1000, in another part in 1500m and the respective heights of 'storey A' would then be 1800m or 2300m. VON KLEBELSBERG therefore concluded that the 'storeys' could be identified from their relational heights. He therefore considered the three 'storeys' as a single altitudinal series and therefore he consequently began to speak of 'series relation', taking it as a rule for identifying 'storeys' in different parts of the Alps.

In southern Africa it naturally seems to be far easier to take up this idea of 'series relation' because of the missing tectonic influences. This is to say that wherever one finds this series of Tertiary and Cretaceous levels (or 'storeys') within the same relational distances of one 'storey' from another and a particular one 'covered' as it were by all the necessary higher ones one can be sure that one deals with the selfsame level - however distant they occur from one another, whether in the southern Cape or in Natal or in eastern Transvaal or in South West Africa.

The present writer did indeed take a close view at all the 'storeys' in all the larger river basins around southern Africa - there was never an exception from the rule.

This finding can naturally only mean that southern Africa must have been a tectonically stable block since the oldest and uppermost storey came to be formed.

A last aspect may be mentioned in this context. KAYSER (1986) noticed that the foot-region of the Great Escarpment usually occurs in the altitudes of the highest levels, especially that of 800m and 1000m. Yet he was not able to explain this phenomenon, because such identical altitudes are totally inexplicable if one clings , as KAYSER did, to the notion of tectonic movements. The fact of similar foot-regions of the Great Escarpment in corresponding altitudes can only be explained if one abstains from this notion of different tectonic up-lifts, warpings and so on.

d) Dating the high-lying planations and their origin

Usually geomorphologists and geologists explained planations as surfaces formed in connection with erosional cycles, either in the sense of DAVIS or in that of W. PENCK. Moreover, such cycles were then connected with surfaces (or cycles) elsewhere. MABBUT, for example, connected his surfaces in the northwestern Cape Province and in the south of South West Africa with those of KING in Natal, in this way connecting his Bushmanland surface with KING's African surface. Doing this is however neither possible nor admissible because the height of the surfaces compared are very different from one another. According to KING and FAIR (1954) most parts of the African surface lie at an altitude of 600m, wheras the Bushmanland surface occurs at 1000m. From the rule of series relation it follows that it must needs be older than the African surface.

The 600m level must be regarded as Upper Cretaceous (see ch. 5). By inference, the Bushmanland level must then be older. MABBUT (1955) however ascribed an early to middle Tertiary age to the Bushmanland level - an age that, by the way, does not comply with the age given by KING to his African surface. MABBUT inferred the age of the Bushmanland level from the higher level above, that of the Namaqualand. To this MABBUT ascribed an Upper Cretaceous age because bones of a contemporary dinosaur had been found in a depression of an outlier of the Namaqualand surface. To use these bones for dating purposes is however objectionable, because they only prove that the surface existed when the bones were fossilised.

ABEL (1959) assumed that his Keetmanshoop level (which is the equivalent of MABBUT's Bushmanland level according to the respective altitudes) is of Upper Cretaceous age, because it lies in front of the Great Escarpment which according to ABEL came into existence in Middle to Upper Cretaceous times. All these datings are however by inference only, direct proofs missing. The same holds for ABEL's dating of the Seeheim terrace (800m) as younger Tertiary. This very young age is in complete contrast to the fact that the level lies well above the uppermost Upper Cretaceous level in 600m.

SPREITZER (1963) started his datings from the calcified Kuiseb conglomerates, which lie at an altitude of 900 - 1000m. SPREITZER (1963, 342 - 343) took them for young Tertiary deposits, again without direct proof. The nature of the conglomerates is comparable to the gravels and sands mentioned by BEETZ and GEVERS in connection with the levels in 300m, 400m - 600m. Then they cannot be younger Tertiary. According to OLLIER (1978) the conglomerates lie on the Namib Unconformity Surface. This surface is situated at an altitude of 1000m (at least in the region where the conglomerates appear) and the conglomerates themselves are covered by the Tsondab Sandstone Formation. This formation is generally believed to be early Tertiary (TRUSWELL 1977, WARD 1984). The conglomerates must then be older, at least Upper Cretaceous, perhaps even older. OLLIER (1978, 173) thought the Namib Unconformity Surface to be of a pre-Upper Cretaceous age. Though this is a guess only, it is in good accordance with our view of the age of this planation surface. Since so far no fossils have been found in the conglomerates so that it is impossible to give their exact age one can safely assume however that they are far older than what KORN and MARTIN (1937) believed, i.e. a connection of these gravels with pluvial pleistocene periods (p. 459). Such an age may safely be given to many other conglomerates that are not covered by the Tsondab Sandstones or which were deposited in the Kuiseb canyon after the Tsondab Formation (WARD 1982).

WARD (1984) proposed another name for the Kuiseb conglomerates, i.e. Karpfenkliff conglomerates, after a certain locality within the Kuiseb area. According to him the bulk of these conglomerates have to be fixed to the Middle Miocene. Such an age is however not in accordance with OLLIER (1978), on whose work WARD expressly based his argument (1984, 456). According to WARD himself the Karpfenkliff conglomerate lies on top of the Tsondab Formation which, with its arenaceous quartz sands, is the oldest deposit on the unconformity surface. The sandstones must therefore be taken as a witness for an early proto-Kuiseb river in the forelands of the escarpment and in an altitude of roughly 1000m ab. s-l. This proto-Kuiseb valley was - according to WARD - very shallow and broad (up to 20km), apparently without any defined channel. This finding is by the way similar to what BEETZ and GEVERS had found out before about the proto-rivers connected with the 300m and the 400 - 600m levels. The observations of all these authors moreover suggest that the coast must have been rather near - or, to put it in an other way, that the valley bottom was not much above the coastal level of the time.

Since the 800 and 1000m levels lie above the one in 600m, to be dated as Upper Cretaceous, our view is that the two higher levels must certainly be older than these and that the Tsondab Sandstone and the Karpfenkliff (or Kuiseb) conglomerates are as old as the two levels in question. They are at least Cretaceous. DINGLE et al. (1983, 227) mentioned Cretaceous deposits from Angola (at Kangnas) at a height of 1000m ab. s-l. (By the way, the Tsondab Sandstone Formation so far is without any datable fossils.)

WARD (1982) used for his datings the assumptions by KORN and MARTIN (1937) and KING (1972) which however are based on assumed Late Tertiary up-lifts. This age of the up-lifts is nothing but a deduction from the 'young' incision of the valleys. WARD (1984, 457), starting from this, went on to say: "Therefore the deposits predating the incision are considered to be of a Tertiary age."

The flat and level nature of the 800m and 1000m levels together with the fact that they also cut across different rocks, faults and folds show that they are not subject to any structural control. This - together with their very horizontality - implies that these surfaces must have an origin similar to that of the lower coastal terraces (or planations). This means that they also are thalassoplains in the sense of ch. 5 g. As was shown above the Karpfenkliff conglomerates were deposited in the neighbourhood of a shallow coast.

This coast belonged to a proto-Atlantic that had begun to invade the depression simultaneously being formed ahead of the rifting between South America and southern Africa. The Lower Cretaceous Pomona deposits on the coast of the Namib desert can be taken as a proof for the proto-Atlantic. Thus the main requirement for the forming of a thalassoplain in this region was in existence in the Lower Cretaceous.

Though the surfaces in question do cut across the folds of the hard quartzites they nevertheless are best developed either on sediments with a horizontal bedding or on not resistant folded rocks as for instance the Bokkeveld and Dwyka shales or tillites. Here the surfaces form deep embayments and even inter-montane basins. The very flatness of the levels in these basins have led some geomorphologists to the assumption that these flat levels originated within the water table of large lakes. This assumption was arrived at by KUNTZ as early as 1912. Such an idea shows that connecting the origins of these surfaces with sea levels is not so absurd after all.

It is moreover remarkable that - in the cross sections given by MARTIN (1968, 125) - the flattenings exactly occur
(1) in the altitudes of 800m and 1000m (and, characteristically, in the levels of 400 to 600m as well) and
(2) especially across the non-resistant Dwyka tillites.

With regard to sea level fluctuations in the Lower Cretaceous FLEMMING and ROBERTS (1973) as well as HAYS and PITMAN III (1973) showed that indeed several fluctuations happened during that age.

The question then is not so much that of the existence of such fluctuations but rather that of how far up the fluctuations reached and whether the then existing continents became flooded by a sea rising as high as 800m or even 1000m above the present sea level, i.e. well above even the known great transgressions of the Upper Cretaceous. There is however no evidence to support such an assumption, neither on land nor on the shelves, the fact being that rifting had just begun and plate tectonics could not yet play such a dominant role for the alterations within the ocean basins as must have been the case during the Upper Cretaceous. We are therefore forced to conclude that the 800m and 1000m thalassoplains became formed in the context of only minor sea level fluctuations. If we today find them however in the present altitudes they can have reached this position only by a twofold up-lift of about 200m each: one up-lift after the forming of the plain in 1000m, another after the forming of the one in 800m.

From the example of the 600m thalassoplain it has become clear that the corresponding eustatic sea level rise must have been a rather sustained one, i.e. lasting over several million years. Since the widths of the levels in 800m and in 1000m are even far more pronounced than the width of the 600m level it must be assumed therefore that the corresponding sea level rise must have been an even more sustained one as in the case of the 600m level.

The datings of the two levels as of a Lower Cretaceous age can be underpinned by findings to be dealt with in a later chapter on the peneplain in 1200m ab. s-l (and higher).

Since they cut across the Valanginian faults they must be younger than these. WINDLEY (1984) indeed referred to an Albian eustatic sea level rise (ca. 100 my. BP).

The discussion about the 800m and 1000m planations can be summed up as follows:

(1) They possess a completely level and horizontal nature.
(2) They have very large extensions.
(3) They developed in the neighbourhood of the sea level in connection with transgressional periods.
(4) They must be assumed to be of a later Lower Cretaceous age.
(5) Two up-lifts must have occurred, each after the forming of each level and each in the order of 200m.

e) Consequences for the evolution of the Great Escarpment

As has been stated above, the Lower Cretaceous planations reach the very foot of the Great Escarpment rather often (see Fig. 9.2). The consequences of this observation for the evolution of the escarpment are the following ones:

(1) The foot of the escarpment must have been formed within an area influenced by these transgressions and thalassoplains.
(2) The age of the Great Escarpment must then needs be as old as these transgressions, i.e. Lower Cretaceous.
(3) The present-day position of the escarpment must have been reached as early as this age. Scarp retreat on as large a scale as KING wanted it to be must be completely ruled out, at least within the last 100 my.
(4) It is not so much the escarpment that actively receded but the coastal forelands that were passively and gradually enlarged in front of the escarpment.
(5) Southern Africa must have been a high-lying block since the Middle Cretaceous.
(6) River incision must at least be as old - it is not at all only Tertiary (WARD) or even Late Tertiary (KING).
(7) Since the planations (partly as 'terraces') follow the drainage systems right back to the escarpment itself these drainage systems must also be as old as the Lower Cretaceous. They are probably even older than that.

f) The problem of inter-montane basins

Inter-montane basins from different areas all over the world have roused the interest of geomorphologists. We have had the opportunity to refer to such basins in connection with the thalassoplains. Some general inferences can be drawn from the occurrence of such basins in southern Africa. Since, however, such discussion is of no further avail to the main theme itself (i.e. the evolution of the Great Escarpment) such a discussion is not dealt with here, but in Appendix D.

g) Comparison with Hövermann's findings

For the coastal parts of South West Africa HÖVERMANN (1978) is of the opinion that there is only one peneplain, rising with gentle ascent from the coast towards the interior, as shown in his Fig. 2. Nevertheless, in his Fig. 1 (represented here as Fig. 6.3) he comes very close to our view of a stepped nature and, indeed, in similar altitudinal positions. The flattened nature of his Fig. 2 is, in our view, due to the fact that the overlying sands and fine gravel blot out most of the relevant geomorphological features.

Fig. 6.3: Profile through the Swakop-Kuiseb-Gap from the coast near Swakopmund towards the interior according to HÖVERMANN (1978)

Profil durch die Randstufenlücke von der Küste bei Swakopmund ins Innere nach HÖVERMANN (1978)

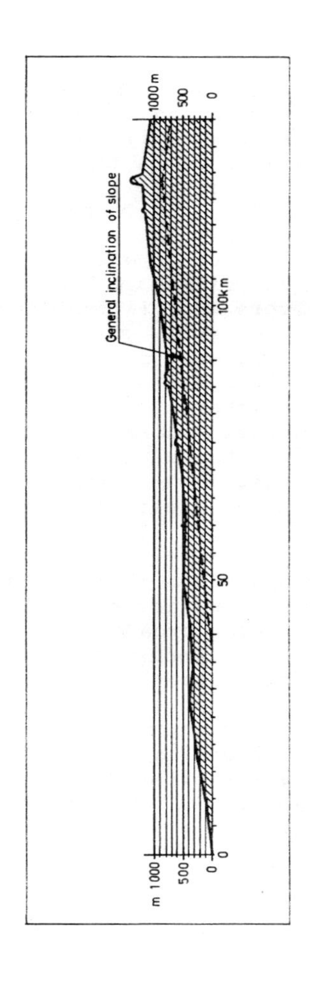

7. DURICRUSTS AND THEIR SIGNIFICANCE FOR THE EVOLUTION OF THE ESCARPMENT FORELANDS

Three kinds of duricrusts are to be observed in the region covered by the present inquiry: i.e. laterites, silcretes and calcretes.

a) Laterites

Laterites occur in Natal and in the northern Transkei. Their altitudes vary between 400m and roughly 1000m ab. s-l. (See for instance the cross sections in MAUD 1968, here presented as Fig. 7.1). There do not seem to occur any fossil laterites below the altitude of 400m ab. s-l, though traces of latisols do show up (cf. MAUD 1968; HELMGREN and BUTZER, 1977 - the latter for the southern Cape).

The author was able to visit many of the places with fossil laterites in 1983, kindly guided there by Dr. MAUD who certainly is the person who knows the Natal laterites best. The crusts are rather thick (2 - 4 m) and have partly been weathered again from above turning the upper layers into thick red soils. Originally the crusts must have been far thicker than today, since some of the red soils have a considerable thickness of their own (1 - 2m). Because of the thickness of this top the weathering process forming it must have lasted for a rather long time. This in itself makes clear that the fossil laterites must be of a rather old age.

All these places with crusts of fossil red soils are however only relics of a cover that originally must have extended over a broader surface in Natal and the Transkei. Where they occur today they have been preserved only on interfluves that have so far not been reached by erosion.

The occurrence of the fossil laterites seems to depend (according to the author's observations in Natal and the Transkei) on two factors: the one is constituted by the type of rock, the other by the planatory level on which they rest.

As to the first factor (rocks): the laterites and red soils only occur on the felspathic TMS. As to the second factor (levels): all of them are bound to those in the altitudes between 400m and 1000m (cf. the profiles: Fig. 6.2).

The farther north one travels the more laterites and red soils can be also found on the crystalline basement and the Karoo beds (Dwyka, Ecca), but only in altitudes of 900m to 1000m. In this altitude and in northern Natal (east of the Pipklipberg, west of the small town of Pongola) the laterites form the broadest stretch that the author has come to know in all Natal.

Two things may be gathered from this description (see also foto 51):

(1) The farther to the south or southwest the more the laterites are dependant for their existence on a favourable rock-type. Thus the laterites in most parts of Natal and in the Transkei seem to have been nothing but outposts of a northeastern laterite formation. This then is to say that the climate suited for their forming can in these areas only have been a marginal one.
This fossil situation can well be compared with the recent situation in Natal - with red soils occurring only on dolerites, i.e. again on a favourable rock-type in marginal climatic conditions (see ch. 16). (Today the boundary of active deep red weathering of all rocks runs along a line - from east to west - that is situated in the north of Mbabane in Swaziland. See Fig. 16.10.)
(2) The laterites are bound to the transgressional levels of the Cretaceous.
This observation makes sense when compared with the findings by VALETON (1983). From her world-wide comparative research VALETON arrives at the following conclusions.
(a) The lateritic duricrusts, especially those with a high alumina content, were formed on coastal plains near the sea where a high groundwater table was guaranteed. VALETON therefore speaks of groundwater laterites.
(b) On the Gondwana continent all lateritic duricrusts are of a pre-Eocene age, because since the Eocene many crusts in India, Australia and southern Africa can be shown to have been widely destroyed, new ones not to be formed again since then. (Cf. the remarks on the deep weathering of the laterites in Natal.)

These conclusions must be regarded as further evidence for the milieu in which the coastal planations became formed and for the age of the thalassoplains, i.e. the ones from 400m on and higher. (VALETON seems to think that very warm and humid conditions continued from the Upper Cretaceous into the Eocene. Some paleontologists, as for instance ERBEN in Bonn, do not share this view. ERBEN, from evidence found in the northern and southern forelands of the Pyrenees as well as in southern China, drew attention to the fact that a marked world-wide change took place from palm-forests in the Upper Cretaceous to conifer-forests in the Eocene. This period of change seems to have lasted for about 3 to 4 million years, before warmer conditions set in again. Both in length and in the marked lowering of temperatures the period can be considered to be the fore-runner of a development that ultimately resulted in the Great Ice Age. After a warming-up in the Eocene laterite forming processes were renewed on a world-wide scale. Thus Upper Cretaceous laterites and later Tertiary ones must be distinguished.

b) Silcretes

Silcretes could be observed by the author especially in the Cape Province where they occur on all levels between the one in 300m and the one in 800m. Regarding the geological periods these observations mean that silcretes were formed between the Middle (or late Lower) Cretaceous and the Miocene. The area where silcretes appear stretches from East London in the southeast towards Calvinia in the northwest. (Fotos: 47, 48, 20, 31.)

The silcretes are obviously bound to the thalassoplains; the heights in which they occur are the same ones as the heights of the planations. There are no silcretes in between these heights. Furthermore, they are obviously of the same age as the laterites in the northeast, at least where the Cretaceous levels are concerned. On the whole, the silcretes are far better conserved than the laterites and they also must have formed a continuous cover.

As was mentioned for the laterites, this cover was destroyed over larger areas by younger erosion. Thus today they are only preserved on interfluves or as seperate mesas and tafelbergs, especially in the Cape Fold basins and in the inter-montane basins in the north or east of the fold belt.

Fig. 7.1: Laterites in Natal according to MAUD (1970)

Lateritvorkommen in Natal nach MAUD (1970)

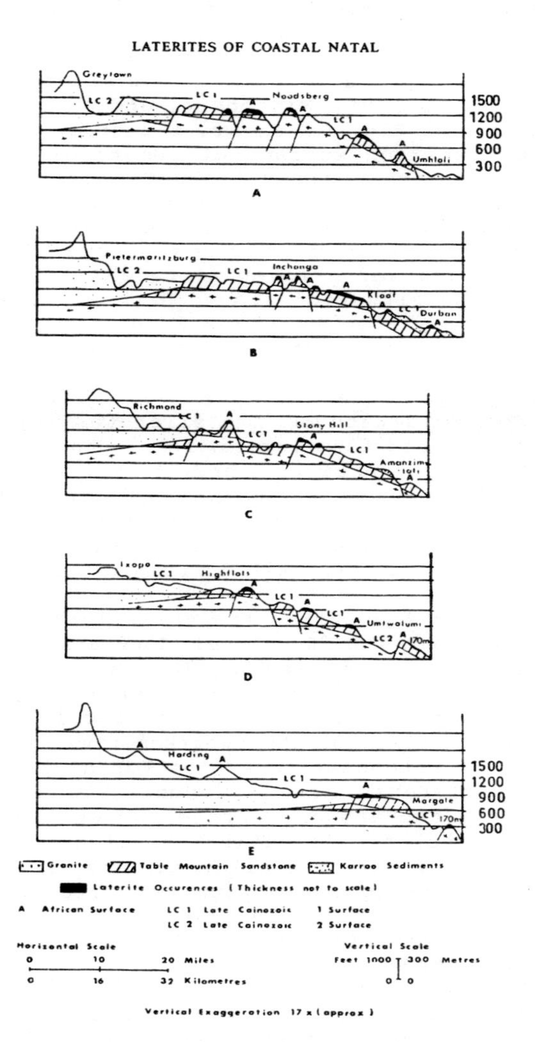

It should be stressed once more that everywhere their altitudes are the same ones as those of the planations.

As the cross section in Fig. 7.2 shows the silcretes of the 600m level cap the Lower Cretaceous sediments of the Glenconnor Panhandle as well as the neighbouring heights of the Cape ranges. In the ranges, the silcretes both occur on the quartzites (TMS, Witteberg) and on the shales (Bokkeveld, Dwyka). As Fig. 7.2 shows, the crust is obviously bound to the planation surface of identical height - and with it cutting across all different rocks, even including the Lower Cretaceous conglomerates and sandstones. By this latter fact it is once more demonstrated that the 600m planation must be younger than the Lower Cretaceous. A similar conclusion was arrived at by FRANKEL and KENT (1937, 35) from the situation around Grahamstown. Both authors made references also to the silcretes in the lower altitudes (1934, 4). (See moreover Fig. 7.3)

The silcretes are however not of the same quality in the different altitudes. The silcretes higher up (i.e. those on the 500m and 600m levels) form a hard and thick crust. Their thickness is between 1 and 3 - 4m. Beneath the crusts there appears a very thick horizon in which the rocks are kaolinised. This kaolinisation horizon is sometimes as thick as 6m. Altogether, silcretes showing the mentioned proprieties are here called silcretes A.

Different from these are the silcretes B. They appear together with the 300m level. They are partly developed on the Alexandria Formation as MARKER and SWEETING (1983) have pointed out, but on all other rocks as well. A typical situation for silcretes B is for instance the one in the southern basin of the Swartkop river in the west and southwest of Uitenhage (north of Porth Elizabeth). (See fotos.) The silcretes here frequently appear on the Bokkeveld shales, but are only preserved on the interfluves of the many rivulets dissecting the slopes of the fold range. On the interfluves the silcretes are always bound to the 300m level, capping the sloping down interfluves where they flatten in correspondence to the level. In the erosional cuts between the interfluves silcretes are naturally no more to be seen today.

Their easy erosion is due to the fact that silcretes B do not consist of a continuous hard crust as is the case with

Fig. 7.2: A cross-section through the Cape Fold Belt and the Glenconnor Panhandle

Profil durch die Kappketten und das Glenconnor Panhandle

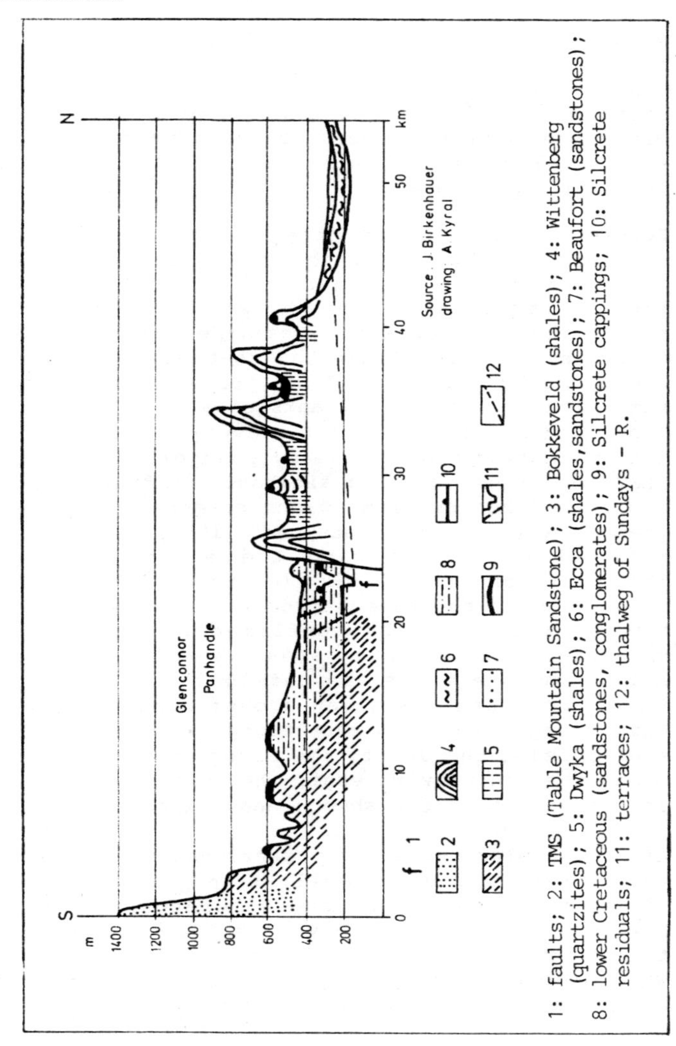

silcretes A. Silcretes B usually consist only of a layer of brown cobbles, with a thickness of between 10 and 50cm. The weathering horizon beneath the cobbles is also by far not as thick as the one under silcretes A. Silcretes B are sometimes referred to in the literature as 'gravels'. This is the case because they seem to consist of pebbles. The pebbles are however of no fluvial origin. They are quartzitic cobbles only partly rounded. Their brown colour is due to the content of iron oxide, having become mingled with the siliciumoxide in the silicifying process in which the cobbles were formed. Because of their weaker nature silcretes B do usually not form mesas and tafelbergs, though these do occur in some places.

Reference to the silcretes, mesas and tafelbergs of the inter-montane basins in the Cape Fold Belt of the southern Cape Province is also made by LENZ (1953). LENZ worked on the Tertiary remnants of the western Little Karoo between the Gourits and Breede rivers. He dealt minutely with the silcretes (and conglomerates that are sometimes baked into the crust by the silicifying process), usually standing out very clearly along the flanks of all the ranges in his area. Today the silcretes here (as everywhere else) are cut into long tafelbergs and mesas by younger erosion. According to LENZ the mesas occur in altitudes of 800m, 600m, 500m, 430m and 300m (thus confirming our observations); though some of them "however, appear stepped" (1953, 209) most of them are regarded as sloping towards the centre of the basin in which they occur or even sloping (as it were: being tilted) along the strike of the ranges from east to west - a feature that however could not be confirmed by the present author's observations in the same area. Wherever LENZ gave the exact heights of the mesas and tafelbergs always the same altitudes are stated as given above: along the Langeberg, the Couga Berge, the Warmwatersberg, the Grootvlakte, or along the Buffels, the Groot, the Slang and the Gourits rivers or along the Klein Swartberge, the Rooisberg, the Gamkaberg, or in the basins of Oudtshoorn and Laingsburg.

All these places expressly mentioned by LENZ were studied by the present writer as well, with special reference to the altitudes of the mesas and silcretes. In no place was he however able to notice any marked and continuous sloping or tilting or fall in the heights of the mesa surfaces; there was moreover a complete indifference to any special location, i.e. along rivers, in the basins, or along the slopes of the ranges. Wherever they occur they are bound to the selfsame surfaces and in perfect consistency with the levels of the thalassoplains and with the rule of series relation.

Only where the mesas (or silcretes, responsible for the steep upper slopes of the mesas) immediately border the slope of a range and from there protrude into the basin can one see a gentle slope. Such a sloping does however not owe its origin to tilting (as LENZ believed) but rather to the fact that it rises in accordance with the local groundwater table just in front of the range.

(Apart from the heights of the mesas mentioned so far it must be mentioned here that there are also mesas in the heights of 700m, 900m and in 1100 to 1200m, i.e. in levels intermediate to those of the thalassoplains. Some of these mesas were also described by LENZ 1953. They are here not dealt with on purpose, because of a systematic treatment in ch. 8.)

For the remnants in 300m LENZ (1953, 225 and 226) accorded a Miocene age, which is in good accordance with our own dating. LENZ also connected these remnants with a marine transgression (227, 232). The remnants higher up are according to LENZ of an older age. Eocene-Oligocene was thought to be appropriate by him.

In doing so he referred to the Enon beds in the hinterland of the Algoa Bay, filling for instance part of the Glenconnor Panhandle. LENZ regarded the Enon beds however as Upper Cretaceous, which, as we know, they are not. Therefore the dating given to the higher remnants by LENZ must be considered as too young.

In referring to the mesas above it has been stated that conglomerates are part of them. As was also pointed out these conglomerates have been baked into the crust. Many observations make it possible to say that this is true for all levels.

Taking the thalassoplain theory as a background this fact can be explained very easily, especially if one remembers what has been said about the observations of BEETZ and GEVERS in South West Africa. There the conglomerates and the nature of the sandstones pointed to the fact that they had been accumulated near the level of a then sea. Since the forming of silcretes is - according to VALETON - a consequence of an equivalent groundwater table in the neighbourhood of a sea level both the occurrence of conglomerates and their becoming silicified within the same horizon makes good sense.

In explaining the vertical differences in the natures of silcretes A and B one has to assume that these are due to a change in climatic conditions from the Cretaceous to the younger Tertiary, i.e. from hot or very warm to less warm, meaning that the hotter a climate is the better conditions prevail for forming a very hard and consistent silcrete with a corresponding kaoline horizon beneath it and also resulting in larger thicknesses.

The horizontal change from the laterites in Natal to the silcretes in the Cape Province on the same levels also indicates a change of climatic conditions, this time however not through time but through space at the same time - a change from more humid conditions under which laterites were formed to more arid conditions under which the silcretes were formed. A similar change from the more humid conditions in Natal to the more arid conditions in the Cape Province can be observed today.

c) Calcretes

Calcretes, also called surface limestone, can be observed on all levels above the one in 1000m whenever climatic conditions are arid enough for the preservation of these surface limestones. The latitude of Calvinia (31.5 degrees south) - or a latitude a little north to it - seems to be the southernmost region where the calcretes could be preserved - at least according to the findings of the present author. (Foto 34.) From there on to the north they become very abundant all over the countryside and stretch from here towards central South West Africa. Though they seem to be especially significant for the 1000m level (cf. the calcified Kuiseb or Karpfenkliff conglomerates) they seem however not to have been formed synchronously with this surface. As HÜSER (1976), BLÜMEL (1981), KORN and MARTIN (1937) have shown the calcretes were formed as late as the Quaternary and are connected by the mentioned authors to the Pleistocene pluvials.

Fig. 7.3: Profile through the Langkloofbasin in the southern Cape Province

Längsprofil durch das Langkloofbecken in der südlichen Kapprovinz

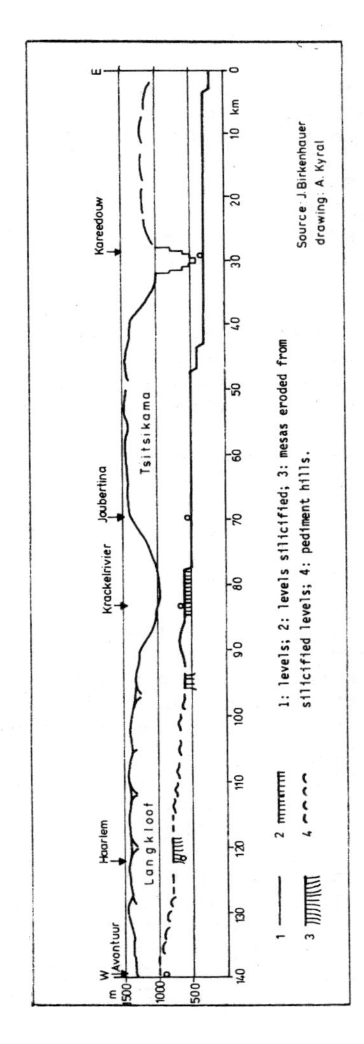

Altogether the calcretes have no bearing on the evolution of the Great Escarpment; therefore they are not dealt with here in any more detail. (Readers interested in such details should turn to the authors mentioned.)

d) Duricrust provinces and their significance

The description of the duricrusts as given above indicates that they occur in distinct areas, or provinces rather, of their own. (Fig. 7.4).

Starting with the laterites one notices that they begin to appear in the northern Transkei in very few and small patches, become a little more frequent in central Natal and only form larger areas of continuous crusts in northern Natal. Moreover, the farther south they occur the more they are dependant on particular rock types (i.e. sandstones with a high felspathic content), the farther north they are the less rock nature makes itself felt. This kind of distribution is apparently due to the prevailing climatic conditions, meaning: the warmer and the more humid climates are the more these climates are favourable to the forming of laterites. If such an assumption is true (besides the findings by VALETON mentioned above of the laterites being formed in the neighbourhood of a then sea level) one has to deduce that in Upper Cretaceous times a spatial change from the more humid to the more arid must have occurred, in a similar way as it does today.

Silcretes are particularly abundant within the basins between the Cape ranges and in the basins immediately to the north of these ranges (Fig. 7.3). It is remarkable however that there are no silcretes on the 1000m level, on the Nama shales and north of the latitude of Calvinia. Comparing this distribution with today's climate one might be able to say that though silcretes need surroundings more arid than those of the laterites they obviously cannot be formed where conditions become too arid.

Another type of quartzitic cover occurs north of the Doring-Olifants-river region and stretches from there into South West Africa. In this province slopes, flats and hills are very often littered with loose fragments of milkquartzes. The fragments form continuous covers, spreading serir-like (Foto 50) over slopes and hills. They usually show a thick-

Fig. 7.4: Duricrust regions and related phenomena in southern Africa

Hartkrustenregionen und verwandte Pänomene im südlichen Afrika

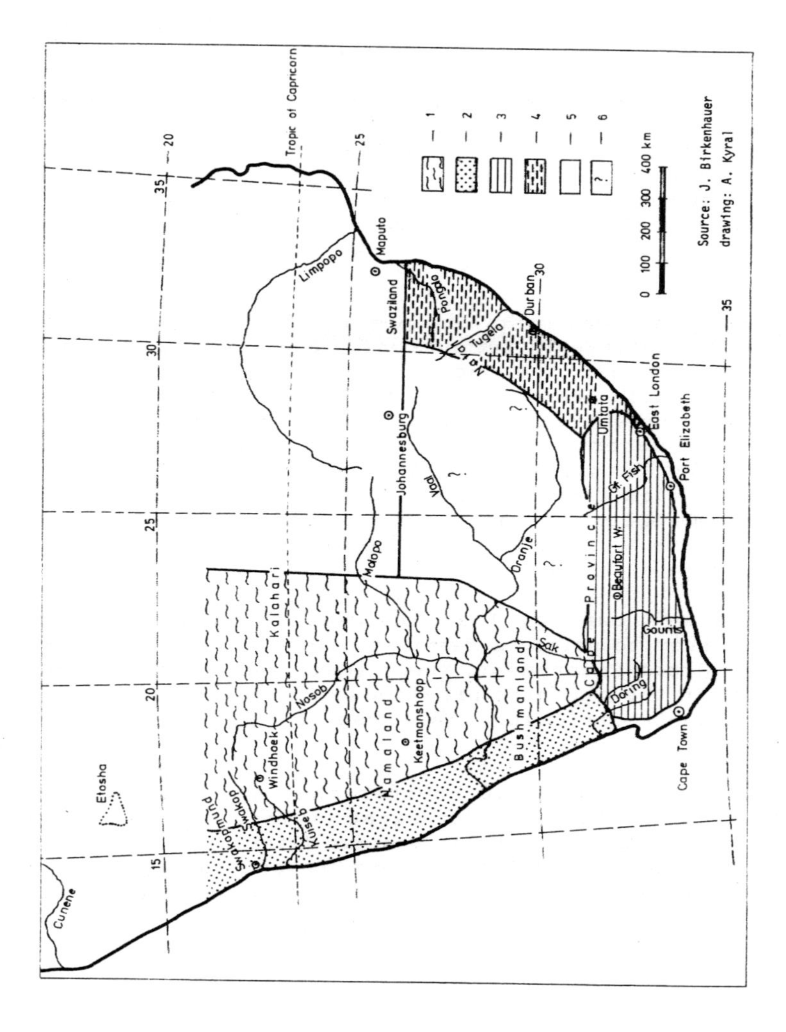

1, calcretes; 2, milk quartz cover (serir); 3, silcretes; 4, laterites and fossil latosols; 5, no phenomena; 6, no observations by another.

ness of up to 50cm. These covers therefore possess enough silica for any process of silicification. Yet nothing of such a process nor any kind of cementation of the fine grit into breccias or the like can be observed. It must therefore be concluded that climatic conditions are obviously too arid today and are not able to provide the necessary humidity for the silification process (quite apart from the fact that the serir-like cover does not lie in the neighbourhood of a sea level and a corresponding groundwater table).

Taking all these observations together they lead to the conclusion that silcretes cannot be used as witnesses for arid or even desert conditions as was formerly believed by geomorphologists and geologists (cf. SCHWARZBACH 1959). Instead, they must be taken as witnesses for a semi-arid (or even semi-humid) though hot climate. Both temperatures and precipitation had to be sufficient to guarantee the solution of silica.

That such conditions must have prevailed in southern Africa when the silcretes were formed can be demonstrated from two more observations: the one is the observed rise of the groundwater table against the face of the quartzitic ranges (though in themselves dry because of the vertical cracks draining the water away but at the same time providing it for processes outside); the other one consists of the fact that the conglomerates and the pebbles and the gravel which later became silicified had to be steadily transported from higher parts - which could only be managed with a certain water supply.

Altogether one can say that the duricrusts as climatic witnesses seem to point to distinct paleo-climatic provinces along the Great Escarpment. These again seem to correspond roughly with the climatic provinces of today - or at least not very much differing. This aspect will be taken up systematically in ch. 16.

It should be pointed out however that the silification process, though reduced, went right on into Miocene times (i.e. down to the 300m level), whereas laterisation seems to have come to an end with the Upper Cretaceous. A climatic shift for the regions concerned is therefore indicated since then.

e) The formation of silcretes and its significance

As has been pointed out apart from semi-arid (or semi-humid even) conditions silicification seems to be furthered by the groundwater table in the subsoil since an equivalent water-table in the subsoil is necessary for providing sufficient moisture throughout the year. Such a watertable will especially provided in the neighbourhood of a sea level. VALETON (1983, 434) seems to have been the first to have pointed out that duricrusts (both silcretes and laterites) should be seen in connection with such a condition and, what is more, with reference to a world-wide distribution. Within such a frame silcretes and laterites are nothing but equivalent groundwater formations, their difference being due only to differing climatic conditions.

According to BUSCHE (1982,1983) the widely distributed silcretes of northern Africa must have developed in broad and flat basins with an insufficient drainage. This naturally means that they were also formed in or near a respective groundwater table.

According to DOUGLAS (1982, 42) both forms occur in low energy flood plains and in deltas, i.e. near a respective groundwater table within tropical or subtropical coastal regions where large tracts are water-logged either all year round or at least over a season.

GEVERS (1936, 78), in talking about the Lower Cretaceous Pomona formation with its quartzites, stated that these quartzites owe their origin to an intensely silicified surface that in itself went together with an intensive kaolinisation below this surface. As we have seen, such horizons of kaolinisation - and rather thick ones at that - can very often be observed beneath the crusts in the southern and eastern Cape Province. (The samples with the large kaolin content talked about when dealing with the 'Grahamstown peneplain'were taken from these horizons, by HAGEDORN and BRUNOTTE as well as by the present writer.)

LENZ (1953, 232) also concluded from his research that "a seasonnally fluctuating groundwater table was produced. Under these conditions colloidal silica and iron oxides were leached from the underlying formations and cemented the overlying deposits" (which had become accumulated on these flood plains). As early as 1937 FRANKEL and KENT (in dealing with the silcretes of the Grahamstown peneplain) arrived at the similar conclusion (i.e. a consolidation of crusts occurs under the condition of a fluctuating watertable in the subsoil).

When LENZ (1953) - as has been stated above - referred to deposits asccumulated on floodplains he by that especially meant the accumulation of conglomerates. Such conglomerates are indeed very prominent and characteristic for all levels; yet not all of them were silicified. Conglomerates are very characteristic for the 300m level of the Uplands-Ruggens-Plateau in the southern Cape or along the Keurboom valley near Plettenberg Bay or in the Worcester Basin. They are further on very marked on the levels in 500m, 600m and 800m. Along these levels thick layers of such conglomerates or gravels can be observed. The very thickness of these layers seem to have furthered the process of silicification - or it was at least been made easier.

f) The geomorphic significance of the duricrusts

The significance of the laterites and silcretes can easily be gathered from what has been said above. It may be summed up with the following statements.

(1) Duricrusts must be taken as a further evidence for the forming of the horizontal surfaces in the neighbourhood of a sea level.
(2) They can supply further evidence for the dating of the levels (i.e. the laterites in Natal as Upper Cretaceous and the silcretes of the Cape Province as Post-Lower-Cretaceous).
(3) They moreover must be taken as evidence for the horizontality of the levels since the groundwater tables (at least partly) responsible for the forming of the duricrusts must also have stretched along in height-constant altitudes. Otherwise the long stretches of corresponding crusts could not have been formed.
(4) They provide a further clue for the question of how the thalassoplains developed. A closer inspection shows that the plains were developed especially well in the larger sinuous embayments in front of the escarpment as well as in the inter-montane basins, where a high groundwater table is demonstrated by the mesas and silcretes. In such low-energy basins and embayments the running of all water must naturally have slackened much - and the more

so even when the sea level rose and the sea advanced inland. The geomorphic situation therefore here meets the requirements necessary according to BIRD (see ch. 5g) for the development of planation surfaces in the neighbourhood of the sea.

g) The formation of thalassoplains

At this stage of the treatise we are at last able to sum up the seven conditions that so far have been found out as being necessary for the formation of thalassoplains. These conditions seem to be the following ones:

- the neighbourhood of a sea level;
- the connection with flood plains of low energy;
- the connection with river flats,lagoons, spreading and meandering rivers in shallow and broad channels on and along such plains;
- the existence of a corresponding groundwater table;
- the spreading of fluvial sands and gravels near sea level;
- the lateral erosion of deeply weathered rock near sea level by sheet floods and by meandering rivers on the near-coastal flood plains;
- the laterisation or silicification made possible in these circumstances (favourable climatic conditions provided).

8. HIGHER INTERMEDIARY LEVELS

In travelling along the inter-montane basins between the Cape ranges the author repeatedly observed flattish sections in an intermediary position between the main levels. The very flat parts of the long lower sections of the slopes within these intermediary levels are perfectly blended into one another. Thus the necessary requirement talked about when KING's pediplanation theory was discussed in a former part of this treatise is indeed present here. These flattish sections must be regarded as true pediments in the full sense.

Now, such pediments can be observed in the average altitudes of 700m, 900m and 1100m. It is moreover very remarkable that they occur as it were rhythmically between and above the mayor upper planations.

Many examples can confirm this general description, i.e. in the basins of Joubertina and Haarlem (700m), in the very large embayment within the Great Escarpment of Aberdeen, Beaufort West and De Rust (900m), on the southern side of the Orange river between Pofadder and Kakamaas (900m) - with very flat wide hill tops at that and very flat slopes (less than 2 degrees), or again in the south of Vioolsdrif (900m) or again over vast stretches of the Bushmanland and the wide regions of Keetmanshoop, Mariental, Koes and Rehoboth in South West Africa (1100m).

It is especially in the region mentioned last that pediments are most pronounced in heights between 1000 and 1100m.

Inspecting all these places and areas in a closer way one comes to notice that a pronounced occurrence of the intermediary levels is obviously dependant on the type of rock. For the consistent pediment levels can only be made out where shists and shales (of whatever age) abound; there are no such forms at all on granites, granite-gneisses, dolerites and quartzites. Instead, in these rocks inselbergs abound, often deeply weathered into mere ruins, sometimes into heaps of débris only.

These different rock types can thus be classified either as 'pediment-afined' (shales, schists) or as 'inselberg-afined' (granites etc.).

The thus very remarkable affinity of two different landforms to two differing rock types reveals that the intermediary levels as such possess a nature very different from that of the thalassoplains where the proprieties of rocks do not play any role at all.

All this goes to say that the intermediary levels are structurally controlled whereas the thalassoplains are completely independent of any such control (as we have seen before). Thus the petrographic nature (or 'petrovariance'; German: Petrovarianz) makes itself very strongly felt with regard to these two forms.

Continuing the inspection one soon notices that, in every case, each higher intermediary level is referred to the thalassoplain below; for each thalassoplain below is surrounded either by a belt of pediments or by a belt of inselbergs dependent on the rock types.

If the rocks change over from one type to the other then the outlook of the surrounding belt also varies. Accordingly the belts are circular or semi-circular following the embayments of the escarpment or the inter-montane basins. The thalassoplains may be said to be 'framed in' by these belts - sometimes with steeper slopes, sometimes with flatter ones, again dependent on the type of rocks. These belts can be well compared to ramps of roads or fortifications as they lead up from one lower plain to the next higher one above. Therefore these belts may be called 'framing-in ramps' (German: 'Rampengürtel'). The higher up these 'framing-in ramps' are the broader and wider and flatter they become.

The same is true of the inselberg-belts: the higher up they occur the more the inselbergs are worn down to mere rubble heaps and the broader the belts become.

The broadest and most flattened-out ramps as well as the most worn down inselberg fringes are found above the highest of the thalassoplains, i.e. the one in 1000m ab. s-l, which is in accordance with the fact that it is the oldest of them all. (Cf. Fig. 8.1.)

Such observations can only be interpreted as being a consequence of the length of time through which a thalassoplained surface existed thus forming the long and stable base level for all denudative washing-off processes on all the neighbouring slopes above the level.

The regular interlinkage of the landforms described is obviously due to a particular chain of events. This chain consists of the following elements:

(1) the forming of a thalassoplain as a base level,
(2) the forming of pediments above it by washing-off processes (in German: 'Spülpedimente'),
(3) the forming of continuous belts - either of inselberg fringes or of ramp belts by these same processes but leading to differing landforms because of the rock types.

Consequently each fringe or belt can be dated: each belonging to the thalassoplain that formed its base level. Ramps and inselbergs above the 1000m level for instance must then be later Lower Cretaceous. The length of time during which the denudational processes since then could have worked on the rocks thus explains why the landforms above the 1000m level show such marked worn-down aspects as they do.

From this it may be assumed that ramp belts and inselberg fringes higher than 1100m must even be older than the ones correlating to the 1000m level.

One can compare these datings with the attempt by RUST (1970) to ascribe a certain age to the inselbergs which were studied by him in an altitude between 1000 and 1120m in central South West Africa. He dated them as older Tertiary, with the reservation that their forming could well have begun as early as post-Karoo times, i.e. after the dolerite intrusions, since some of the inselbergs consist of worn down dolerites. The dolerite intrusion ended, as we have seen, in the late Upper Jurassic so that the forming of these inselbergs may well be as old as the age given.

As will be seen from ch. 9 such an old age cannot be justified. Yet even a Lower Cretaceous age for the inselberg fringes is indeed a very old one - which is certainly a very surprising conclusion.

One last thing may be dealt with: the spatial distribution of the intermediary levels. Starting from the east, they first come into appearance in a very marked way - as the

Fig. 8.1: Landforms in the Pofadder area in the nothwestern Cape Province

Das Relief im Gebiet von Pofadder in der nordwestlichen Kapprovinz

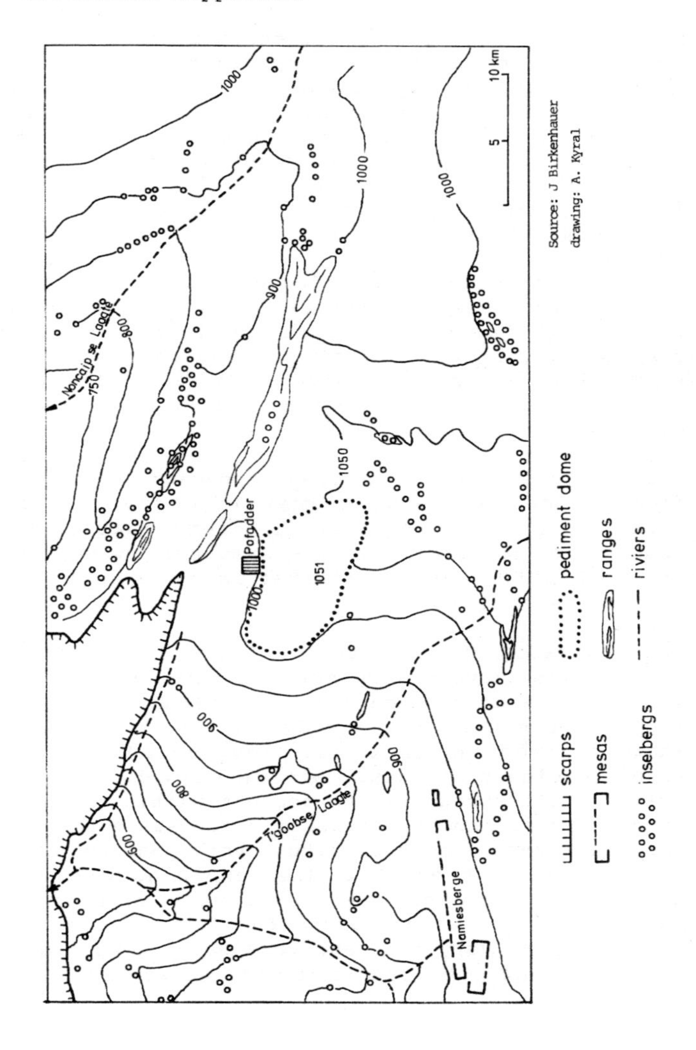

earlier description has shown - in the southern Cape Province. Going on from there to the north or northwest resp. they become more and more pronounced. Such a distribution is naturally in accordance with the growth of aridity in the same direction, both today as well as in the past.

Similarly, pediplanatory processes begin to make themselves felt in a very pronounced way in the same direction. The forming of real pediments begins, as also MABBUT had shown (1955). This again is moreover compliant with what LOUIS and FISCHER state in their handbook (1980). Their statement as well as the observations are contradictory to KING's assumption about the taking place of pediplanatory processes in Natal (both today and in the past) (KING 1972; KING and FAIR 1954; FAIR 1947).

It must altogether be concluded that pediment and inselberg forming processes can steadily go on through the ages once a base level has been established (by the forming of a thalassoplain, for instance) remaining as a common base since then. Under such circumstances the fringes and belts can become wider and wider - as long as suitable climatic conditions prevail.

Within these southern African landscapes it is indeed justified to use the term of what in German has been called 'traditionale Weiterbildung', a term that might be translated as a 'forming-on in the same morphodynamic tradition'. It was BREMER (1965) who first used the term in explaining planations in Australia.

It is because of these tradited on-going processes that the intermediary levels very gradually blend into the main base levels below them. Such a gradual blending can especially be found in the Orange river basin (cf. Fig. 12.3), but only where sedimentary rocks are prevalent. Here, in connection with the oldest of these main base levels, the one in 1000m, relief differences are the poorest within the whole research area. This is the case because processes here could continue through a very long time. Thus very poor relief differences could become established. This explains why pans and vleis begin to appear everywhere over very wide areas. They are particularly prominent along the Sak river course. (Driving through these very flat and very monotonous plains for hours and hours one gets the idea that it is these plains which really should be called the true 'African surface' - and not the one in 600m as was the opinion of KING.)

It is exactly these monotonous levels in 1000m (and in 800m) which - together with the corresponding intermediate levels - are meant by OBST and KAYSER (1949) when they again and again enumerated the occurrence of pronounced flattenings in the altitudes between 800 and 1200m. It is these levels - that together with the intermediate belts and fringes - constitute the real background to what was called the 'oberes Randstufenniveau' ('upper escarpment level') by these two writers and again by KAYSER (1986). Bearing this in mind it is indeed easy to agree with OBST and KAYSER on these two points: the altitude of the 'upper escarpment level' and its prominence just in front of the Great Escarpment. (It is however not possible to agree with them on the age they attributed to the 'oberes Randstufenniveau'. The age given was that of Mio-Pliocene (1949, 238), whereas it really is late Lower Cretaceous.)

9. HIGHEST SURFACES

a) Occurrence

As has already been mentioned in ch. 6, there can be noticed some more levels above the 1000m level and its associated belts and fringes that cut for instance across the folds in the Witteberg and TMS quartzites of the Cape Fold Belt. Nevertheless, within the Cape Fold Belt traces of these higher surfaces are rather scarce, owing to the deep dissection of the mountains by the valleys. The farther north one comes however and the broader the higher elevations tend to become - as is the case for instance in southern and central South West Africa - the more prevalent these highest surfaces become, in such a way that they themselves and their altitudes can easily be fixed. For these highest surfaces two distinct levels can be made out: one above 1200m and another one above 1800m ab. s-l. (Fotos: 32 - 35; 16, 30, 43.)

The lower one is here called the P 2 surface, the upper one the P 1 surface.
The P 2 surface may extend over tens and even hundreds of kilometres and slopes up from 1200m usually to 1500m and several times even up to 1700m.
The P 1 surface forms remnants only across the highest mountainous areas. It slopes up from 1800m to 2000m.
Wherever the two surfaces occur they both possess the same four and very distinct characteristics. These characteristics are:

(1) They are very smooth and flat, though not horizontally level. In contrast to the thalassoplains they possess a very gradual gradient with degrees of 0,5 to 1.
(2) They include very flat shield-like pediment-domes (in German: 'Schildinselberge') of a sometimes rather extended nature (several kilometres in width or diameter).
(3) At their fringes a very pronounced pedimentation goes on as must have been the case throughout the ages since the surfaces were formed. These pediplanatory processes especially concern the embayments. Where, with its embayment, the lower surface reaches with its embayments into the higher one these embayments show a steeper gradient than the one given in (1). The gradient may here even reach 6 degrees, without a loss in flatness and smoothness. A pronounced scarp separates the lower

surface from the one above. The scarp usually occurs at an altitude of 1500m ab. s-1, but sometimes as high up as 1700m.
As was described for the intermediate levels before Inselberg fringes and ramp belts surround the surface,the same processes still forming them. the frings and belts around the p_2-surface are even more prononced than those of the lower levels. This observation may be taken as a first indication that this surface must even be older than the two lower planations.

(4) Both surfaces bevel across all structures (rocks, faults, folds) (though their belts and fringes of course are structurally controlled).

From the description and the characteristics given the conclusion follows that these two surfaces are indeed true peneplains - be it in the sense of DAVIS or be it in the sense of BÜDEL and LOUIS. This is the reason why for denominating purposes the letter 'P' was chosen ('P' for 'peneplain').

Whereas the higher plain can well be observed in the Khomas and Naukluft highlands as well as in the highlands east of Windhoek or again in the Groot Karasberge in the south of South West Africa, the lower plain stretches all around these highlands. The city of Windhoek is itself situated in one of the embayments of this lower plain. Another prominent embayment is the one that is followed by the road from Rehoboth to Windhoek.

Though the occurrence of the lower plain is especially well developed in the region just referred to it is nevertheless extant also on the eastern margins of southern Africa. Here it can be observed in all the high lying basins of all major river drainage areas. FAIR and KING (1954) described it from northern Natal, where it occurs in the upper reaches of the Pongolo river and in the inter-montane basin of Paulpietersburg (by KING at that time thought to belong to his 'African' surface), the upper basin of the Mooi river and the Tugela, the Mgeni basin, the Himeville-Underberg basin on the upper reaches of the Mzimkulu, the so-called (by KING) Kokstad-Cedarville bevel and similar other 'bevels' in the Transkei.

GEVERS (1936, 68) described the dissected remnants of an oldest and highest peneplain in the Khomas highlands in 1900m to 2000m ab. s-l. He thought that, to the north of these highlands, the level continued over the inselbergs by way of a so-called 'summit-level' (in German: 'Gipfelflur'). According to him, this 'gipfelflur' should be taken as an indication that the peneplain was downwarped to the north. In so far as however only remnants of inselbergs are used for such a procedure it seems very doubtful indeed to reconstruct the occurrence of a peneplain and even more so its down-warping. Even apart from such a doubtful reconstruction all observations are contrary to the assumption of any down-warping.

The surface in 1200 to 1400m was also dealt with by ABEL (1954) and called by him 'Hauptfläche' ('main peneplain'). Accordingly, the surface below it (the one in 1000 to 1100m) was called 'Vorfläche' ('marginal plain'). The higher level in 1800 to 2000m was thought by ABEL (1959) to be part of his main plain. The main plain was described by him as very prominent around and south of Rehoboth - which indeed is the case. According to ABEL the highest parts in 1800 to 2000m are nothing but domed up parts of the main plain, owing to tectonic movements which now form the Windhoek and Khomas highlands.

MABBUT (1955) also referred to two highest surfaces - both of them in the same altitudes as was stated here. The one is the Great Namaqualand surface (1200 to 1700m ab. s-l), the other the Namaqualand and Khomas Highland surface (1800 to 2000m ab. s-l). MABBUT gave a Cretaceous age to both surfaces, in good accordance with our views. The Bullsport Flats south of the Naukluft fault line in the north of Maltahöhe (in South West Africa) was taken by him to be part of the lower plain - which indeed it is.

Remarkably, KAYSER (1986, 3, 13, 47, 53) also described surfaces in similar altitudes in the eastern Transvaal. An upper surface according to him can be found in the highlands south of Pietersburg (1700 to 1800m ab. s-l with a flat culmination in 2000 to 2100m). The summits of the Barberton mountains farther south are part of the same level. KAYSER regarded this upper highland level altogether as a geomorphologically intact and genuine peneplain bearing all signs of an old landscape in the sense of DAVIS. The level is cut all over the formations and rocks from the Transvaal series to the Karoo series. In spite of its altitude KAYSER (1986, 3) gave its age as Miocene only. According to KAYSER the lower surface - by him called 'Hochlandrandniveau' (in English: 'highland marginal level') - forms embayments within the upper one similarly to our observations mentioned above. Examples for such embayments can be found in the Barberton mountains, Swaziland and northern Natal or the adjoining parts of Transvaal resp., as for instance the embayment that is followed by the railway track from Piet Retief to Ermelo: in altitudes of 1300 to 1400m, excavated from Ecca beds above the old granite. With this example the embayments in northern and central Natal can be connected, lying in the same altitudes and forming parts of our lower peneplain. The inter-montane basin of Lydenburg in 1200 to 1300m, the granite highlands of Pietersburg in 1300 to 1500m and areas in Swaziland with similar heights in 1400 to 1550m are - according to KAYSER (1986, 53) further examples of this surface.

Having compared the findings of other authors with our own observations there seems to be no doubt about the existence of two highest surfaces as genuine peneplains (or 'Rumpfflächen' in German): the P 1 surface in altitudes between 1800 and 2000m, the P 2 surface in altitudes between 1200 and 1700m.

It is indeed remarkable that so far all authors are in agreement with each other about the heights of these two peneplains and about their existence. Fundamentally, as one should realise, such agreement really does not leave much room for any assumed tiltings and down- or up-warpings of southern Africa.

A further point of agreement, at least between KAYSER and the present author, is the fact that both regard the two surfaces as the only genuine peneplains (in the sense of DAVIS or BÜDEL, 1957, and LOUIS 1964).

The particular conditions under which these peneplains were formed can still today be studied within the larger embayments mentioned above. A particularly good example is the embayment followed by the road from Rehoboth to Windhoek (South West Africa). (See Fig. 9.1.)

Within this embayment the gradient gradually steepens to 6 degrees and then, with a sharp knick-point, breaks away towards the steeper slope (or even scarp) above it. In the embayment exists an active groundwater table just beneath the softly sloping surface as one can see from the fact that the embayment is well wooded (or bushed) in spite of longer droughts and missing streams. (The same is the case with all similar embayments.) The groundwater table thus indicated continuously reaches up to the knick-point where, upon reaching it, it is immediately disrupted.

The gradient risings both of the watertable and of the embayment surface can by the way be well compared to what has been stated about the face of the silcretes in the Cape ranges (cf. ch. 7.e).

The observations indicate that planation (and denudation) of the surface were (and are) directed by the groundwater table. Perhaps even the whole peneplain was formed in connection with a corresponding groundwater table. If this assumption is safe it must be concluded further that the plain-in-forming must have remained relatively near to a base level - and that moreover for a rather long time in which the wide stretches of the plain could be formed.

Planatory processes seem to go on even today as the gradual deepening of the embayments towards the slopes suggest, in spite of the fact that the peneplains must have been uplifted several times and for several hundred metres at that. Thus the continuous widening and deepening of the embayments is another case of 'traditionale Weiterbildung'.

From the fact that each peneplain finds its lower end in the same altitude one has to assume that each plain originally began to be formed along a base level common to the whole plain. For example, the altitude where the lower peneplain sets in is, all around southern Africa, always found in the present height of aproximately 1200m ab. s-l. Such common base level can best be visualised in the neighbourhood of a presumed sea-level. If such a sea-level remained at the same height for a longer period (as may well be assumed in comparison with conditions regarding the thalassoplains) it may moreover well have served for thoroughly establishing the base level for all denudational and erosional working done on the peneplain.

Fig. 9.1: Profile of the Rehoboth embayment within the Windhoek Mountains area, southwest Africa

Profil durch die Rumpfflächenbucht von Rehoboth im Bergland von Windhoek, Südwestafrika

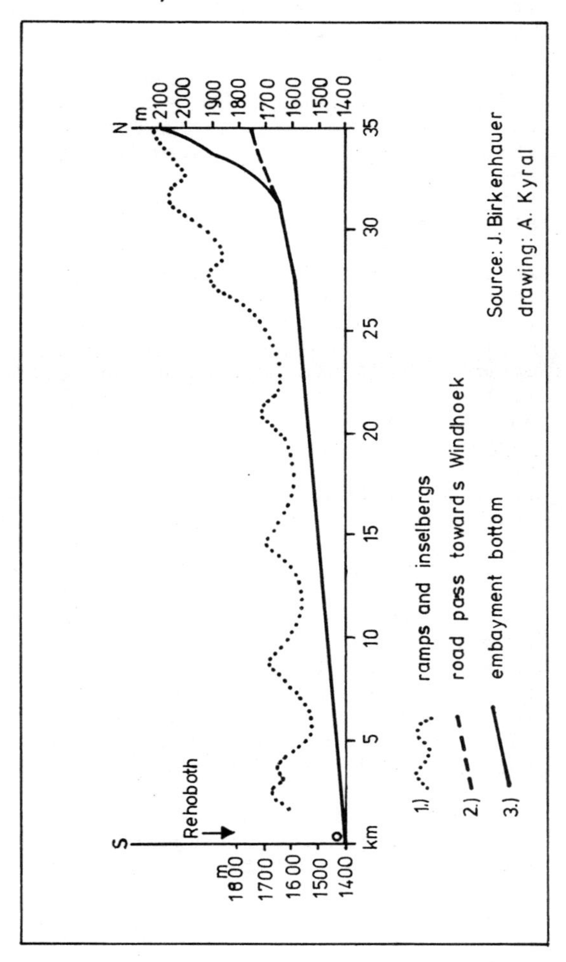

b) Age of origin

As the peneplains P 1 and P 2 lie above the 1000m level they must be older than this level, i.e. older than late Lower Cretaceous. Such an older age is not only suggested by their height positions but directly indicated from the fact that the respective embayments (together with the ramp belts and inselberg fringes) are entrenched into each higher surface. The age of the peneplains - at least that of the P 2 plain - can moreover be established from the following observation. The P 2 plain must be younger than the Valanginian faultings and the Jurassic dolerites because the plain cuts across these features - as can clearly be seen for instance in the Klein and Groot Karasberge.

Another indication for the age in which the P 2 plain was formed can be gained from the following constellation. The gigantic dolerite inselbergs of the Omatakos between Otjiwarongo and Okahandja (northern South West Africa) tower above the P 2 plain (here at an altitude of 1400 to 1450m) with a sudden and steep rise of 800m (the summits are 2268 and 2175m resp.). The dolerites themselves intrude through Etjo-beds and partly lie on top of them. This means that the dolerites are younger than the sedimentary Etjo-beds. Because of their higher resistancies the dolerites have protected the Etjo-beds from being removed by the planatory processes prevailing all around the dolerite pinnacles of the Omatakos. The Etjo-beds have indeed been removed completely everywhere else, though once they must have spread over the whole region as can be shown from other remnants of them at the Otjikuberg, 6km to the east of the Omatakos, and from the large Etjo mesa on which the Etjo farms lie (22km to the northwest). In between these remnants the plain rests in a distinctly lower level than the beds, showing that in the forming of the plain the beds (and some of the dolerites topping them) were removed.

Since the Etjo beds can be dated as late Jurassic and the dolerites intruding through them as early Lower Cretaceous the peneplain must be younger than the Etjo beds and the dolerites.

Similarly, the Erongo mountains (their magmas equally of a Lower Cretaceous age) tower above the plain, undercut by it. The Erongo magmatites were dated as 120 my BP (WINDLEY

Fig. 9.2 a: Schematic cross-section of the main geomorphic features in southwest Africa

Schematisiertes Profil der geomorphologischen Grundzüge im südwestlichen Afrika

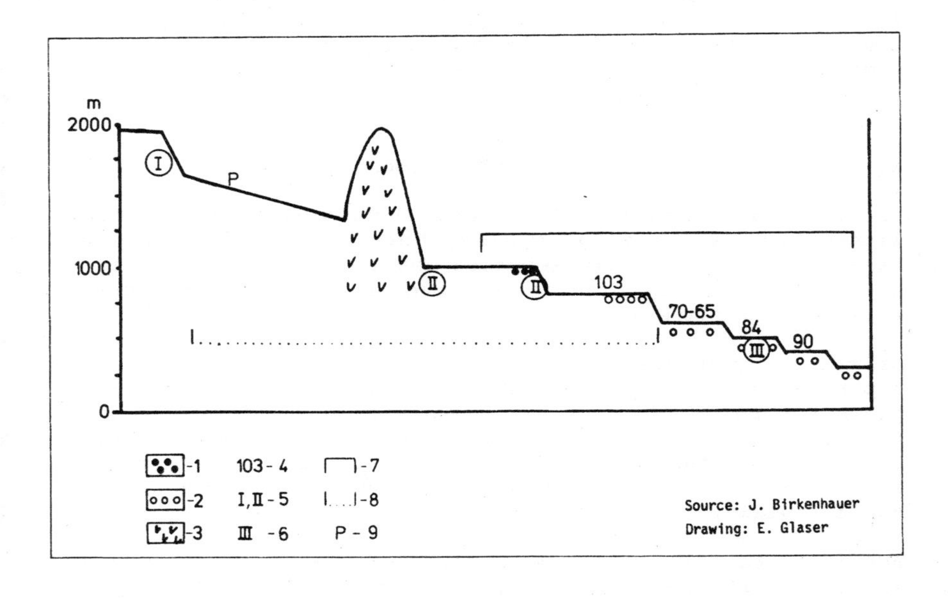

1 = Kuiseb-Conglomerates
2 = Silcretes, conglomerates (Cape Province only)
3 = Volcanites
4 = Age in million years after Hancock and Kauffman 1979; Windley 1984
5 = Broad pediments, strongly eroded
6 = Scarps, ramps, younger inselbergs
7 = Thalassoplains
8 = Lower Cretaceous planations
9 = Peneplain (in the sense of Büdel and Louis)

1984), i.e. Barremian. The undercutting P 2 plain must then at least be post-Barremian.

Such a dating can moreover be regarded as a confirmation of the view given above about the age of the 1000m thalasso-plain as post-Valanginian. It now can be safely assumed that it is even younger than post-Barremian since the post-Barremian P 2 plain became dissected and embayed from below, i.e. from the 1000m level. An Albian age (100 my BP), founded on the separate evidence of sea level rise at that time (as was already mentioned above), fits very well with the conclusions arrived at for the P 2 plain.

c) Tectonic and geomorphic implications

The two main peneplains (P 1 and P 2) must have been formed within two main erosion cycles, both of them presumably based on two distinctly separate sea levels (height difference: 600m), as the discussion above of how the lower peneplain was formed has shown.

If this assumption is correct southern Africa must have experienced an up-lift of 600m after the forming of the P 1 plain and before the developing of the P 2 plain. Another uplift of 200m must have occurred between the forming of the P 2 plain and that of the 1000m thalassoplain. A further up-lift of again 200m preceded the forming of the 800m thalassoplain.

According to WINDLEY (1984) quite a series of up-lifts did indeed occur between the late Jurassic and the early Cretaceous.

Throughout this age of up-lift (as it may aptly be called) southern Africa must have reacted as a tectonically stable block since no tiltings whatever (and the like) can be observed in the surfaces between 800 and 2000m, a fact again demonstrated by the constant base level for each surface all around southern Africa.

If this conclusion is correct it is indeed possible then to connect any given surface of the same altitude in whatever region it may occur and whatever the distance is between these regions. Under these conditions the very heights can be used for connecting equivalent surfaces. In the same way and for the same reasons it is moreover possible to integrate the findings of other scholars in so far these are related to particular heights.

In attributing a Lower Cretaceous age to the P 2 surface (in fact: post-Barremian) the higher P 1 surface may then probably be of an end-Jurassic age.

Starting with this assumption one can go on to say that the P 1 peneplain was formed before the rifting of Gondwanaland. Thus it would really be the original Gondwana surface if one uses the terminology introduced by KING. (In the terminology of KAYSER it is the 'Binnenhochland',or 'highland level' in English.)

Adopting KING's term, the P 2 surface would then be equivalent to his 'marginal Gondwana-level', later on renamed by him and FAIR (1954) as 'post-Gondwana'. This name was retained by KING till 1972. In 1982 KING had its name changed to 'Kretacic'. (KAYSER's name by the way for this level was 'Hochlandrandniveau' or 'highland marginal level' in English, both in 1949 and 1986.)

All such reasoning is naturally rather highly speculative but it can be founded on one or the other observation. As we have seen above, the P 2 surface became formed when rifting had already begun. Therefore the assumption about the higher P 1 level of being the true 'Gondwana level' may not be far from the geological truth. If all this is the case the P 2 level then cannot be taken as the 'marginal Gondwana level', since actually it really is the first 'African' surface proper. KING however has called it the 'moorland level' instead since 1982. (This level by the way corresponds to KAYSER's 'oberes Randstufenniveau' or, in English, 'higher escarpment level'.)

Whatever these niceties of name-giving and name-comparisons are: the one and only truly 'African' landscapes with very broad and wide extensions are those of the P 2 plain as well as the two uppermost thalassoplains (including, of course, the similarly widespread intermediary surfaces). These truly 'African' surfaces were all formed during the second half of the Lower Cretaceous.

d) Classifying names

At this stage of the treatise it seems to be advisable to sum up the terminology used by the different authors by way of a table (see Table 9.1). The explanations given above should be taken as a preliminary explanation of the table. Included in the table are the heights and datings as given by the different authors so that the may easily be compared.

e) Age and development of the Great Escarpment

From the discussion it can be gathered that the Great Escarpment was subject to a development consisting of two separate stages: the preparatory 'marginal Gondwana' stage (in the sense of KING 1972) and the more or less final 'African' stage.

In the first stage, probably Barremian in age, the lower P 2 peneplain became bevelled into the escarpment, at least partly (northern Natal, central South West Africa), since here this surface immediately connects to the escarpment proper.

In the final stage - consisting of sub-stages - the higher thalassoplains, especially the ones in 800 and in 1000m, undercut the escarpment, as is demonstrated from their proximity to the scarp (see Fig. 1.1). This undercutting process must have taken place during middle Cretaceous times.

One can altogether sum up that, because of the rather old age of these processes, the Great Escarpment itself has indeed to be regarded as a very old macro-geomorphic erosional form - shaped at that over a larger part of the earth's continental crust.

A particular and rather young 'scarp-retreat' can however not be verified.

Fig. 9.2 b: Cross section from the ocean towards the highlands along the Swakop river in the central part of southwestern Africa

Profil vom Atlantik in das zentrale Hochland des südwestlichen Afrika im Gebiet des Swakop-Flusses

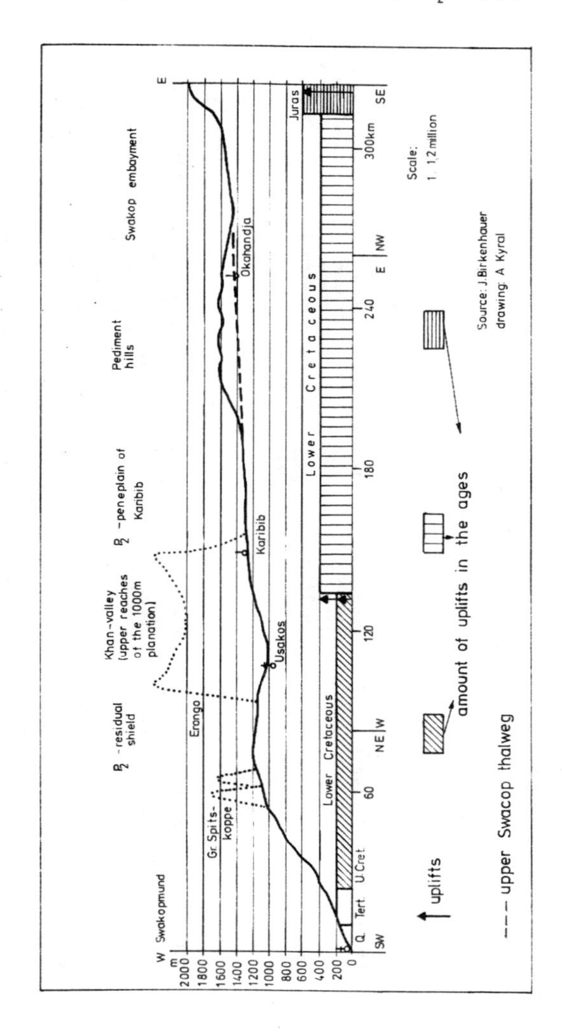

Table 9.1: Altitudes, ages and terms for surfaces in southern Africa according to different authors

Höhen, Alter und Bezeichnungen der Niveaus im südlichen Afrika nach verschiedenen Autoren

Altitudes (in m ab. s-1)	King: 1972 and earlier	King 1982	Kayser 1949, 1986	Mabbutt 1955	Birkenhauer
1800 - 2000	Gondwana	Gondwana landscape late Jurassic	Binnenhochland (highland level) Late Cretaceous to early Tertiary	Khomas Highland surface Cretaceous	P 1 - peneplain Late Jurassic
1500 - 1800					PIS I (pediplanatory intermediary surf.) Early Cretaceous
1200 - 1500	marginal Gondwana, post-Gondwana	Kretacic landscape Lower Cretaceous	Hochlandrandniveau (highland marginal level) Eocene	Namaqua Highland surface Great Namaqualand plains, Kaap peneplain Cretaceous	P 2 peneplain Early Cretaceous
1100 - 1200					PIS II Early Cretaceous
1000	African older Tertiary to Miocene	moorland landscape late Cretaceous to early Cenozoic	oberes Randstufenniveau (higher escarpment level) Mio-Pliocene	Bushmanland surface early-middle Tertiary	T 1 (Thalassoplain 1) Lower late Cretaceous
900					PIS III late lower Cretaceous
800				2600 feet plane Upper Tertiary Lower Pleistocene	T 2 late Cretaceous
700					PIS IV
400 - 600		rolling landscape Mio-Plio-	unteres Randstufenniveau	2000 feet plane Upper Tertiary	Upper Cretaceous T 3 a: 600

10. PEDIPLANATION

In dealing with the intermediary levels their origin was brought - perhaps in a little cursory way - into connection with planatory processes caused by the development of pediments. Pedimentation - or pediplanation - must therefore be dealt with in a more systematic way.

As we have seen, KING (1972, 1982) and FAIR (1947) attributed the so-called retreat of the Great Escarpment in Natal at least partly to pediplanation. Yet such a process could not be observed anywhere in Natal - neither today nor for former ages. (Cf. ch. 3d)

Coming from Natal and the Transkei and travelling west, one has to go for about 400km west of East London (as the crow flies) in order to reach regions where true pediments show up as marked features indeed and true pediplanatory processes are still going on, eventually leading to pediments and even totally pediplained landscapes. (Fotos: 1, 43, 44.) Going in the direction as described above, the first region where the forms and the processes can be noticed to a larger extent is the high-lying inter-montane basin of Willowmore, 200km in the northwest of Port Elizabeth. In this basin very broad and very flat pediments can be observed, both below (in 700m) and above (in 900m) the 800m thalassoplain. These very broad and flat pediments are here entrenched into the steep slopes of the Cape ranges.

Travelling north and leaving the last fold range behind one in the south (i.e. the Droekloofberge, Boesmansportberg, Grootrivierberge) an overwhelmingly flat and wide countryside in the level of the 800m thalassoplain opens up. Very smoothly and with very low, scarcely perceptible gradients the 800m level begins to change into that of the 900m intermediary level. Over tens of kilometres the level consists of nothing but flat pediments that intersect and blend into a vast pediment plain. Coming from Natal the plain is the first real and grand example of what pediplanation can bring about.

The heights of the plain are however not completely level as the thalassoplains are. Sometimes so-called spitskoppies still rest on the flat pediments, that form domes around the koppies thus indicating that the rocks (that here consist of shales and grey-wackes) must once have reached higher up, but have become pedi-plained away. The surface is everywhere littered with débris and fragments of the shales, lying on top of a very shallow 'soil' of yellowish-brownish sands (just a little loamy). The 'soil' is indeed so shallow that one hesitates to call this very thin horizon immediately on top of the unweathered rocks a real soil. Underneath this pseudo-soil, in about 5cm, the rocks appear, completely unweathered and their strata obliquely cut by the pediment. The débris (together with quartz fragments) weathers into loamy sands. For it is these sands (and not the débris) that are carried away by occasional sheet floods, transported by them to the very shallow and rather broad 'valleys' that because of their very shallowness can hardly be called 'valleys'. The German term 'Flachmuldental' can aptly be applied to these 'valleys'. They are filled up with nothing but these sands.

In time, the weathering process going on on the pediment domes gradually destroys even the central spitskoppies so that after a while only a flat knoll within the dome denotes the place of a former koppie. Very often, even these knolls have disappeared so that the pediment dome has now received the form of a very flat shield. Again, the German term 'Schildinselberg' can aptly be used for describing such a very flat shield.

Near Aberdeen and Beaufort West, for example, the 800m level as well as the pediplained intermediary level in 900m (with the process of pediplanation still continuing) reach back into the dolerite sills of the Great Escarpment in large triangular embayments (German: 'Dreiecksbuchten'). The surface of these embayments is still going on to be flattened out by the planatory processes described above. All stages of weathering can be noticed: from the inselbergs still prominent in the immediate fringes of the embayments on towards the spitskoppies and knolls at the 'mouth' of the embayments towards the southern plain.

The Aberdeen and Beaufort West embayments can therefore serve as good examples for how the pediplanatory processes work in general; a particular 'chain' of landforms is produced in this process wherever it is spreading steadily: from (a) the entrenchment of a new base level (here the 800m thalassoplain) towards the gradual dissolution of the hinterland areas into (b) an inselberg fringe or flattish ramp (depending on the rock types: see ch. 8), (c) the shaping of these inselbergs into spitskoppies and knolls, (d) the forming of flat shields which finally (e) intersect into a gently rolling plain.

This generalisation drawn from the example of the Aberdeen and Beaufort West embayments can similarly be observed in all other instances from here towards the north or northwest of South West Africa - wherever climatic conditions are suitable (i.e. semi-arid and warm or even hot).

Since the process has gone on for a considerably longer time on each higher level it is not at all surprising that the highest intermediate level, that above the 1000m plain, has grown very wide with extensions of tens and tens and even hundreds of kilometres in some places. Often the areas on the fringes have widely and totally been consumed into the pediplain, especially on shales.

In the same way, it is not at all remarkable that many parts of the highlands have been consumed by the pediplanatory embayment forming process starting from the even older P 2 plain.

Another example for the process was studied by MABBUT (1957) in the Steinkopf area of Little Namakwaland (south of the Orange in the northwestern Cape) with crystalline rocks. He described similar forms and pointed out the thin detrital horizon, the unweathered rock below, the hill wash that sweeps away the scree fragments. The embayments, smaller or larger, were aptly described by him as the 'vanguards' of pediment forming.

Summing up this chapter the morphodynamic consequences drawn from it seem to be the following four ones:

(1) Pediments obviously do only develop where subtropical or even tropical climatic conditions are semi-arid or even arid, but do certainly not do so in the sub-tropical or tropical humid conditions of, for instance, Natal or Swaziland. Especially favourable conditions prevail in the lee of the Cape ranges in the Little Karoo and eastward of the Namib desert.
(2) These subtropical (or tropical) semi-arid and arid conditions must have been prevalent in all these areas for a very long time: a fact to be concluded from the very

pronounced nature of the pediplains. The time-span can be estimated to be at least 100 my (since the Barremian or Albian for the 800m and 1000m levels) and even more (120 my) for the P 2 plain and its adjacent parts.

(3) Pediplanation does not result in the forming of a horizontally completely level plain but in a gently undulating landscape with flat swells.

(4) It does not lead to a peneplain proper in the sense of DAVIS or BÜDEL or LOUIS or VON WISSMANN (1955).

11. MODEL FOR THE EVOLUTION OF THE GREAT ESCARPMENT AND ITS THALASSOPLAINED FORELANDS

At this stage of the treatise it is at last possible to propose a model of how the Great Escarpment and its forelands have evolved during a time of approximately 150 my.

The model is represented in the form of a generalisid profile for which Fig. 9.2 can serve as an example.
The profile shows several features:

(1) the age of the two upper peneplains;
(2) the age of the thalassoplains and the eustatic sea level rises leading to their forming;
(3) the interconnection (at least partly) with the age of up-lifts;
(4) the long age of standstill after the age of up-lifts with the long opportunity for the erosional and denudational onslaught;
(5) the repeated advance of the forelands against the sea as a consequence partly of up-lifts and especially of the fall of the sea level.

How much southern Africa may have risen totally by the successive up-lifts is a question that can only be answered if one knows where the sea level lay before the onset of the Cretaceous transgressions. VAIL et al. (1977) and HANCOCK & KAUFFMAN (1979) apparently thought the sea level before the Upper Cretaceous to be more or less comparable to that of today. If this assumption is correct and if we moreover assume that the P 1 peneplain in 1800 to 2000m ab. s-l was equally formed in the neighbourhood of a sea level we must come to the conclusion that the total amount of up-lift since the Lower Cretaceous is around 1800m.

According to OBST and KAYSER (1949) - see Table 3.1 - the up-lifts equally amount to a total of 1800 to 2000m though altogether taking place in a far younger time (i.e. since the late Cretaceous).
KING (1972) - see Table 3.2 - however estimated a total of up to 3200m. In reaching such an estimation he does not seem to have taken into account the thickness of the dolerite sheets intruding the layers of the Karoo system.
Since peneplains and especially thalassoplains did play a prominent part in the evolution of the Great Escarpment it seems advisable to sum up also the six main characteristics

for the development of the thalassoplains. These characteristics are:

(1) Ingression of the sea with a slowly but constantly rising level over a considerable time, lasting for at least several million years;
(2) equivalent rise of the groundwater table, being dammed up towards the interior by the raised sea level;
(3) the slackening up of water running down towards the newly formed base level (forming of a 'Staubereich');
(4) forming of huge flood plains with shallow and wide river channels and a horizontal spreading of sand and gravel along and over the flood plains, gradually accompanied by the silicification of the sediments;
(5) simultaneous decomposition of rocks in the hinterlands;
(6) quick removal and wash-out of débris on the flood plains.

Another characteristic of the thalassoplains not mentioned before can in this context also be drawn attention to. The lower and younger, not so wide thalassoplains here always referred to by their average heights of 200 and 300m actually lie in altitudes between 170 and 220m and, respectively, in 270 and 330m. This means that these plains possess a certain gradient towards the sea. Similarly, the plain in 400m slopes down from 430 to 400m. The upper and wider plains however do not have any gradient at all.

Perhaps one has to assume that the upper thalassoplains originally posessed similar gradients which have become flattened out by a slight epirogenetic tilting towards the interior which however did not extend towards the coastal plains. Another explanation might be that the time-span needed for erosional and denudational processes to wipe out any gradient was sufficient for the higher plains but not for the lower ones.

12. THE EVOLUTION OF THE RIVER DRAINAGE SYSTEMS ALONG THE COURSE OF THE GREAT ESCARPMENT

The planations with their horizontally constant altitudes provide an excellent instrument for the reconstruction of the river drainage systems - a point that has been mentioned before but must now be made use of in a more systematic way. It may be remembered from ch. 5 and 6 that the coastal levels turn inland into the river catchment areas where they run along in the way of terraces, but always remaining at the same altitudes. (This phenomenon is, by the way, very similar to the 'trough surfaces' ('Trogflächen') in the Rhenish Massif. Cf. BIRKENHAUER 1973.)

Some other authors, notably OBST and KAYSER (1949) and KAYSER (1986) described terraces which however - according to them - show a continuous drop down and along the the river thalwegs. If this is the case, the terraces must then naturally be regarded as fluvial ones. But according to observations of the present writer the opinion of OBST and KAYSER could not be confirmed. Not a single fluvial terrace accompanying a river thalweg along longer stretches could be found, not within the whole coastal arch from northern Natal to South West Africa. Rather, all 'terraces' really always do remain in the same altitudes - and as such form another good proof for the rule of series relation.

'Terraces' are most impressive naturally along the great valleys, for instance of the Pongolo, the Tugela, the Mkomaas, the Great Fish, the Sundays, the Groot-Gamka-Gourits, the Breede, the Olifant-Doring-Tankwa, the Orange and the Fish rivers.

Along the thalwegs of these greater rivers (but also along those of the lesser valleys) a certain phenomenon is connected with the 'terraces' (Fig. 12. 4; Fotos: 31, 3).
This is the following phenomenon: Wherever the terraces come to an end upriver, they swing round and traverse the valley within its bottom. At all these places the valley bottom (a) markedly flattens out over a longer stretch of the thalweg and (b) broadens, (c) forming thereby a smaller or larger, flat valley basin.

Similarly, the scarps (or slopes) between the 'terraces' curve across the thalweg. Where this happens the upriver thalweg begins to enter a steeper relief, thus forming a

Fig. 12.1: Profiles of some river thalwegs

Ausgewählte Flußlängsprofile

a) Pongolo

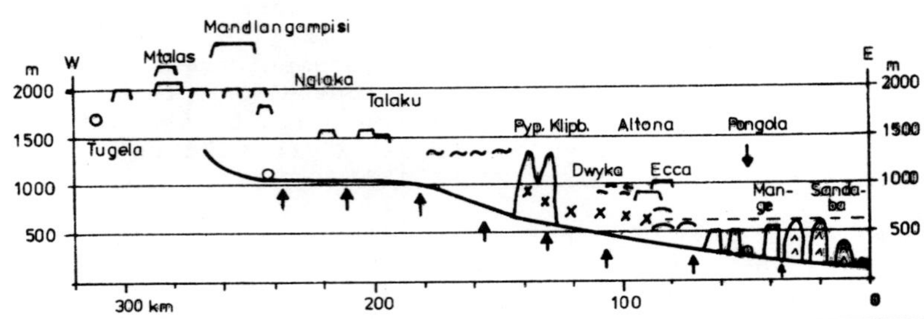

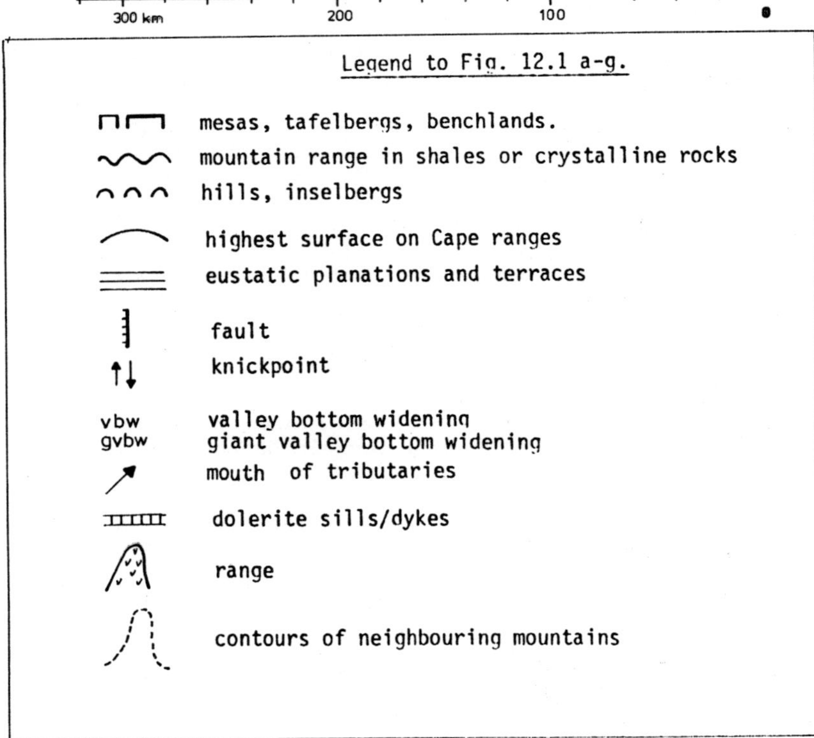

Fig. 12.1: b) Mzimkulu

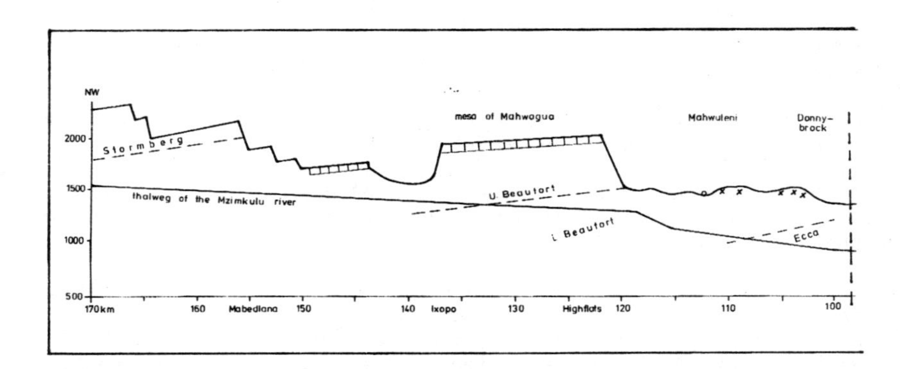

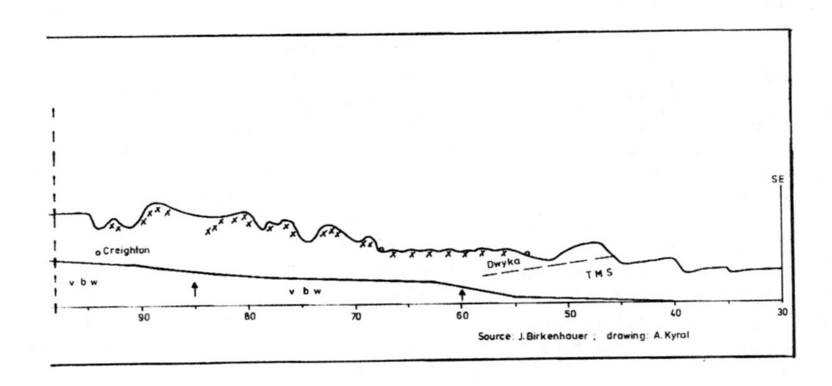

'close' to the valley basin below. Therefore these closes might be called 'valley bottom closes' (German: 'Talbodenschlüsse'; cf. BIRKENHAUER 1973). The 'valley bottom closes' describe an oval around the basins below. Moreover, wherever the thalweg crosses such a 'close' the valley gradient naturally always steepens perceptibly: a characteristic knick-point is produced. Higher up, towards the next crossing of a 'terrace' over a valley bottom the thalweg flattens out again and, similarly, at the next crossing of a slope or scarp, the next knickpoint appears.

In this way there are knickpoints along the whole thalweg wherever it passes over from one flattish valley basin to the next one above.

The phenomena described is a regular one at whatever river and river system one has a look. It is exactly these knickpoints in these situations that do occur most regularly along the thalwegs of all rivers. There are no other regular knickpoints. (The occurence of these knickpoints is, by the way, in perfect agreement with the rule of series relation.)

A comparison with profiles and cross-sections drawn by other authors shows that knickpoints, valley bottom widenings and planation crossings occur in the suitable heights. Such a finding is the more valuable since other authors drew their figures without knowing about the possible implications for the problems in question in this treatise. One figure is represented in RUST (1987), showing a profile along the Hoanib in northern Namibia, here as Fig. 12.1.f.; another figure is taken from WATSON et al. (1985), here as Fig. 12.1.g. The profile from RUST is in remarkable agreement with our notion, the one from WATSON et al. a little vaguer. This may perhaps be explained by the network of altitudinal points used by the latter authors, which is not as dense as the minute drawing by RUST.

To illustrate this striking regularity some examples are given from the major rivers along the coastal arch. (Fig. 12.1.a - e).

Taking up the 'terraces' once more: they also occur within the same altitudes wherever a river breaches a mountain range by way of a gorge or a poort or a deeply incised valley. Travelling towards such valleys (unknown before) one can apply the rule of series relation and predict where and

Fig. 12.1: c) Dwyka - Gamka - Gourits

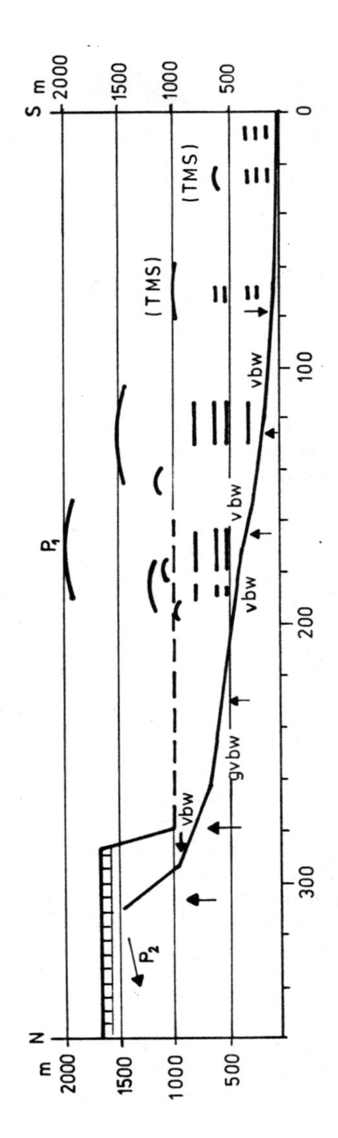

in which altitudes the 'terraces' must occur. It is this predicting faculty that really makes it a true rule.

Along the breachings the 'terraces' are naturally not as broad as the coastal plains, but, though narrower and narrower, they are always there, distinct, however resistant the rocks are that were cut through by the river or were plained over by the terrace-forming processes. That this is the case even with the hard and folded TMS and Witteberg quartzites is a very striking fact.

Another striking fact is that the coastal planations - especially those in 600, 800 and 1000m ab. s-l - do reach back through these often very narrow gorges or poorts into the interior behind the Cape ranges (Fotos: 23 - 25). KING and FAIR (1954), for instance, also drew attention to the fact that their 'African' level - here meaning the 600m one - swings right back to the base of the Great Escarpment along the valleys, a special example being the basin of the Tugela and its major tributaries. (See Fig. 5.1 and 5.2.)

In the cases where the 'terraces' do not reach back behind the ranges they however make themselves felt - either within the ranges or the basins between them - in the widening of the valley bottoms (as was described above).

In the eastern Transvaal this phenomenon was expressly referred to by KAISER (1986, 33, 43, 44) - though he was not able to explain it - in dealing with the thalwegs of the Olifants, Blyde, Sabie and Crocodile rivers. He also mentioned that - below these valley bottom widenings - the levels accompany the thalwegs as more or less broad terraces. These terraces were called by KAYSER the 'lower main terraces' (in German: 'untere Hauptterrassen') when talking about the terraces in 400 to 600m ab. s-l. and 'upper main terraces' or 'terrace' ('obere Hauptterrasse') when talking about the 900 to 1000m level. In calling these levels along the valleys 'main' terraces KAYSER was really confirming the very prominent nature of the levels. According to KAYSER the same features can be recognised everywhere in Natal and also in the eastern Cape.

In the eastern Transvaal KAYSER connected the 'lower main terrace' with river gravels lying on top of the Lebombo range near the coast. In spite of this it should however been born in mind that the levels are no true fluvial ter-

Fig. 12.1 d: Fish river

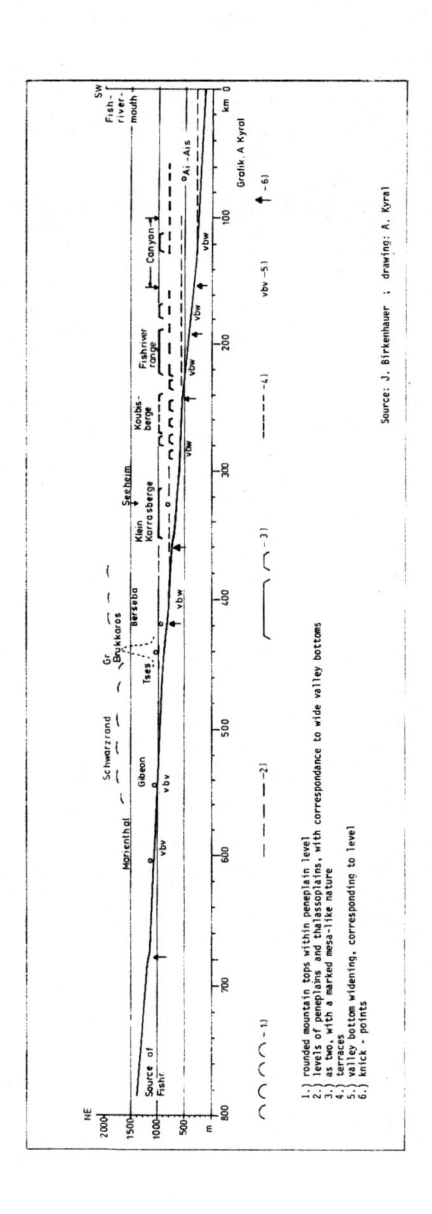

races with a certain gradient, as was pointed out above. This being so they must be considered as features that are completely independent of rivers and can neither have been subject to their lateral erosion nor to any gradient-oriented sedimentations by the rivers.

Special attendance has been given by various writers to the complicated drainage pattern of the Cape Fold Belt. The first writer to occupy himself with its evolution was ROGERS (1903). Dealing especially with the Gourits river system ROGERS held the longitudinal valleys to be even prior to the Lower Cretaceous sediments.

He therefore thought an antecedent course of the Gamka river (and other rivers) - by using transverse valleys through the ranges - as highly improbable. He envisaged the later river drainage system to have developed through headstream erosion starting from the longitudinal basins. Some rivers, as for instance the Buffels river - that probably flowed through Garcia's Pass in the north of Riversdale - became captured in this way by the Gamka. Most drainage systems later on became newly superimposed on the Lower Cretaceous sediments that buried the mountains; it was only then that the poorts were developed through headward erosion.

LENZ (1953) also thought that any evolutionary reconstruction had to fall back to the Cretaceous sediments which had been filled into the longitudinal basins in connection with an advance of the sea. After that an up-lift should have happened so that the erosional power of the rivers could be increased for a headward erosion of a new drainage system, in this way eventually breaching the ranges of the Aasvoelberg, the Rooiberg, the Gamka Hills and the Swartberge. In the basins flood plains were formed thus preparing the way towards silicification. Then a further period of up-lift must have happened according to LENZ so that the Groot river breached the mountains to the north. In the Mio-Pliocene the sea advanced once more and as far north as the Aasvoelberg, followed by a next period of up-lift so that the rivers were enabled to cut through the Mio-Pliocene deposits. It was then only that the Langeberg was breached and the drainage lines as they are today were finally established.

FABRY (1924) was also opposed to the idea of super-imposition and, before LENZ (1953) and even TALYAARD (1949),

Fig.12.1: e) Swakop

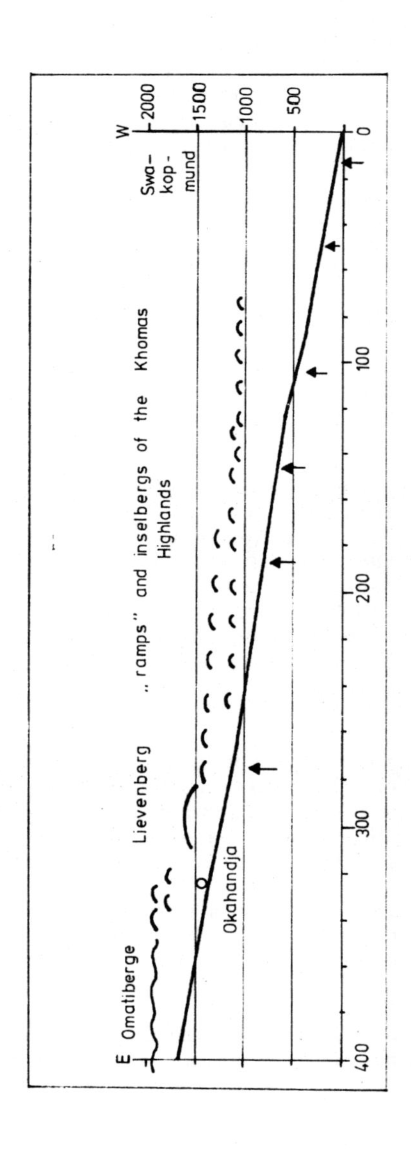

suggested headway erosion (after the sedimentation of the Lower Cretaceous beds) as the only way to explain the poorts. He saw today's main divides as the consequence of this headward erosion.

Apart from the Cape Fold Belt there has not been done much work with regard to the largest river area in the whole study area, viz. the Orange system. HELMGREN (1979) was more or less mainly concerned with its alluvial history only. ABEL (1959) considered the Orange valley as a very ancient form because, analogous to the Kuiseb-Swakop Gap in the north, the Orange valley uses a 'gap' within the continental margin and the escarpment of a comparative age.

JAEGER (1932) dealt with the Konkip and Fish rivers as the northern tributaries of the Orange. He interpreted their upper reaches as having been inherited from the peneplain above. The surfaces were excavated later. The Fish river canyon itself should be of Quaternary age.

Similarly, OBST (1937) thought the Orange to be a rather young feature coming into existence only after and with the (mainly Tertiary) downwarping of the continental margin. Other examples for a similar process should be the Kunene, the Great Fish, the Great Kei rivers or the Olifants-Doring-Tankwa system. According to OBST all these rivers were 'robbers'. The Orange, for example, by headward erosion, incorporated one drainage system of the interior after another into its own system. Kunene and Tugela were not as successful in sawing back through the escarpment.

Of the two, the Kunene did however reach back into the interior, there tapping older rivers and their valleys which once had flown to the Etosha pan. BEETZ (1950) attributed a strong headward erosion to the Kunene in doing this tapping and though the erosion to be contemporary with the upwarping of the continental rise during the Tertiary.

In comparison to the findings of former authors it can now be attempted to describe the results arrived at by the present writer are somewhat different.

The results for the drainage systems in the Cape Fold Belt and its hinterlands are summarised in Fig. 12. 2. This is possible because all the cross sections for all breaching

Fig. 12.1: f) Hoanib

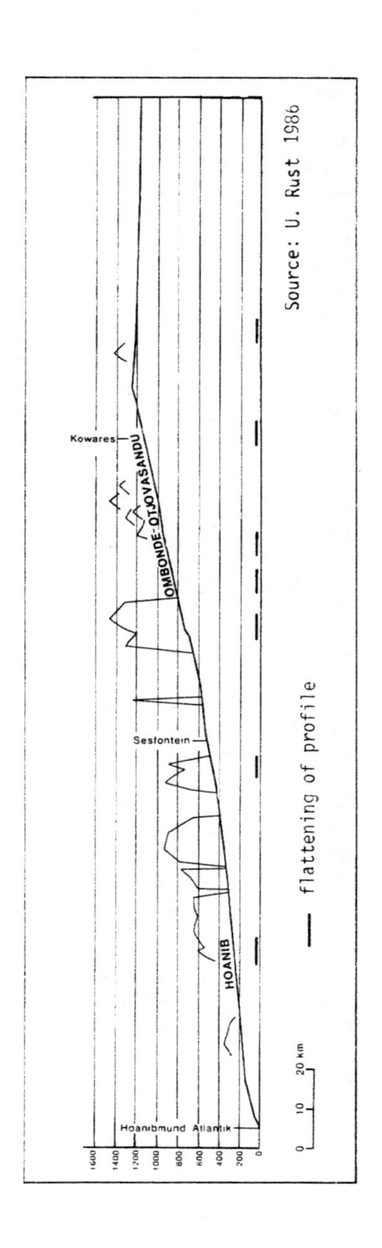

valleys revealed that all of them had indeed had a very similar evolution.

A detailed description of each drainage system is not needed, because the general picture is clear. This picture is constituted by the fact that all larger drainage systems have found their outlets through valleys which - proved by the equivalent 'terraces' - must have existed since the Lower Cretaceous and - again by proof of the equivalent 'terraces' - were formed in connection with the thalasso-plains - on and along the alley-like extensions of these through the Cape ranges. Since after the regression the thalassoplains must have served as a new surface for the newly emerging river systems these rivers must indeed be partly regarded as being superimposed (in the way of ROGERS).

The thalassoplains themselves - on following a sea level rise - however, could only intrude backwards through the ranges into the interior so far if there existed thalwegs prior to the transgression and if these thalwegs had been brought down deeply enough by erosion to become flooded. Only then would the thalassoplain (or the transgression resp.) be able to 'conquer' the (pre-existing) valleys and passes and thus pave the way for the later drainage routes.

In this way and step by step the poorts with their graded profiles came into existence, from the highest and oldest ingression level in 1000m ab. s-l down to the one in 300m.

Because of the narrow poorts as the only outlets through the ranges the drainage level seems to have become fixed in quite a lot of cases since at least the later Lower Creta-ceous. This naturally goes not to say that the drainage systems have been completely handed down since then in the way they exist today. This can be seen from some of the passes which are dry today (for instance, Garcia's Pass and Kareedouw Pass),yet show all the levels running through them, complete even with silcrete ledges and mesas. As everywhere else, such phenomena can appear only where the passes had become lowered by fluvial incision before. Only then was it possible for each lower transgression to pass through them. Streams might have passed through them again after regression.

Fig. 12.1: g) Cross profile showing the relation of the Lowveld terraces to past and present relief
Profil durch die Lowfeld-Niveaus im Vergleich des heutigen Reliefs mit dem Vorläuferrelief

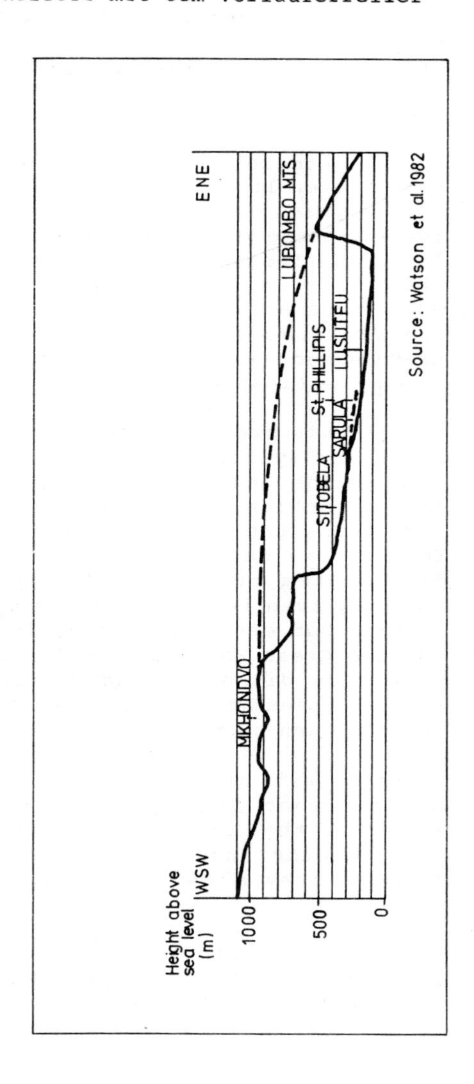

As Fig. 12. 2 clearly shows the main drainage systems - Sundays, Groot, Gourits and their various tributaries - must have used the same valley paths ever since the Lower Cretaceous. This fact means, altogether, that it is indeed not necessary to seek rescue with complicated assumptions of ever renewed headway breaching, as for instance LENZ (1953) was forced to do. The fact moreover means that the drainage system is strikingly old: Lower Cretaceous or possibly even late Jurassic.

What a predecessor of this very old drainage system would have looked like is naturally open to conjecture. Yet what of the great obstacle the Cape ranges must originally have formed, there once must have been a predecessor system that later and lastingly became overthrown as a consequence of the one chain of events that really could be the cause for such fundamental change: the large trans- and ingressions of the Cretaceous. The present writer's guess is that the predecessor system was developed in a sub-sequential trough along the Great Escarpment.

Within this trough there may however not have existed just one river system, but two, each of the two huge enough in itself. The one system seems to have drained the eastern trough section. This system is centred on the Great Fish. Even today the Great Fish still circumvents the Cape Fold Belt in the east. If the assumption is correct the Great Fish system must then be regarded as one of the oldest valleys, even in world-wide comparison. Though all this is nothing but mere conjecture, it would well explain the fact why the Great Escarpment - far into the interior atthat is so deeply dissected and severed into tablelands, mesas, spits-koppies and the like exactly where it has been slit up and cut open by the Great Fish.

Starting with this idea of a subsequential system centering on the Great Fish other pieces fall into place as, for instance, the markedly long eastern (or southeastern) directions of the Sundays and the Groot (or Sout) rivers - as long as they flow along this conjectural subsequential trough.

The other old and large system seems to have drained the northwestern section of the trough with a proto-Doring and a proto-Tankwa as its headstreams, which still run along the escarpment in a northerly (or northwesterly) direction. The

Fig. 12.2: Passes and breaching valleys in the Cape Fold Belt

Durchbruchstäler und Pässe in den Kap-Ketten

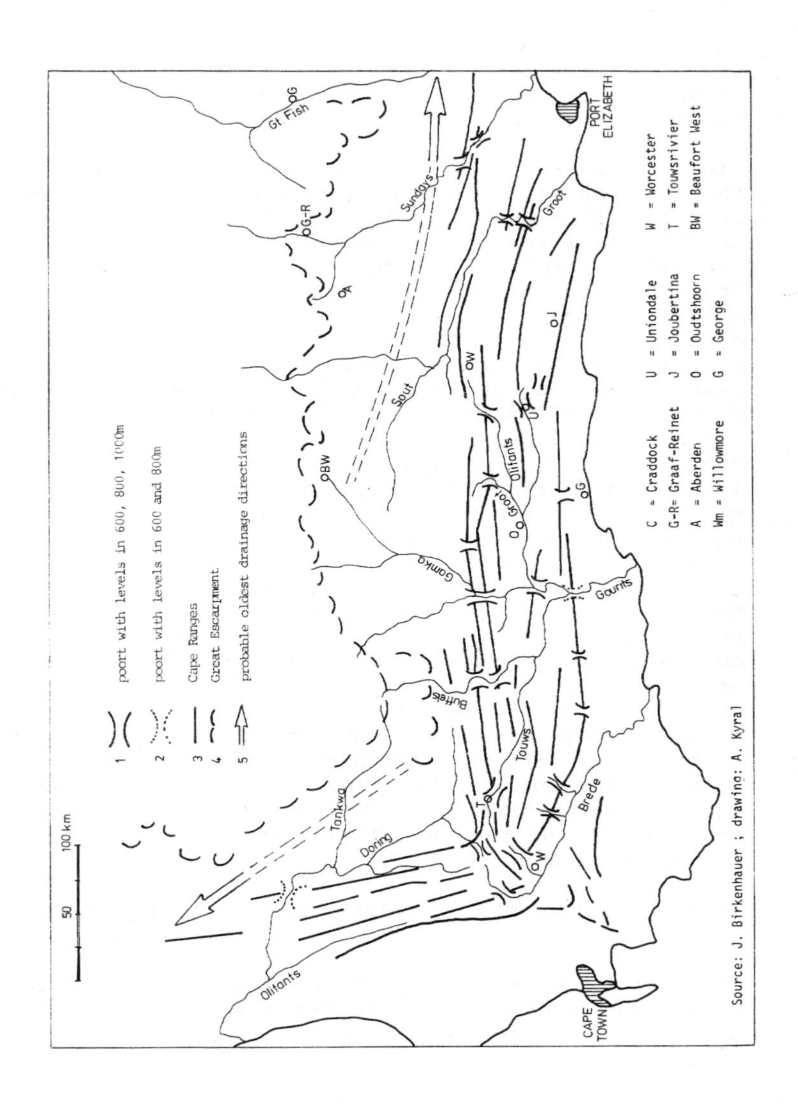

northwestern drainage system may later have been cut up by the Olifants system, i.e. since the Lower Cretaceous. The whole northwestern proto-system may well have centred on the Orange as its main stream.

The divide between these two proto-drainage systems could well have been the main divide of today: from the Komsberg escarpment in the north to the Witberge west of Laingsburg in the south.

Along the Orange itself it is a very remarkable fact indeed that all the Lower Cretaceous levels extend very far inland. In the neighbourhood of Upington, 500km inland from the sea as the crow flies the Orange valley bottom still lies at an altitude of only 800m ab. s-l and from there it needs another 250km (as the crow flies) to reach the next higher valley bottom in an altitude of 1000m (in the neighbourhood of Prieska). Where the tributaries - i.e. Hartebeest and Sak to the south and Konkiep and Fish to the north - reach the same levels they also flow in basins, very broadly opening up in these altitudes with very low gradients of the thalwegs. The combined thalwegs of the Hartebeest and the Sak need ca. 300km for a very gentle drop from an altitude of 1000m to that of 900m. Taking this into consideration it is not remarkable (as it otherwise would) why the valley bottom consists of a chain of very shallow pans and vleis.

From the description given of the Orange system it has become clear that the two Lower Cretaceous thalassoplains penetrated very deeply indeed into the interior of the continent. This observation leads on to the assumption that an equivalent depression must have existed prior to the thalassoplains so that the depression (very wide at that) could be made use of by the deep marine ingressions. The depression itself must furthermore have served as an outlet of the proto-Orange. If these conclusions are correct the Orange river system must also be a very ancient feature in the drainage of southern Africa (to be compared only with the Great Fish system).

HAUGHTON and FROMMURZE (1928), GEVERS (1933), KNETSCH (1940) and ABEL (1959) have all come to a similar conclusion.

The contour-lines of the 800 and 1000m thalassoplains are given in Fig. 12.3. They illustrate what has been described so far. The occurrence of the 800 and 1000m levels (together with their equivalent intermediate levels) over such vast distances is indeed understandable only if the corresponding ingressions were able to enter a widely opened up countryside within a preexisting great depression.

Similarly, the lower (i.e. Upper Cretaceous) ingressions penetrated deeply along the main valleys. The 600m level can serve as a very good example for this (Fig. 12.3). Because of the large extent of up-lift between the Lower and Upper Cretaceous (see ch. 10), valleys were formed that were particularly deeply incised and rather narrow. It becomes understandable therefore that only smallish 'terraces' could be formed when the ingressions penetrated into the steep and narrow valleys.

Some consequences can be witnessed in the Fish river canyon and the Orange gorge below the Augrabies Falls.

The Augrabies Falls with a total height of 200m (between an altitude of 800m above and that of 600m below) is a direct consequence of the fact that the Orange had to cross the scarp between these two levels. The drop in the thalweg - and together with it the falls - exactly occur where the 600m valley bottom close crosses the vally. Gorge, falls and canyon: they can altogether be explained by using a model for their evolution as the one proposed here.

As to the Fish river canyon, one can safely say that it owes its existence largely - or indeed its first appearance at least - to the same drop from the higher valley bottom in 800m to the lower one in 600m. Later on the canyon became accentuated by the deep ingression of the Tertiary 300m level. This level, as a consequence of the equivalent base sea level, invaded the canyon as far up as the main view point of the canyon (in the north of Ai-Ais). Near this place, in the crystalline basement, the 300m level forms the valley bottom close.

From all circumstances described above, one feels justified to assume that the Orange drainage basin existed before the earliest transgressions in the Cretaceous and must therefore be regarded as a feature of the Gondwana surface. This assumption is corroborated by MARTIN (1984) who also gave an Upper Jurassic age to the Orange basin.

Thus, even before the opening-up of the south Atlantic ocean the Gondwana continent seems to have been drained to the west, at least for this portion of the continent. If the attempts to re-set the continents on both sides of the Atlantic as parts of the old continent are correct (cf. HAYS and PITMAN III, 1973; HERZ, 1977; SIEVERS, 1983) the Orange basin must have been situated off the basin of the Rio de la Plata. It may well have been that both drainage systems formed a single huge interior drainage system of the old continent, similar to the great central basin of North America today.

Though all this is certainly guess-work only it is nevertheless not futile. It helps to visualize the larger early landform-units. An understanding of them may come in useful for throwing some light on the paleoclimatic conditions and their larger topographical framework. They will have to be discussed in a later chapter (see ch. 16).

From the asymmetric distribution of the terrigenous sediments on the shelf in the west of the mouths of the Orange and the Olifants DINGLE and HENDEY (1984) came to the conclusion that the two rivers used their present exits in late Cretaceous times, but that they had a common exit (with the Orange crossing the escarpment by way of the Calvinia wind gap) in the Tertiary up to the Upper Miocene. The alterations between the exits are, according to the authors, due to periods of low sea level (distinctly far lower than today as the existence of the Cape Canyon shows) which promoted river capture adjacent to the western escarpment. While all sediments on land are missing and the conclusions can only be called tentative ones (the authors frequently use the term "suggest") the paper can be taken as another indication that the Orange and Olifants catchment areas should be regarded as a single super-system. The 'gaps' reconstructed by the authors can as easily be explained by the different thalassoplanations, preparing gateways for the later headward river capture. (The paper by MC CARTHY et al. 1985, made use of the findings by DINGLE and HENDEY but are not pertinent to our research. Moreover they use data from MABBUT, 1985, that, according to our findings, must be revised.)

The knowledge gained from the Orange system and its evolution may be applied to the explanation BEETZ (1950) gave about the complicated pattern of the Kunene river. The

Fig. 12.3: The main geomorphic features of the western Orange basin
Die geomorphologischen Grundzüge des westlichen Oranjebeckens

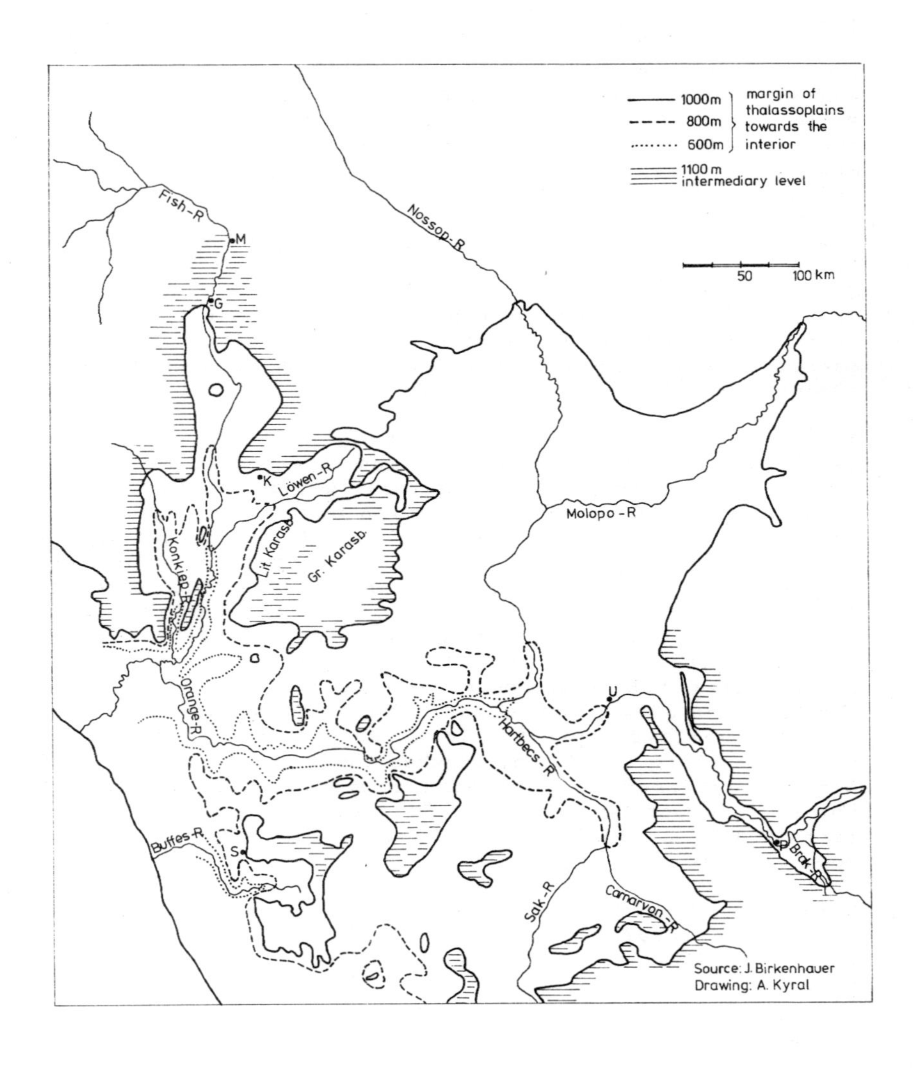

river, near Eriksons Drift, Humbe and Fort Rocades, flows in a valley bottom at an altitude of 1000m. The bottom, near these places and upriver from them, is very wide, for many kilometres. This in itself is very characteristic for valleys in this level as we have seen in the Orange basin. Then, 90km downriver from Fort Rocades, the Ruacana Falls begin, just where the river leaves the 1000m level and drops down to the one in 800m. It is a situation comparable to the Augrabies Falls, the only difference being that the Ruacana Falls are one 'storey' higher up. Characteristically again, the 1000m level accompanies the valley towards the Ehombo mountains, through which the river then breaks in a canyon and gorge, 60km long, before the valley (again very characteristically) widens when its bottom reaches the 800m level. Where this again drops off to the 600m level the Montenegro Falls set in.

Thus, falls and gorges and canyons are nothing but the usual knick-points between the levels where the thalwegs cross the scarps between the levels - only in a more dramatically accentuated way.

It is well possible as BEETZ thought that the Kunene truncated an older drainage that was directed to the Etosha pan, as quite a host of streams still going into that direction suggest. If so, the truncation must have made itself felt first on the 1000m level and then become fixed by way of the gorges downriver. In itself the truncating process must have followed along pre-existing valleys that served as 'alleys' for the ingressions, first for the one in 1000m, then for the lower ones. It may well be, as BEETZ showed, that the Coroca river in the northwest served as such a first valley, but later became discarded as a consequence of the 600m and 300m ingressions. For BEETZ pointed out that some old valleys cross the lower Kunene from southeast to northwest where the Kunene reaches the coastal plains. These old valleys seem to have been tributaries of the Coroca before they were completely filled up with sands - characteristically as high as the 300m level. After that they became disused.

Using the transgressions and the truncations initiated by them as an interpretative framework the 'mystery of the Kunene' as BEETZ called it can be understood in a far easier way.

Fig. 12.4: Valley bottom closes in 200m and 300m in the Oudsthoorn basin (southern Cape Province)
Talbodenschlüsse in 200 und 300m im Becken von Oudtshoorn (südliche Kapprovinz)

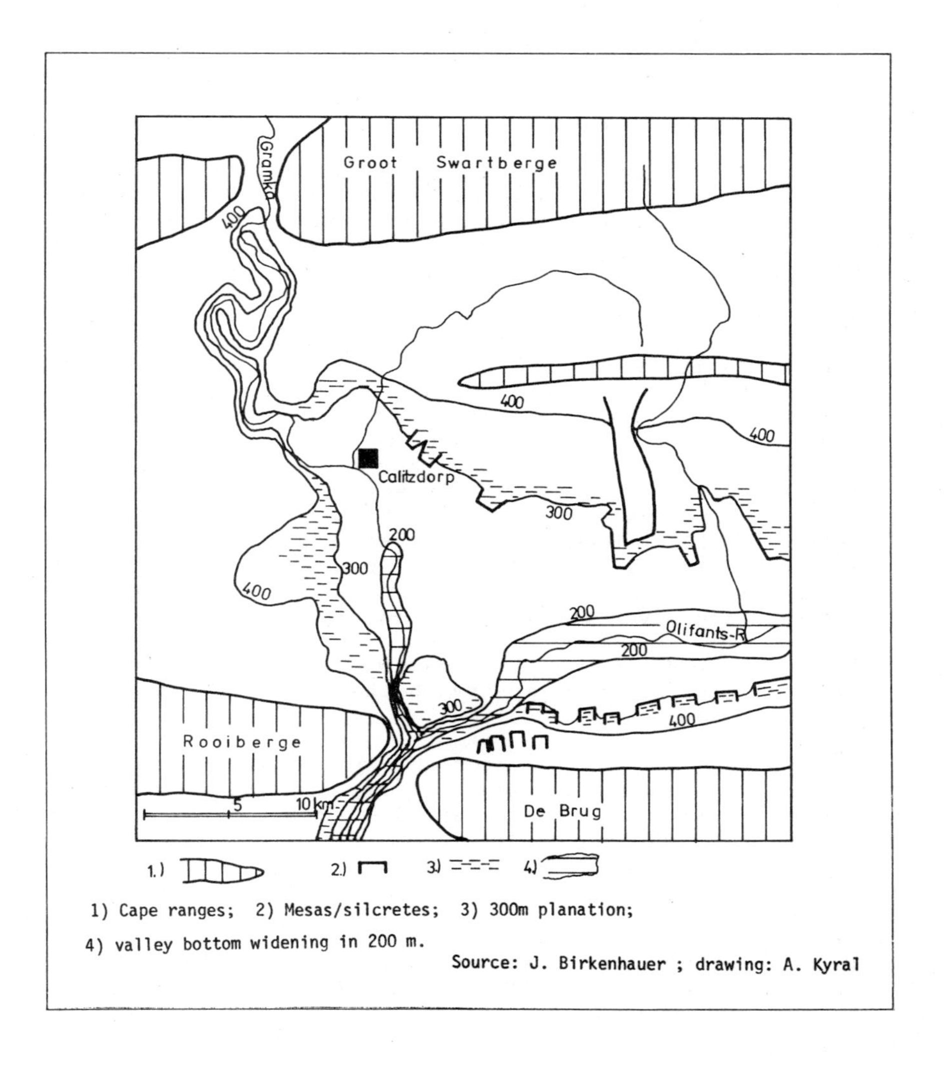

1) Cape ranges; 2) Mesas/silcretes; 3) 300m planation;
4) valley bottom widening in 200 m.

Source: J. Birkenhauer ; drawing: A. Kyral

Also, what for JAEGER (1923, 1930, 1932) was a big stumbling stone, i.e. that the higher inselberg chains in front of the Great Escarpment are crossed through by the rivers towards the coast, can be explained in more elegant a way. JAEGER, in explaining the breachings, had to resort to a first period of up-lift, followed by a complete down-lift - and yet had to admit that even such an assumption was not able to explain all of his observations, with the additional difficulty that the postulated down-lift could not be proved by other observations.

Turning our attention now to the eastern side of southern Africa we there find that the valley course of the Olifants river has often been payed attention to because of its intricate course, its cuttings, breachings and rebreachings. DAVIS, following VON RICHTHOFEN, considered the valley to be superimposed on a sedimentary infill that reached as high up as the present peneplain. This - significantly - is equivalent in height to the thalassoplains in 800m and 1000m. For KAYSER (1986, 13) the present peneplain cuts across all different rock-beds and therefore must be regarded as the original surface from which the river became incised down gradually in successive periods of up-lift. Though KAYSER did not expressly state it as thus his assumption leads on to the notion of antecedence. Anyway, the valleys and basins - here as everywhere else - must have been in existence before the 1000m level became formed. The penetration of the level as deep inland as it does is possible only because it could pass along other valleys and basins, i.c. the whole pre-existing river drainage systems.

In the Olifants valley the lower thalassoplains also reach back into the drainage system, both as 'terraces' and river bottom widenings.

As a consequence of the later transgressions, thalwegs different from the earlier ones may have become superimposed through other gaps (newly lowered and newly developed) thus leaving the older poorts as wind gaps (similar to the development of the poorts and passes in the Cape Fold Belt). The indeed intricate river courses can thereby be eventually explained.

Finally, with regard to the breaching valleys of the Lebombo range (cf. KAYSER 1986, 70), the following observations can be made along the Pongolo or Mkuze, for instance: all 'terraces' can be seen, either above the valleys or along them, both the higher ones in 400 to 600m, as well as the lower ones (300 and 200m). From these facts it can be gathered that the valleys were handed down (as it were) from the higher levels to the lower ones with each trans- and regression. But whereas KAYSER (1949, 144, 145) explained the valleys as being superimposed from the 200m peneplain of the Low Veld, the existence of the higher 'terraces' points to the conclusion that all the breaching valleys must be older than the 200m level. Superimposition seems to be feigned only. Moreover, the 200m level was at least partly removed by peneplanations, for which the 100m level served as a base level. (See Appendix E.) The latter circumstance was also mentioned by KAYSER (1949, 145). It is however contradictory to his own argument.

WELLINGTON (955, 49) also used the concept of superimposition. Though we think his concept wrong, WELLINGTON's argument may be given here in full text, for completeness' sake. "In the eastern portion of the watershed the rivers rising on the Ecca surface are being imposed on the older denuded surface to the north. Thus the Olifants river flows from the 'Karoo watershed' across the pre-Karoo Formations margining the Bushveld basin and so to the eastern lowveld. The Komati has a somewhat similar course. The key to the character of the plateau edge in this part of South Africa appears to be the recognition of the origin in the eastern Transvaal drainage on the Karoo surface, and, secondly, the effect of the great Lebombo flexure and subsidence in the mid-Mesozoic, which laid open to plateau drainage the steep slopes for the newly formed Mocambique Channel.... The present course of the Great Escarpment seems to be largely the effect of the headward erosion of post Lebombo subsidence streams, cutting back into the flexed and probably faulted margin of the plateau. The two processes of superimposition and headward erosion have apparently gone on together, the result being the receding escarpment of the Transvaal Drakensberg ... transsected by super-imposed rivers."

Summing up, it does seem to the present writer that the development of the river drainage systems in the forelands of the Great Escarpment (and also in the drainage areas well behind it) everywhere around the coastal arch makes altogether better sense if considered within the framework of the transgressions and thalassoplains in the Cretaceous and in the Tertiary. Moreover, this framework alone makes it possible to understand how the planations were enabled to cut back to the very foot of the escarpment, and even through the Cape Fold Belt, by the express use of pre-existing drainage systems.

In this way, together with the embayments formed by thalassoplanation along the drainage systems, it is finally possible to explain the sinuous nature of the escarpment in the easiest way.

13. THE NATURE OF THE GREAT ESCARPMENT

It seems advisable at this stage of the treatise to review the previous findings with regard to the main question of the inquiry: the evolution of the Great Escarpment and what the nature of the escarpment is.

So far it should have emerged that the escarpment was subject to three main conditions. These three conditions seem to have been the necessary ones. Therefore they are called 'formatives', and it is under the influence of these 'formatives' that the Great Escarpment has come into existence as it stands today. The three 'formatives' are:

(1) The three main tectonic depressions of southern Africa which pre-figure the strike of the sedimentary layers as well as that of the dolerite sheets. A very good example is presented by the circumferential strike of the Karoo series in the southern Orange basin.
(2) The presence of layered rocks (including the dolerite sheets) that allow the forming of cuestas.
(3) The erosional and denudational attack from the drainage basins against the higher fringes surrounding them.

These three 'formatives' are essential for defining what the Great Escarpment is. Bearing these three essentials in mind it must be pointed out that the Great Escarpment in a proper amd true sense should be spoken of only when and where all these three essentials occur simultaneously. Where one of them is missing, the Great Excarpment in any proper sense is missing also. This goes on to say that no other mountainous area along the coastal arch of southern Africa should be regarded as being a constitutive part of the Great Escarpment proper as long as one of these three essentials is missing - even if such a mountainous area runs, more or less roughly, in the strike of the Great Escarpment proper, as is for instance the case with the isolated horst of the Khomas highland block.

In the following paragraphs the two latter essentials or 'formatives' must be dealt with in some more detail. For the first one (as being rather self-evident) it here only needs to be pointed out that the structuring of southern Africa into three tectonically distinct basins is certainly not a trait that is owed to the breaking up of Gondwanaland but a trait that already existed before the breaking up (cf. OLLIER 1985).

Formative 2

In Natal, in the Transkei and in the eastern and northwestern Cape Province the Great Escarpment is clearly bound to the most resistant rocks. These are especially the dolerite sheets. Wherever these sheets are sufficiently thick the Great Escarpment presents an unbroken and continuous front. Such a front is the case over a very long stretch of distance indeed: from the Hantamsberg in the northwest (near Calvinia) down to the Komsberg Eskarpment, then right on to the Natal Drakensberg, and on from there into northern Natal, where the Ecca quartzites intersect with the dolerites. This uninterrupted stretch amounts to more than 1500km (the embayments not counted in).

Now, in spite of the fact that dolerites occur in many more places north of the line formed by the Hantamsberg in the west and the basin of Piet Retief in the east, the dolerites however do not appear as sheets north of this line, but only as isolated cones or dykes, sometimes as swarms of these. Wherever this is the case, weathering, denudation and erosion has only led to the forming of inselbergs, but nowhere to the forming of a cuesta.

Along the stretch mentioned above no other stratified rocks cause such towering cuestas as the dolerites do. If cuestas do appear on such other rocks they are only minor and fully subordinate to the dolerites, as, for instance, the middle Ecca quartzites in Natal, or the Beaufort sandstones in the Little Karoo.

This observation obviously means that under prevailing climatic conditions other rocks are not resistant enough to bear cuestas, and through them, the Great Escarpment. (From what has been said before, it is equally clear that such climatic conditions must have prevailed throughout the geological past for a considerable time.)

In the northwest of the Hantamsberge (with which the unbroken stretch towards the south and east begins) there occurs a considerable and conspicuous gap within the Great Escarpment. This gap is coincident with the disappearance of the dolerite sills. And when the escarpment re-appears, after the gap in the northwest, it suddenly is made up by the sandstones and even greywackes of the Ecca and Dwyka series - and farther to the northwest by the sandstones, greywackes and limestones of the Nama system. (See Table 13. 1.)

It is from this gap then that these and similar rocks substitute the dolerite sills for the forming of prominent cuestas - a consequence obviously of altered climatic conditions (arid and semi-arid).

It should be furthermore pointed out that the Great Escarpment is not just one single cuesta and not formed out and upon just the same rock layer. As the benches of rock change along the arch of the escarpment, so change the cuestas. Thus the Great Escarpment must be considered to be rather a chain of cuestas (German: 'Schichtstufen' or better: 'Schichtungsstufe'). As such, the Great Escarpment is probably the longest 'cuesta chain' (or 'Schichtungsstufe') in the world.

In the northeast of southern Africa, on the thick old quartzites and dolomites of the Transvaal system, similarly successive cuestas are built up - and with them the Great Escarpment.

In the opinion of the present writer it is to be concluded that the Great Escarpment, as a landform, ends where the 'cuesta chain' ends, i.e. where the chain is broken over too long a distance. This is the case, for instance, in the northwest, where, north of the Nosob-Molopo-Fish river basins the Khomas Highlands begin to rise and where, between the Tsarisberge in the south and the Kaokoveld Cuesta in the north a gap appears over a distance of 500km. Thus the Tsarisberge is to be considered as the last real buttress of the Great Escarpment proper towards the northwest. Similarly, in the northeast, there appears a large gap north of the Waterberge towards the Limpopo basin.

Formative 3

We have here to deal with the sinuous nature of the Great Escarpment. This sinuosity can now be explained with respect to the different thalassoplains and their intermediary levels and furthermore with the condition of how far these were able to reach right back to the foot of the Great

Escarpment, to be undercut by these levels. Such an undercutting must have gone on along the pre-existing valley basins, these forming embryonic embayments. Within these embayments the sediment layers must have been efficiently removed. The undercutting process is responsible for one of the great characteristics of the Great Escarpment: i.e. its front, its buttresses, outliers and so on tower above the embayments; it is the removal (German: 'Abräumung' oder 'Ausräumung') of complete sheets by which the plain surfaces of the embayments (within the levels of the thalassoplains of course) were formed. Both processes happened synchronously and went on over considerable time; it was by both processes that the Great Escarpment proper came into existence.

Thus, along its whole course, the Great Escarpment proper (i.e. in the sense of a 'cuesta chain') is a true erosional scarp, modelled as such and in its sinuosity in connection with the thalassoplains - which themselves followed into the large pre-existing river systems of the Orange, the Olifants, the Gourits, the Sundays, the Great Fish, the Great Kei, the Mzimvubu, the Mkomazi, the Tugela, the Mfolosi and the Pongolo rivers, or even farther to the north the Olifants river system of Transvaal.

Table 13.1: Escarpment and rock series in the northwestern Cape Province and in southwestern Africa
Die Gesteine entlang der Randstufe in der nordwestlichen Kapprovinz und im südwestlichen Afrika

Escarpment	Region	Series
Langeberg	Loeriesfontein	Dwyka, Ecca
Koubiskouberge	Loeriesfontein	Ecca
Richtersveld, in the W of Springbok	south of Orange river	Nama, Ecca
Huibplato Klein Karasberge Tirasberge	southern South West Africa	Nama

14. SLOPES, SCARPS AND CUESTAS

a) General remarks

In the preceding chapters the terms 'scarp', 'escarpment', 'cuesta' have been used in order to denote alignments of particular slopes with certain regular features over wider distances. A feature is then called 'regular' here if the landform in question possesses some characteristics that can be generalized as, for instance, comparative heights between base and top, similar altitudes, or a restriction to a particular type of rock. The difference between scarp and escarpment is judged to be one of dimension (escarpment higher than a scarp), the difference between scarp and cuesta one of structural control (which the former must not have, but the latter must). The word 'scarp' has here been used so far in order to denote slope features that - though in continuous alignment - are a hundred or two hundred meters in height. In the context of this book the term 'scarp' has usually been reserved for those slopes that occur between one thalassoplain and the next. Such scarps form the distinct erosional margin between one level and the next.

One of the regular aspects of the thalassoplains - except in the 'low gradient embayments' (German: 'Flachböschungsbuchten') and except the gradings into the intermediate levels - is indeed this pronounced rise from one level to the next with distinct knickpoints at the foot and the top of these scarps.

Though the main object of this chapter is the cuesta-forming process, it is advisable to deal with slopes and pediments first.

b) Slopes and pediments

In the eastern parts of the study area, i.e. eastern Transvaal, Natal, the Transkei, the eastern and southern Cape Province, the nature of the slopes is fundamentally the same at whatever altitude or level they occur (Fotos: 38, 55). This is taken as an indication of the climatic conditions having remained in a humid or semi-humid state during a considerable time, presumably at least since the Upper Cretaceous. This assertion may proved by a profile taken from northern Natal. Here, near Candover, on an interfluve north of the Mkuze river, all planatory levels can be observed - from the one at the bottom in 100m ab. s-l to the one in 600m (Fig. 14.1).

In the parts of this cross-section marked I and II some flat slopes occur that might be judged to be just one slope (Fig. 14.2). This judgement, however, is wrong since each of the flattish levels has to be considered as a separate morphogenetic unit. As reasons for this assertion can be given the following ones:

(a) a distinct slope steepening between one flat section and the next (i.e. between levels II and III, at an altitude of ca. 250m ab. s-l;
(b) an as distinct flattening out of each higher level with angles of 2 to 4 degrees towards the respective upper hill-side;
(c) the dissolution of the lower parts of the levels III and II into rounded hill-tops by erosion from below (cf. Fig. 14. 2; Foto 40).

The observation stated under (c) shows that the higher levels are in the process of being attacked from a rejuvenated erosion and are not flattened out by any on-going pediplanatory process.

Therefore the higher levels are separate and distinct features and - since they are being attacked from below - must certainly be considered as being fossil forms.

The flattish parts moreover agree very well in height with those of the 'marine' platforms or 'terraces'. From this comparison it is clear that these flattish levels can by no means be fossil pediments. According to the common altitudes of the 'marine' levels the 'storeys' to be observed in Fig. 14.1 can be dated (I: Upper Cretaceous; II: Oligo-Miocene; III: Upper Pliocene).

From the profile (Fig. 14.1) it can also be gathered that the farther up a level is situated the more pronounced the results of the slope-forming processes have become. Slope-forming is naturally a function of time. The longer the time-span has been the more pronounced slope-forming can have become.

Fig. 14.1: Profile in northern Zululand (Natal) along the road from Candover to Magudu
Profil im nördlichen Zululand (Natal) entlang der Straße von Candover nach Magudu

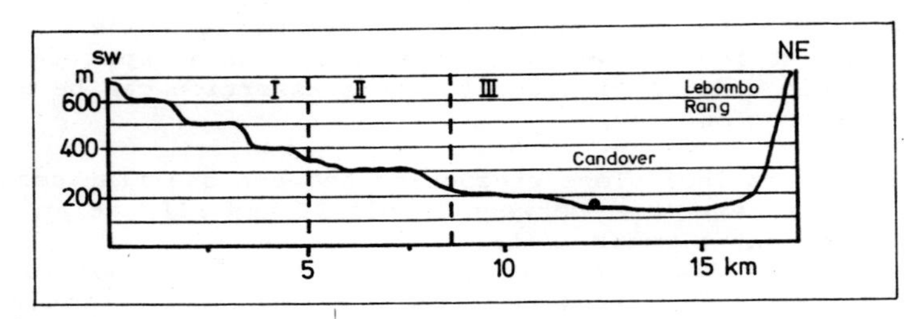

I: Upper Cretaceous levels; II: Oligo-miocene level;
III: Upper Pliocene level.
Rocks along whole profile: volcanics

Source: J. Birkenhauer ; drawing: E. Glaser

Fig. 14.2: Typical slope "garlands" in Natal (schematic) on shales, crystalline rocks, softer sandstones and the passings from one of these rocktypes to the next

Typische Hangsequenzen im Kristallin, in Schiefern, wenig festen Sandsteinen, wie auch auf den Übergängen von einem dieser Gesteine zum anderen

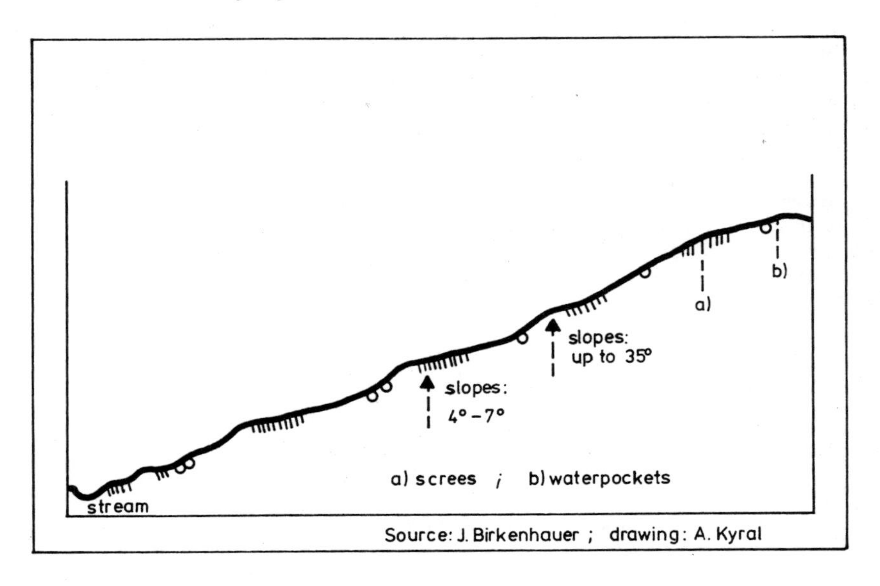

Source: J. Birkenhauer ; drawing: A. Kyral

There is no general change in the outlook of the slope-forms between the lower and younger levels and those higher up. This is another pointer to the conclusion that similar climatic conditions must be assumed for the whole time-span since the Upper Cretaceous. From recent (or sub-recent) conditions one is able to say that humid or sub-humid climates must have prevailed.

The example presented in Fig. 14.1 is taken from an area in Natal where the Drakensberg dolerites are extant. Dolerites are however particularly liable to be formed into rounded hill-tops, a process through which the original level (from which the erosional process started) has become dissected or diffused. The higher up a level is (and the older it then is) the more these rounded hill-tops appear. The farther down - and younger - the level is the less the rounded forms can be observed and the more the level forms are still preserved. The reason for this certainly is that the reshaping processes have set in on the lowest levels rather recently only.

c) Cuestas and related landforms

The cuesta forms are here dealt with together with the related landforms, such as benchlands, mesas, tafelbergs, towering buttresses of layered rocks, outliers and so on. The reason for so doing is that all these forms are restricted to a horizontal or near-horizontal rock-bedding.

In southern Africa, the related landforms usually occur in the forelands of the cuestas proper, i.e. in those areas where the cuestas have been attacked from below with the valleys cutting into their fronts. Much of the very rugged nature of the Drakensberge and their towering buttresses depend on this circumstance, for instance. The same is true about the wide benchlands, mesas, the projecting and interdigitating flat interfluves or the spitskoppies in and up the river basins - of the Tugela, the Great Kei, the Great Fish and their tributary valleys (Fotos: 41, 42, 43, 52).

All these forms owe their existence principally to the same shaping process - however humid or semi-humid or semi-arid the particular regional climatic conditions are today or have been in the past.

Fig. 14.3: Typical slope forms in hard and in shaly rocks (Natal)
Typische Hangformen in harten und schiefrigen Gesteinen (Natal)

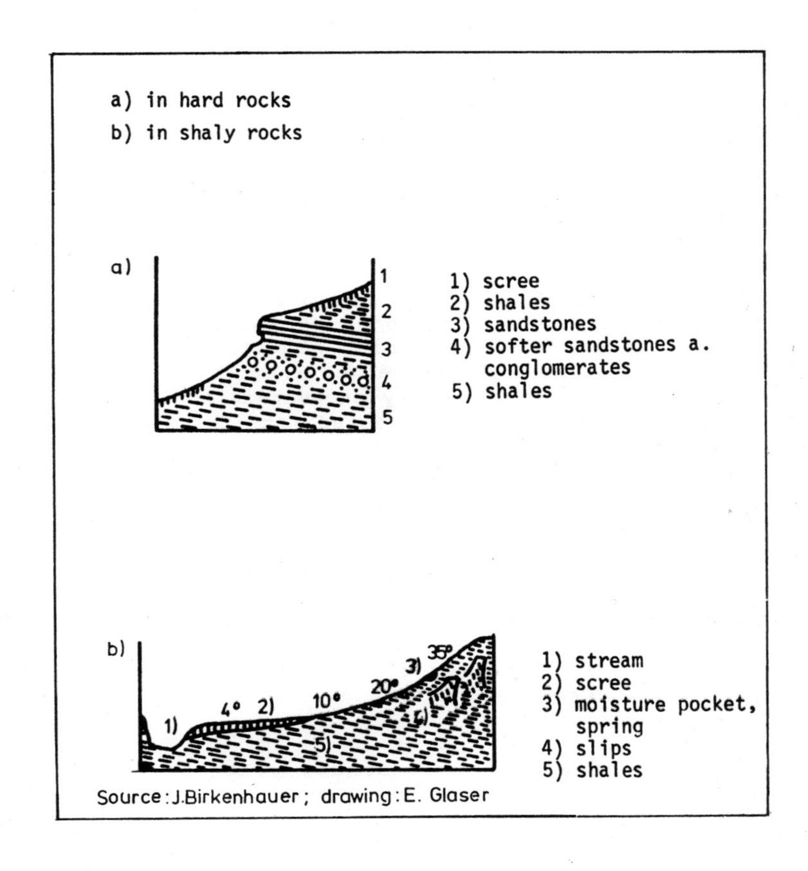

This is a process strongly connected with the water that is drained away through the cracks and joints of the resistant rock-beds. Wherever such percolating water is stopped by a layer of shales or similar rocks underlying the cracked rocks above and is therefore forced to form a local water table and wherever this local sheet-bound table is cut by an outward slope water pockets occur or even springs, forming regular spring lines (German: 'Quellhorizonte'). Even in the rather dry years that happened in Natal between 1978 and 1983 these springs and water-pockets - being fed by the percolating water from above - remained wet so that small streams trickled down from them, continuing the erosion process below each pocket or spring, i.e. the forming of a rill. The fact that even deep gullies have been formed below the pockets and springs indicates that they have 'functioned' in the way described through a considerable time.

Above the pocket or spring another characteristic feature has developed, i.e. the nodge beneath the more resistant layer. The nodges - occurring all along the spring line - have been formed because the wet layer beneath the resistant sheet has been continuously excavated. As a consequence the nodge has continuously retreated into the slope by headward erosion. Eventually, the undercut rock, hanging over the nodge, breaks away along its cracks and joints, sometimes in tables and boulders as huge as houses. (This is especially the case with the Cave Sandstones in the Drakensberge.)

All along the spring lines these processes take place everywhere so that the whole stratum is effectively attacked and gradually forced to move backwards - though not in the way of slope retreat along the whole front and parallel to it, but rather in the way of smaller and larger embayments. Embayments here also serve as the vanguards, comparable to pediplanation.

Wherever harder rocks alternate with shales (or similar rocks from the valley at the bottom right up to the highest crests and buttresses at the top) the process is repeated again and again so that the whole larger slope is being attacked continuously over its total front.

Rills or gullies do not only occur beneath the springs, but also immediately above the nodges, where they are often rather deep. Such rills or gullies reach up to the next

Fig. 14.4: The 'amphitheatre-effect', illustrated from the escarpment landscape near the Sani Pass, Natal (schematic)
Der "Amphitheater-Effekt" am Beispiel der Randstufenlandschaft am Sani-Paß (Natal, schematisch)

1 : dolerites
2 : cave sandstone
3 : excaved amphitheatre
4 : water pocket horizon
5 : small amphitheatre below water horizon
6 : forming of nodges and caves
7 : quartzites or harder sandstones and conglomerates with shales in between

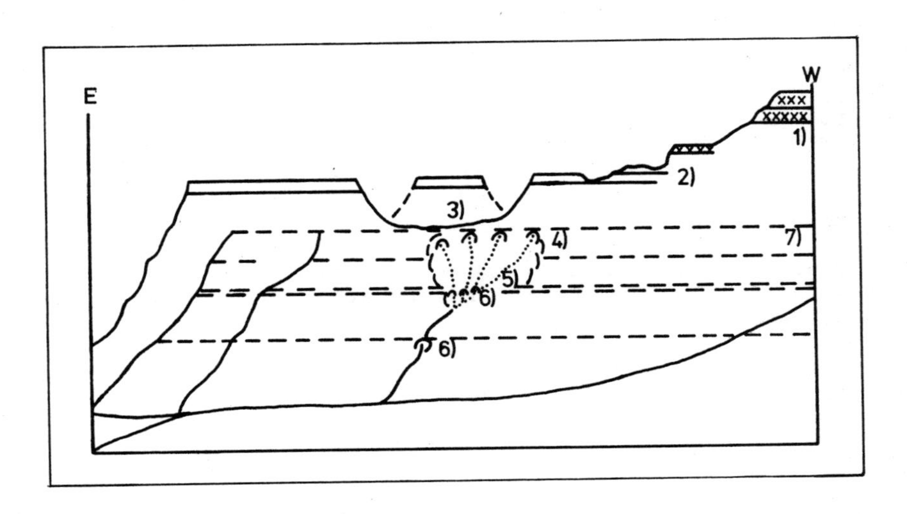

harder stratum and from there to the next one (and so on), i.e. the water supply for such 'rilling action' is fed from each new water table below each of the hard beds. In this way streams (or gullies) develop and 'jump up' as it were from one hard bed beneath to the next higher up (and so on). This may be called the 'jumping-up effect'.

This effect now is multiplied wherever a stream reaches a shaly layer with a broader water table. There immediately other small streams branch off into the slope sections above. The result is that the same over-lying resistant bed becomes attacked from several places not far from one another, all of them having their identical base level. An embayment begins to be formed which gradually forces the bed back. Very frequently the small embayments join into one large embayment that looks like an amphitheatre. It is in these amphitheatres that erosion naturally is fostered on an even larger scale - successively attacking the harder beds higher up. The 'jumping-up effect' thus leads to what may be called the 'amphitheatre-effect'.

By a combination of both effects whole sheets of strata have effectively been removed. Complete removals - generated by these two effects - especially occur on the interfluves - being attacked simultaneously from both valley flanks (or embayments).

Fig. 14.3 and Fig. 14.4 (showing situations from the Sani Pass area in the Natal Drakensberge) are fair examples illustrating the effects.

Yet they are not only typical for this area but for all similar cases , especially where deeply incised valleys abound.

It is exactly these processes founded on the effects described that have formed the Great Escarpment of the Natal Drakensberge between Kokstad in the south and Wuthering Heights in the north and to which the Natal Drakensberge owe their majestic scenery.

Even if there have been older surfaces along the front of the scarp as KING assumes (cf. the description of the foto from the Sani Pass area in KING 1972) they must have become completely obliterated within these 120 my or so by the processes described. All surfaces that can be observed here today are structurally based.

The features resulting from the processes described above seem to be most prominent where two conditions are given: (1) where the strata show a steady incline towards the interior; (2) where such strata have been ripped open by deeply incised valleys. Under such conditions the cuesta-forming processes have never come to a standstill since the first undercutting of the Great Escarpment.

Where the bed structure is however more or less horizontal or only gently undulating as in many parts of central and northern Natal (for instance in the basins of the Tugela, Mkuze, Mfolosi and Pongolo) the scenery is dominated by large benchlands and/or long ranges of tafelbergs. In front of these one can see surfaces that are basined in and which, because of their very flatness, seem to be very old landforms (if one applies the Davisian notion of the erosion cycle). Yet these flat forms are no proofs whatsoever for any cycle, because they are nothing but parts of the benchland and tafelberg regions which all of them have been formed under the same cuesta-forming processes as the ones described above. And here - as indeed everywhere under similar conditions - the processes are going on.

Though these processes can best be studied under the humid conditions of the Natal Drakensberge they are active even under arid conditions, as can be seen along the extensive cuestas of South West Africa or on both sides of the Fish river basin and especially in its canyon. (The cuesta-forming processes are the main causes for the shaping of such a magnificent canyon.)

Naturally, the processes are slowed up the drier a regional climate is, since the water table is not as saturated as in Natal. Therefore water pockets and springs are not as densely spaced in these arid regions as is the case in Natal. Yet, the self-same effects begin to appear (i.e. 'jumping up', 'amphitheatre') wherever a spring and a stream occur. (Foto 43)

d) Slope retreat

The description of the processes leading towards the shaping of cuestas and mesas (and so on) may have suggested the idea that the retreat of the cuestas (and all similar slopes) is rather quick, especially where there is a combination of favourable conditions.

In Natal two favourable conditions seem to have been prevailed (and through very long times at that, it seems). These two conditions are the following ones: (1) humidity; (2) successive rejuvenation.

Yet inspite of these two prevailing conditions the assumption of quick scarp retreat has to be severely doubted in contrast to what KING (1962, 1967, 1972, 1982) has again and again asserted for slope or for scarp retreat.

In dealing with the question it does not seem necessary to go into the details of slope development, slope sections and their angles (and related matter), though the author took many measurements. (A reader interested in these details may turn to another publication: BIRKENHAUER 1986.)

Generally speaking, the steep upper slope sections are not subject to much denudation and as such seem to be rather durable and lasting forms (cf. LOUIS and FISCHER 1979, 145; YOUNG 1961, 130; BREMER 1981, 25). There are two reasons to be given for this: (1) As JESSEN (1936) already noticed in Angola the steeper sections are less prone to deep weathering than the flat ones because the water necessary for the weathering process is rapidly drained away (steepness, clefts and joints helping each other), which is especially the case, by the way, on cuestas; (2) denudation 'snaps' because, on the steeper slope sections, the catchment area becomes smaller and smaller the nearer it is to the upper knick-point (HORTON 1945; BÜDEL, 1977, for instance, referring to it, in German, as 'Abreißen der Denudation an der Arbeitskante').

Conditions favourable towards an approximately exact calculation of slope retreat can be found in southern Natal. Here, near Margate and Palm Beach, going inland on the dirt road, one crosses three successive scarps or slopes (Fig. 14. 6). The first scarp (A) - near kwaNzimakwe - leads up from an altitude of 100m to one in 180m, from where it smoothly runs up towards an altitude of 250m; at this altitude the gradient steepens again. Scarp A and the flattish level above it are all situated on the crystalline basement. The slope of scarp A has a length of ca. 175m reckoned from the knickpoint on top towards where it flattens out below. Scarp A must have been formed in the time between the 100m level below (early Pleistocene) and the 180m to 240m level above (Upper Pliocene). (For the times given: see ch. 5 c.)

The second scarp (B) must have come into existence between the Upper Pliocene and the Miocene (or Oligocene). The slope of scarp B is widest at the northern end of the map section. Measurements were taken from there. The level above scarp B runs smoothly up to an altitude of 345m ab. s-l where the next scarp (C) begins. Scarp C slopes up to the altitude of 400m (and is therefore the lowest of the Upper Cretaceous levels). Scarp C and the level above it were shaped into the TMS, whereas the flat lower section of the slope is still on the crystalline basement. (Cf. Fig. 14. 5 b). The length of the slope amounts altogether to some 375m.

For the length of time during which scarps B and C were formed different assumptions must be made. Assuming scarp B was formed between the Upper Miocene and the Upper Pliocene then ca. 5 my seem to be enough; if one however considers the Oligocene as the upper time base, then 20 my must be considered. Similarly, scarp C may have become formed between the Maastrichtian and the Lower Miocene or between the Maastrichtian and the Lower Pliocene. The respective figures then are 40 my and 58 my.

In assuming that the slopes began to be formed from the distal tip of the flat section and were forced back from there to more or less the present position of the steep upper section one is able to connect slope length and time-span. The results for the different time-spans are given in Table 14. 1.

Table 14.1: Calculations of scarp retreat in southern Natal

Berechnung zur Hangrückverlegung im südlichen Natal

Scarps	Length in m		my.	cm in 1 000 y.
A	175		2	8.8
B	425	a)	20	2.1
		b)	5	8.5
C	375	a)	40	0.9
		b)	58	0.6

From the calculations one may gather the following results. The retreat of scarp A must have been comparatively quick at first, but tarried very much after a while (results for the older and longer slopes). This tarrying seems to be the result of two conditions: the length of the lower flat slope section and the steepness of the upper section. The rock types do not seem to exert much influence as the comparison between scarp B and C shows.

A slope profile with a very flat and long lower section and a rather steep upper one suggests the notion of'ripeness'. Retreat seems to come to a standstill where such a 'ripe' slope profile became formed. The reasons for such a standstill may be twofold: (a) the lower section has become too flat so that not much material can be transported away any longer; (b) the upper section has become so steep that denudation snaps.

The obvious result of these reflections is that there must exist a critical limit for any parallel slope retreat, or indeed for any kind of slope retreat. On the base of quite a number of angle measurements this limit seems to be reached with angles of between 25 to 30 degrees (disrespective of any rock-type, as the measurements also showed) and with slope lengths of approximately 500m.

It should be noticed however that BREMER (1975, 28, annotation 2) repeated on older statement of hers declaring that slope denudation is the stronger the steeper the slope is. BREMER considered this to be a rule for all geomorphological regions, be they in the tropics or in the so-called ectropics.

Fig. 14.5: a) Levels and scarps near Margate (southern Natal)
Nivaus und Stufen bei Margate (Südnatal)

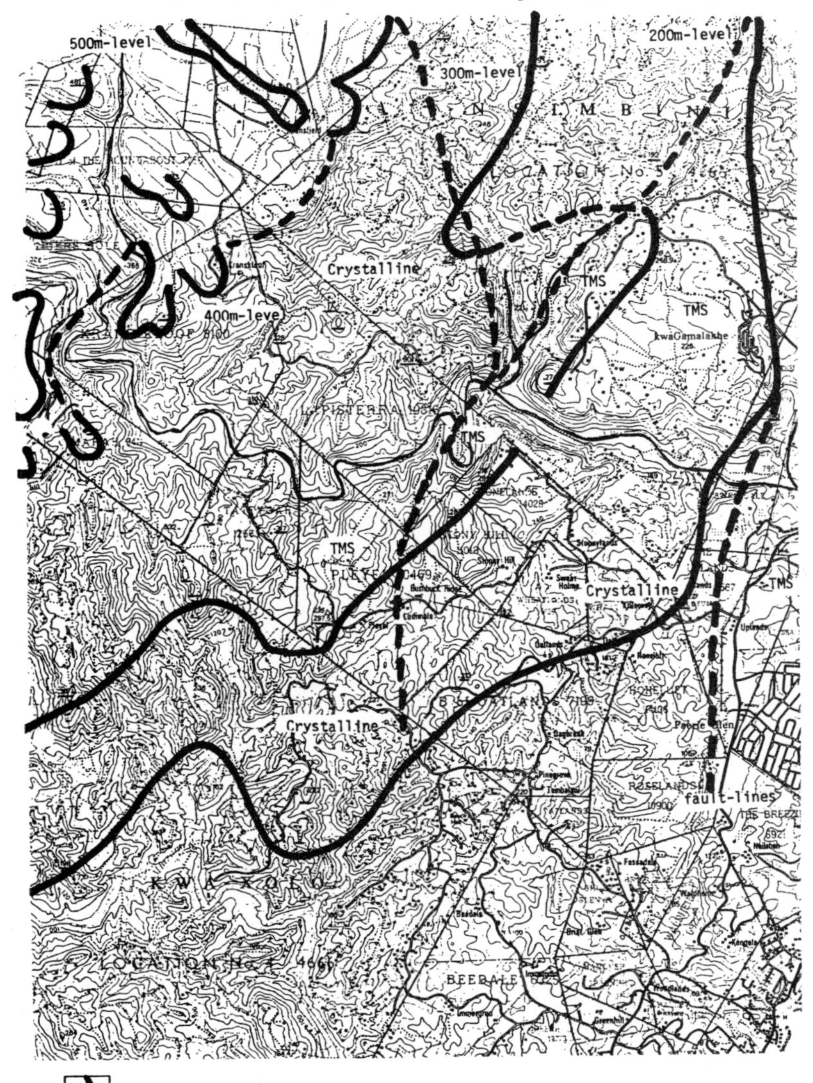

rim of levels

fault - lines

Fig. 14.5: b) TMS-slope near Margate (southern Natal)
TMS-Hang bei Margate (Südnatal)

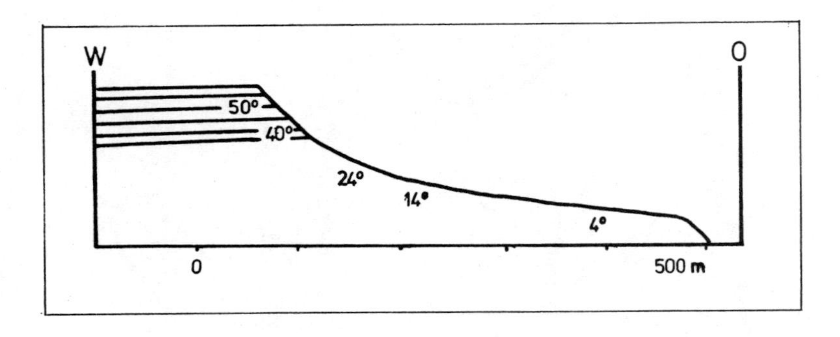

Source: J. Birkenhauer ; drawing: E. Glaser

As can be seen from the cases in Fig. 14. 3 the slope-forming processes have not been rejuvenated there, in spite of the different regressions of the sea level. This observation is a consequence of the fact that the valleys have so far not reached back to the slopes in question.

FAIR (1947) studied slopes in Natal in some detail. He (using the retreat 'imagery') came to certain conclusions which are similar to those of the present writer's. He states for instance: "...hills have retreated beyond the effective reach of lateral planation by large streams" (106), or: "Apart from a narrow zone at the foot of the Great Escarpment the major rivers play almost no part in keeping scarps steep as the latter have retreated beyond the range of planation of such rivers" (110). Other relevant observations are: "...stream spacing is not sufficiently wide to allow for the disappearance of scarps by progressive pedimentation " (113) and "...but continuous regrading does keep retreating scarps higher" (113). In spite of all these observations FAIR remained in favour of active scarp retreat by pedimentation. According to FAIR the process happens like this: "As the supply of débris from above becomes less and less, the talus slope assumes angles of decreasing declivity until it finally disappears". So "...the pediment ... merges into the summit" (113). For all this he did however not give a single example where such a process could be seen in action, nor, where it had been in action before. According to the present writer's observations FAIR would really not have been able to do so, not from anyone place in Natal (or elsewhere in the humid or semi-humid areas).

Altogether FAIR had to concede that "progressive pedimentation of the scarps of structural plains does not occur" (113). In spite of this and the lack of any proof he nevertheless went on to say: "The fact that inselbergs and scarps have retreated considerable distances from their initial positions (though he never proved a single example: J.B.) and have retained (sic! J.B.) their steep sides, is sufficient evidence to postulate a parallel retreat of slopes" (110). Moreover, what FAIR called pediments is really nothing but the waning lower wash-slope section. As he says: "the pediment is essentially a slope of less than five degrees over by far the greater part of its length" (110). (Such a statement is against today's knowledge about the usual angles of pediments, as LOUIS and FISCHER (1979) gave them, but is correct as such if applied to the flat lower sections of slopes, which indeed have, as such (but not as pediments), "a wide occurrence" as FAIR stated on p. 109. FAIR, and perhaps others with him, seem to have taken the lower slope sections as separate entities, calling them slopes and pediments, though to do so is against all geomorphic reason and experience.

With regard to the humid tropics BREMER (1981, 128) observed that even there well drained slopes with angles of more than 45 degrees are very stable. From Ayers Rock in Australia it is known that the slopes there have not shown any changes within 70 my or even more (BREMER 1965; TWIDALE 1978). Thus, together with the observations from southern Africa, it seems cogent to say that once a steep slope has come to be established it has gained a remarkable stability - both in form and position and both in the tropics and subtropics. (Since the subtropics are part of the 'ectropics' BREMER's statement of 1971 (repeated in the footnote of 1975: see above) cannot be held true throughout all of the 'ectropics'.

YAIR and GERSON (1974) came to the opinion that cuestas, even under the very arid conditions in the neighbourhood of the Red Sea, should retreat with a rate of at least 100m - and even up to 2000m - per 1 my.

If one transfers this rate to Natal and tries to envisage what the result would be one should see a flat and extensive coastal plain with a width of at least 10km or of 25 to 30km at the utmost. Yet - in spite of its more favourable humid conditions - such a planation is nowhere to be found in Natal. Probably, GERSON and YAIR interpolated their rates from the time-span when a scarp first comes into existence, when - as we have seen from the discussion about Fig. 14. 6 - slope formation is at its most effective.

The examples for scarp retreat were taken, as we have seen, from the Natal coast. One might wonder now what the retreat of the cuestas higher up in the escarpment might be. Table 14. 2 may give an idea. For the calculations the same basic assumptions were made as for those in Table 14. 1.

As can be gathered from Table 14. 2 the actual scarp cannot be said to have retreated much in spite of the striking span of time that has to be taken into consideration. In contrast to a scarp retreat parallel to itself it is rather the efficiency of what might be called 'embayment retreat' (comparable to the 'amphitheatre effect') that makes itself felt - which is moreover in good proportion to the lengths of the different time-spans.

The very large extent of 'embayment retreat' in the case of the Piet Retief basin seems to be due to the rocks (Ecca shales). As may be remembered the Piet Retief basin was one of the examples mentioned by KAYSER (1986) of the wide basins in front of the escarpment.

e) Further examples of little slope retreat

The fault-lines in the fault belt of Natal can be used as another example of not much slope retreat. For the TMS scarps have not worn back in any considerable way from the fault lines though these are as old as the Lower Cretaceous or the later Jurassic even. Examples for such small amounts of backwearing can be found in areas both with the softer felspathic TMS beds - as for instance in the area of the Eshowe horst in northern central Natal - and with the hard quartzitic TMS - as for instance in the neighbourhood of Port Shepstone and the Oribi Gorge. The boundaries of the horsts are on all sides still formed by the Cretaceous levels (Eshowe) or are beyond the ingressions denoted by the Miocene level (Oribi; Renken near Port Shepstone). (See Fig. 5.3.)

A third example is that of the area west of Pongola in northern Natal. Here the deep fault-line, striking from north to south, separates the crystalline country in the west from the Ecca sandstone country in the east. Along the fault-line the eastern basement was lowered in such a way that the Ecca beds now border the crystalline rocks in the west. The step-faulting here altogether amounts to 400m to 500m. The Cretaceous transgressions reached across the fault-line, but only a little to the west. The Miocene transgression reached only as far as the fault-line itself, stopped there as it were by the high wall of the crystalline rocks towering above the 300m level. Though the 300m level must have served as a base level for erosion at the time of the transgression the crystalline wall was not worn back from the fault-line into the interior. (See Fig. 15.1)

Denudation comes also to a similar abrupt end on the eastern part of the Pongola basin, where, along another fault-line, the Drakensberg lavas steeply rise. No wearing back of the steep slope away from the fault-line can be discerned.

In southern South West Africa the Great Brukaros (90km to the northwest of Keetmanshoop) is another example for the length of time even pedimentation processes need to shape a once very steep hill side into a flatter pediment above the plain. The Great Brukaros (1590m ab. s-1) stands roughly 600m higher than the 1000m thalassoplain encompassing the mountain on all sides.

The thalassoplain (together with its corresponding intermediary level in 1100m) is here bevelled across the Nama rocks which came to be intruded by the carbonatites now forming the Brukaros. By the intrusion the Nama rocks became welled up like a coat around the carbonatites. The latter could be dated and show an age of 84 my BP (MARTIN 1983). The upper flanks of the mountains are still very steep, but, below a certain knick-point, they change into flat pediments. Near the knick-point the angle of the pediment could be measured as having 10 degrees, but farther down it flattens out more and more. The pediment itself has a width of 5000m and can only have been formed within these 80 my or so, pediplanation of the Great Brukaros starting from the intermediate level in 1100m.

As all examples so far have shown steep and high walls they seem indeed to be very slowly attacked and denuded (cf. BIRKENHAUER 1986).

Table 14.2: Cuesta retreat along the Great Escarpment in the Cape Province and in Natal
Ausmaß der Rückverlegung der Stufe in der Kapprovinz und in Natal

Location	Sheet 1:250000	Rivier	Drainage system	Level	Scarp retreat (in km)	Embayment retreat (in km)
Oustasie	Sutherland	Karee	Tankwa	600	0,5-1	2-3
Die Geut	"	Wilgerbos	Gamka	1000	"	1-4
Maritzhoek-Eselkom	"	Waikraal	"	800	"	4-5
Aberdeen	Graaf-Reinet	Kamdebo	Sundays	800	"	8-12
Graaf-R.	"	-	"	"	1	4-10
Pearston	"	Voel	Voel	"	"	4-5
Somerset East	"	-	Little Fish	800 1000	0,5-1	10
Cookhouse	"	-	Great Fish	600	1	3-4
Dejagersdrif	Dundee	Buffels	Buffels	1200	0,5-1	5
Nondweni-	"	Mvundyana	White	1200	"	5
Nqutu	"	-	Mfulosi			
Babanango	"	-	"	1000	"	1-2
Paulpie-	Vryheid	-	Pongolo	1200	"	5-10
tersburg	"	-	"	1000	"	5-7
Piet Retief	"	-	Assegai	1200	"	25-30

15. THE EFFICIENCY OF EROSIONAL SYSTEMS

a) The phenomenon

As we have seen in the preceding chapter, the attack on slopes, scarps and cuestas comes to a standstill or near-standstill after some time, especially if the scarp, slope or cuesta is attacked from the front. This does not mean however that no signs of vigorous erosion can be found. Indeed, vigorous erosion must have taken place - as all the valleys show and, for example, the large and thick deposits on the marine shelf (MARTIN et al. 1982).

Other examples on land are the wide basins of the larger river drainage systems (i.e. Pongolo, Tugela, Sundays). Fig. 15.1 shows a view of the upper Pongolo basin looked at from the east (in the north of Paulpietersburg in northern Natal). The altitudes here are between 1200m in the front and 1400m in the background. The basin shown can also serve as a good example for an inter-montane basin. Fig. 15.2 shows the topographic situation from above. Fig. 15. 3 generalises some features of a section across the Sundays river basin between the northernmost Cape range and the Great Escarpment. For pictorial reasons, the width of the section covers only a strip of 2,5km - yet what huge masses of rocks must have been carried away from this one stretch only becomes rather clear. How truly gigantic the volumes become if the whole basin is considered can easily be imagined; there certainly must have been carried away tremendous volumes and volumes of rocks, if one takes into account all basins.

West of the fault-line near Pongola (cf. Fig. 15. 2), in the middle parts of the river course of the Pongolo, the crystalline basement has become completely bared by erosion and denudation, all rocks on top of the basement having been carried away: both the Dwyka and Ecca beds, together with a thickness of several hundred metres. Today the height of the uncovered crystalline basement stands at 600m, the exact altitude of the 600m transgression. It must be assumed therefore that thalassoplanation must have played a role in the baring of the basement. Towards the interior, in the west of the second major fault-line (cf. Fig. 15. 2), the basement came to be lowered only to an altitude of 1200 m, though over wide stretches. This altitude again corresponds with a planatory level, from which the escarpment has become dissected into mesas and isolated tafelbergs, as Fig 15. 1 shows.

In the catchment areas of the Tugela and its tributaries (esp. the Blood and Buffels) very flat basins with very low gradients (less than 2 degress, often only 0.5 degree) appear, in an altitude similar to the situation referred to above. One could think these basins to be inter-montane ones in the sense of CREDNER and BREMER (1975). Yet they are nothing but landscapes in which a very wide removal of rocks has taken place. Landscapes where such wide removal occursed seem to form a type of their own and might be called 'landscapes of wide removal' (in German: 'Ausraumlandschaften'). Such typical landscapes seem to be bound to situations in which the strata remain horizontal or near-horizontal, as is the case with the basins in the upper Tugela.

It is especially in these 'landscapes of wide removal' that the Great Escarpment has become deeper embayed than everywhere else. Its very marked sinuosity exactly in these areas must be attributed to such landscape types. The layers, though widely removed, are still to be found on the interfluves, as mesas. Nowhere are such large and so many mesas to be seen as on the interfluves in the vicinities of such 'landscapes of wide removal' with their low strata gradients.

Though not on as grand a style the Sundays river basin can serve as another example. Here, along the stretch shown in Fig. 15. 3, on an area of about 100 sq. km, a volume of roughly 71 cub. km must have been removed, equivalent to about 1.6 billion tons.

b) Cases and processes

One wonders by what processes all these masses have been removed and excavated.

From the observations within the basins and on the interfluves one is able to find out about three or four processes that, being combined, seem to be responsible.

Perhaps the easiest case to understand is that of the interfluves, especially where these are built up by rocks of

Fig. 15. 1: The basin of the upper Pongolo (northern Natal), seen from the east (schematic)
Das Becken am oberen Pongolo (Nordnatal), von Osten gesehen (schematisch)

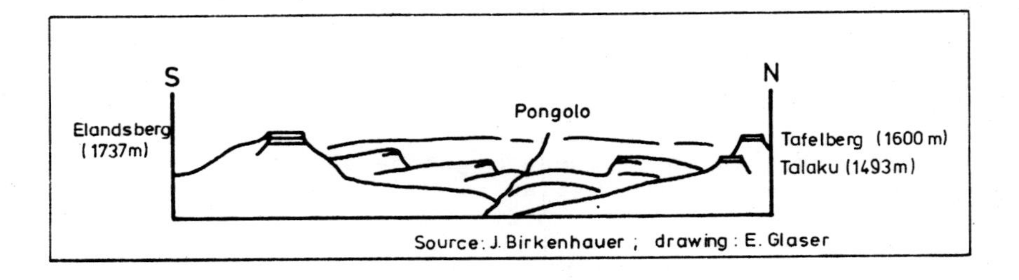

different resistancies and different hydromorphic qualities. Particular examples are shales or schists or granites or gneisses below, dolerites or sandstones or quartzites on top, as can be seen, for instance, in the Valley of a Thousand Hills in the neighbourhood of Durban (Fig. 15. 4). The removal of the TMS on top of the granite-gneiss is here based on the same process that was already described in ch. 14 b - a process which might be called the 'cuesta process', in which spring erosion, the 'jumping up effect' along horizons of springs - in Fig. 15.4 between the granite-gneiss and the TMS - all play their role.

The second case and process might be called the 'wearing down process'. It is similar to the 'cuesta process', but differs in two respects: (a) it concerns far wider areas and (b) in positions not as raised as the one on the interfluves of the basins and benchlands. The wearing down process usually takes place when layers of different rock resistancies alternate with one another as with sandstones or dolerites or quartzites on shales (Ecca series in northern Natal, Beaufort series in the Transkei, Beaufort on Ecca shales in the Sundays river region; cf. Fig. 15. 3) or with shales on the crystalline basement (as Ecca and Dwyka on granite-gneisses in central and southern Natal). In these cases, due to the 'cuesta process', only horizontally expanded, the hard rock cover is eventually worn down and removed. Often, tafelbergs and even spitskoppies only are the last remnants of the once coherent layer. (See Fig. 15.5)

It may be pointed out that it is exactly landscapes underlying such influences that were studied by FAIR (1947) and then cited by him as examples for effective pedimentation (which, however, as we have seen, they are not, viewed above from a close inspection of what goes on on the slopes, but now also from what the wearing down process teaches).

The horizontally expanded wearing down process seems to be rather efficient. Especially on crystalline rocks and on Lower Ecca shales whole sheets of layers have been completely removed across large areas thus completely baring the crystalline basement (as one can judge from the example of the upper Pongolo basin: cf. Fig. 15. 2). (See also: Fotos 52, 53.)

Fig.15.2: Geomorphological features of the upper Pongolo basin (northern Natal)
Geomorphologische Strukturen des oberen Pongolobeckens (Nordnatal)

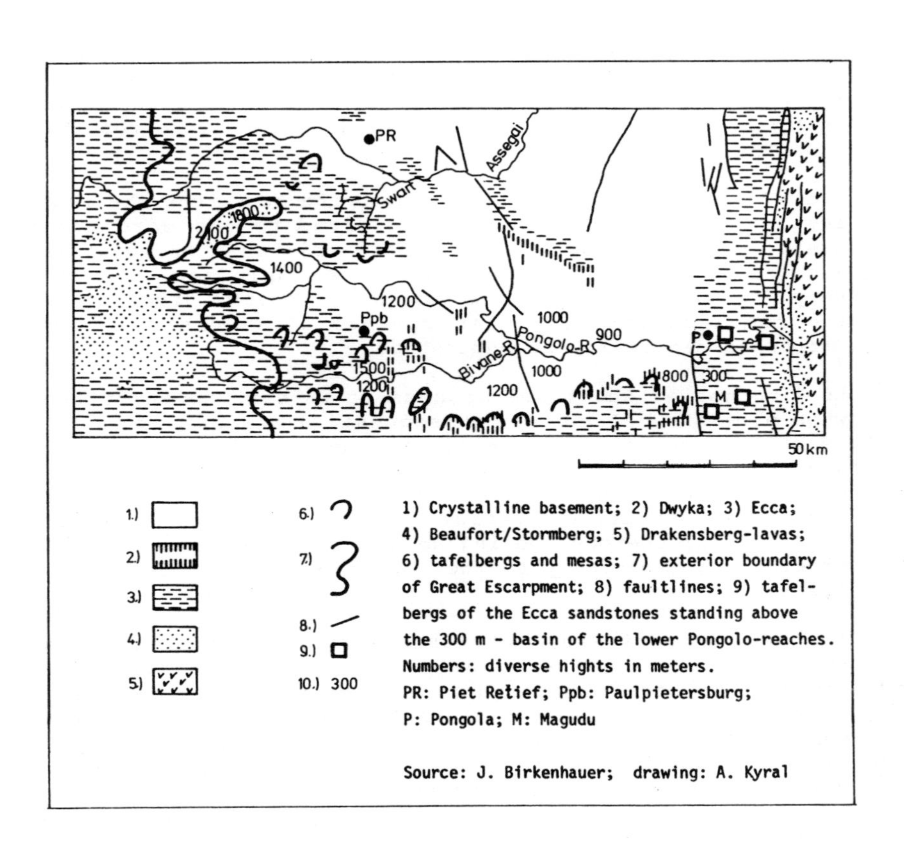

The process moreover can serve to give a first understanding and explanation of what can be observed on Fig. 15. 1. The triangular embayment (Fotos: 1, 2, 27; German: 'Dreiecksbucht') to be noticed here can be understood by way of the wearing down process as seen as the horizontally expanded cuesta process. Triangles of scarps are naturally formed where the basin becomes narrower because it is cut back into the escarpment proper. Such triangular embayments appear wherever alternating rock layers occur along a major river.

The fact that the wearing down process in these embayments is indeed of large importance can be gathered from a comparison between direct scarp retreat (with little results) and what may be termed 'embayment retreat' (with large results). (Cf. Table 14. 1.)

From Table 14. 2 there follows also the fact that the wearing down process is not only reserved to the more humid areas, but has gone on in the more arid ones as well. The Fish river basin is on outstanding example for this in a very dry area. It is only by the wearing down process that the complete destroyal and removal of layers over vast stretches can really be understood.

Looking now at the third process it is best to start from what might be called the 'subsequential effect'. As such it is still another variation of the cuesta effect or process. Wherever a subsequential valley branches off from the main obsequential valley, both incised in a hard top layer, this effect will occur. The incising streams chisel through the hard layers, thereby opening up new flanks of the layers to erosion and especially to the amphitheatre effect. (See Fig. 15. 5.)

In the north of Paulpietersburg (in northern Natal) the Swartrivier is a rather small example only. Nevertheless, towards the stream bed and on both sides of it, the tablelands have completely vanished. Inselbergs as the only remnants have been left over only on the outside of the river basin. In many cases even these inselbergs (consisting of dolerites and/or Ecca quartzites) have so far been reduced by the cuesta effect that only conically shaped hills - the spitskoppies often mentioned before - have remained. What can be observed along the Swartrivier can in the same way be

Fig. 15. 3: Removal levels and amount of removal by planations in the upper Sundays river area (southern Cape Province)
Ausraum-Niveaus und Beträge des Ausraums durch Verebnung im Gebiet des oberen Sonntagsflusses (südliche Kapprovinz)

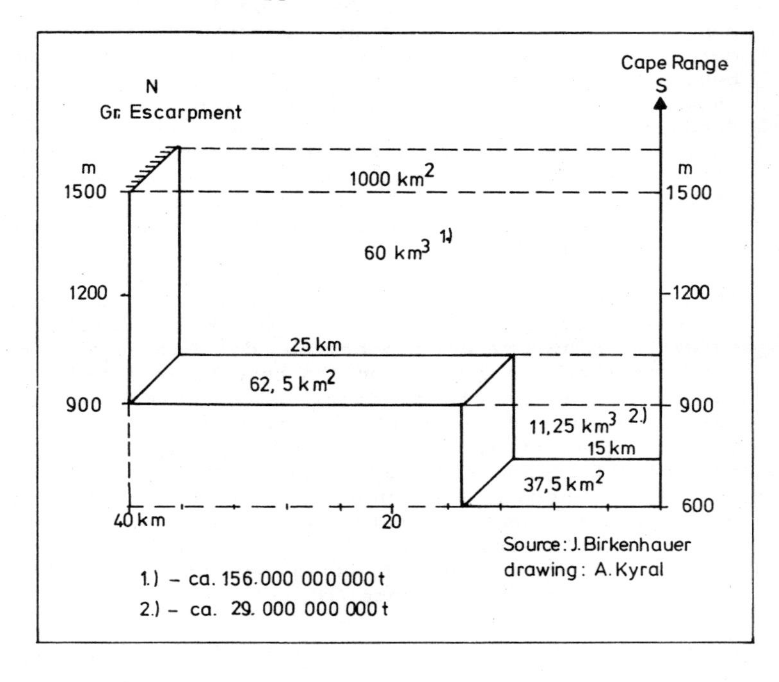

noticed in the basins of the southern Cape - only on a far wider and grander scale. (Cf. Fig. 15. 3.)

In the case of all subsequential valleys removal takes place both from the sides of the cuesta (or benchlands or mesas) and from behind them. Therefore this kind of removal may be called 'removal from the side and behind' (German: 'Abräumung von der Seite und von hinten her'). It is naturally the more effective the more subsequential valleys are developed behind the front-line of a cuesta or a mesa. Fig. 15. 6 shows the generalized picture, taken from the Great Fish and Sundays river basins. Here, each newly cut-in subsequential valley opens up new lines for the erosional and denudational attacks. In this way the process jumps as it were over from one valley to the next. In this way subsequential systems begin to develop as a very effective means of scarp removal. (Cf. Fig. 15. 6.)

It is exactly in those regions where such subsequential systems occur multifold that the Great Escarpment has most effectively become decomposed and dissected; all stages of decomposition and dissection can here be seen and studied: For instance, inselberg fringes and spitskoppies in all stages are found along the margins of the subsequential valleys. Their flat bottoms have beem completely destroyed, the valleys become wider and wider towards the cuesta front, forming triangular embayments. Along these embayments the escarpment has become completely decomposed as a consequence of the total destroyal of all layers forming the escarpment. (In German: 'völlige Auflösung der Stufe als Folge der Zerstörung der Schichtstapel'.)

Eventually, this process of total decomposition may even lead to the forming of a peneplain (or near-peneplain) as is partly suggested by the example of the Tugela basins (described above).

As may be gathered from the description of these processes of rock removal, such peneplains however and their inselberg or spitskoppie fringes do not at all owe their existence to any kind of tropical morphodynamics in the sense of BÜDEL (1957), i.e. they are not the result of a 'double front of planation' (German: 'doppelte Einebnungsfläche').

The fourth and last process or effect is related to the thalassoplains and their intermediary levels. The wearing

Fig. 15.4: Mesas in the Valley of a Thousand Hills (schematic; central Natal)
Mesabildung im Gebiet des Valley of a Thousand Hills (schematisch, Mittelnatal)

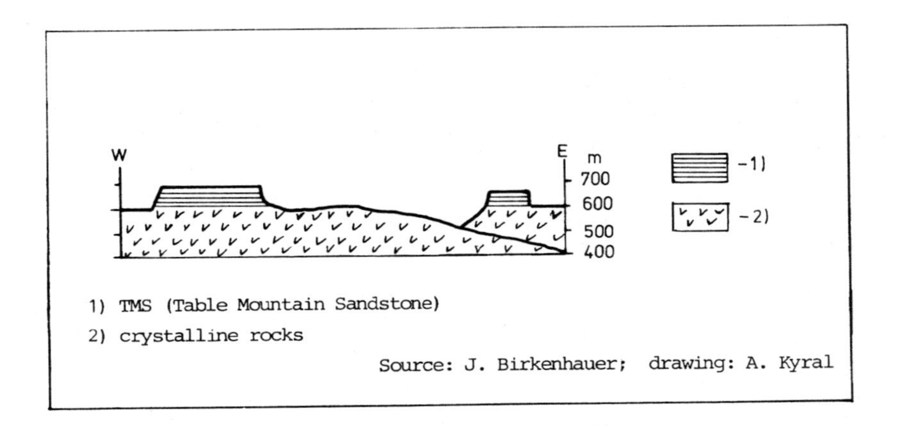

down effect as well as the subsequential effect are both most prominent in the neighbourhood of these levels - whether they were triggered off on the level of the developing plain or be it that eventually the pediplanatory processes have taken over. It must be inferred that the thalassoplains are a constitutive element in one way or other. It even seems that the neighbouring thalassoplain level has played the most desisive part in connection with the processes dealt with in this chapter in the case of the Orange basin and its tributaries and in the basins between the Cape ranges and the escarpment. (For the Orange see Fig. 12. 3)

For it is here that the levels gain their greatest extensions and it is equally here that the planations digitate the strongest (by way of subsequential systems) into and right round the escarpment. To understand this phenomenon a comparison helps perhaps, a comparison with smaller catchment areas. From a comparison of these one can safely say that the smaller, steeper and shorter a catchment area is - even when the base level is very low and the gradient of the thalweg is rather large - the less is its denudational efficiency over its overall area, and vice versa. Transferring this rule to the thalassoplains one is enabled to say that the levelling extent must be a function of the largeness of the area over which a particular marine ingression made itself felt.

Looking for a common 'denominator' as it were one has to think of the water table (i.e. groundwater table). And indeed, in comparing all the four processes or effects described above, one notices that all of them are related to a water table. It is the water table - either beneath resistant layers or along the level of the thalassoplains - that alone extends over a wider area and at the same time draws from a very and continuous catchment area (German: 'Einzugsgebiet'). (It is to be remembered that the denudational force 'snaps' in too small catchment areas: cf. ch. 14 c.)

A wide and continuous water table must therefore be regarded as the one factor that seems to be the most necessary in helping towards the removal of rocks over wider areas, at least in southern Africa (and there over more than 120 my).

Fig. 15.5: The subsequential valley of the Swartrivier (from the northwest) and the removal of different dolerite sills, dissolving the escarpment (schematic; northern Natal, near Paulpietersburg)

Das Subsequenztal des Swartrivier von NW und die Ausräumung verschiedener Doleritlagen mit Auflösung der Randstufe in Tafelberge und Mesas (schematisch: nördliches Natal, nordwestlich Paulpietersburg)

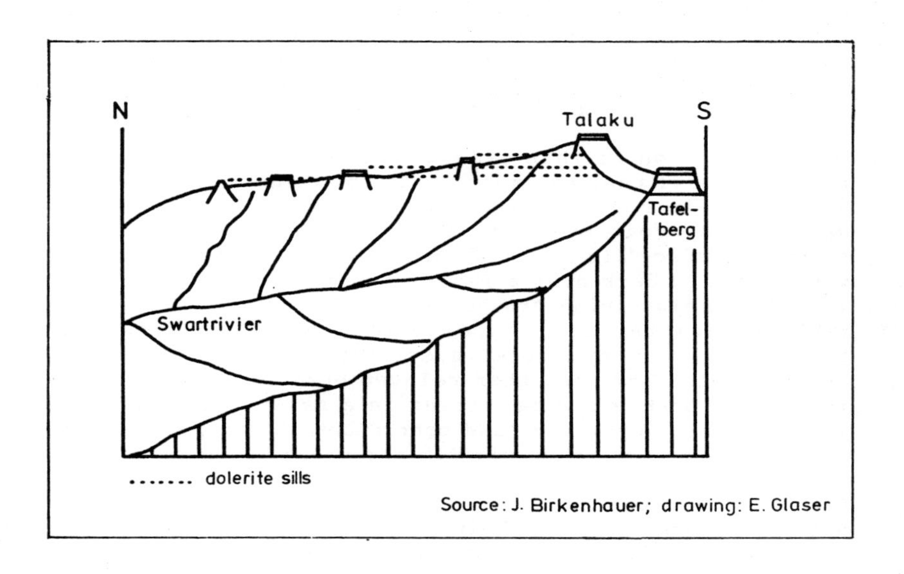

Removal (German: 'Abräumung') has in this respect therefore to be regarded as a function of the size of a water table catchment area.

This function can be written like this: r = f (c) (r meaning removal and c catchment area).

As we shall see in the next chapter (ch. 16) this function has to be seen in close connection with a supplementary process.

The removal of layers over wider areas is certainly not only a function of a sufficient catchment area but also of a sufficient time-span in which it can make itself felt so that the processes of removal can become effective, more or less continuously. This aspect shall be dealt with in the next part of this chapter.

Before going on to this aspect it seems to be advisable to remind oneself that the main agent for removal is indeed the water that is held available by way of what might best be called 'reservoir surface' (German: 'Staufläche'). By 'reservoir surface' is meant an area in which (and over which) water is held back over a larger stretch of land in connection with a high-lying ground water table. Such a high-lying table can either be the result of the water flow being blocked up by impermeable layers (as we have seen from the cuesta effect and the other related ones) or by a sustained ingression of the sea resulting in a raised level which in itself forms a permanent and continuous though perhaps shallow and efficient reservoir for the processes described.

Such a 'reservoir surface' (or 'Staufläche') is certainly not only caused by water being blocked up from downward, but also one of precipitation. For in a humid climate the reservoir will be better sustained than in an arid one. This means that processes of the described kind will be more active in humid climates than in arid ones, where they certainly will become rather delayed or perhaps will be even completely terminated. The catchment areas of the Assegai, Pongolo and Tugela rivers are examples for more humid conditions, the river basins of the Cape Province for more arid ones. This explains why the processes described can still be seen going on on the eastern side of southern Africa, whereas on its western side they seem to have become fossil ones. This also goes to say that recent and subrecent condi-

Fig. 15.6: Subsequential valley systems in the eastern Cape Province (schematic)
Subsequente Talsysteme in der östlichen Kapprovinz (schematisch)

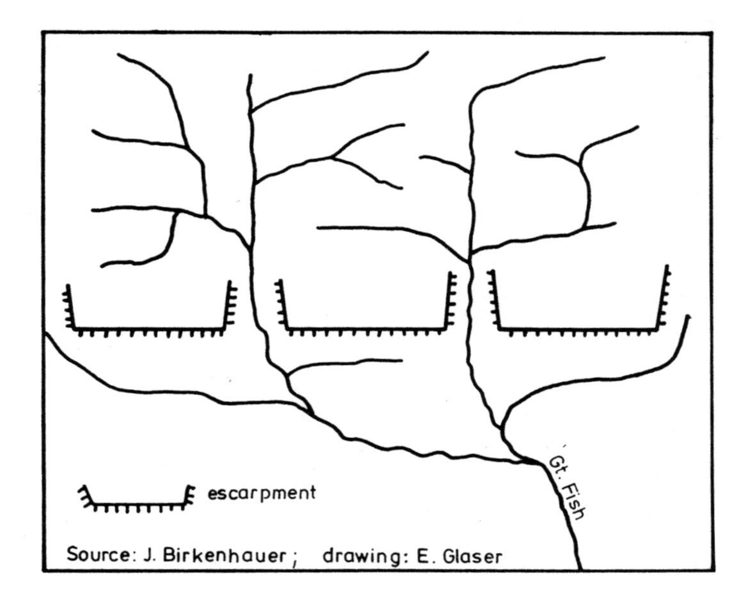

tions in the eastern basins furnish the key for what once must have happened also in the western ones. Or if they do occur as one can observe here and there, their efficiency must be very reduced. The same consequence would apply to the past if very arid conditions were prevalent then, though the length of the time-span (and it is indeed very long for the upper levels) may smooth over some of the slowing-up results caused by a deficient 'reservoir surface'.

Comparing, finally, all the landscapes with rock removal over wider areas (German: 'Ausraumlandschaften') one immediately notices that those with the widest extent of removal became formed during Cretaceous times, both in the east and in the west. Since the middle of the Tertiary (at least since then) the processes of removal seem to have slowed up or have even come to a standstill or near-standstill. A comparison of the widths of the different levels in Fig. 15. 2 and 15. 3 will lead to such a conclusion.

It was also in connection with these removing processes in stratified rocks that the mesas and benchlands came into existence everywhere in the basins along the escarpment. Several conditions which are only secondary with regard to the extent of the phenomena have not been discussed so far. This will be done however in the next part of this chapter.

c) Intensity and extent of erosion and denudation

The question here is what can be made out about the speed (or different lengths of time-span) with which removal processes take place. Some examples will be discussed in the following paragraphs.

The first example is taken from the Swakop and Khan canyons in central South West Africa. In the neighbourhood of the canyons an intense dissection can be observed. Tributary dissecting valleys form real badlands. From this dissected area it is possible to deduce how long it has taken linear erosion to cut into the former plateau-like platforms, if one looks at the lower thalwegs of both valleys. Going upriver along the Swakop and starting at its mouth one sees dissection begin where the base level is formed by the 100m 'terrace'. This 'terrace' is itself incised in the 200m level. Between the two levels the dissected parts have a width of roughly 1km on both sides of the valley. Since the lower level was formed in the Lower Pleistocene and the one higher up not earlier than the Upper Pliocene the time-span for the dissection between the two levels may not excede 1my. Dissection becomes more pronounced where the 200m level serves as the base level for the (Miocene) level in 300 m. Dissection grows markedly where the Miocene level serves as the base for the Upper Cretaceous planations. These have been turned into badlands over a width of 5 to 7 km on both sides of the valley. The time-span needed to bring out the badlands here may be calculated to be between 40 and 58 my.

Other examples are taken from Natal and the eastern Cape Province. The sites of the examples are shown in Fig. 15. 7. The sites were chosen under the follwoing respects: they should differ in landforms, in rocks and in ages. (It may be pointed out that here, as in the example from the Swakop and Khan, the same favourable circumstances prevail, namely the tectonic stability of southern Africa since the Valanginian, and the possibility to use the levels as guiding lines.)

The situations of the different examples are shortly described as follows.

Site 1: Interfluve between Stanger and Kranskop. On TMS (cf. Fig. 6. 2). Two time-spans are possible: (a) the interval between the forming of the 400m level and that of the 800m level (= 22 my; cf. VAIL et al. 1977; HANCOCK and KAUFFMAN 1979; Fig. 5. 6; 5. 7); (b) the interval between the 600 m level and the one in 800m (= 7 my; HANCOCK and KAUFFMAN 1979; Fig. 5. 7).

Site 2: Sundays river area (cf. Fig. 15. 3). On Ecca and Dwyka shales. Interval: between 800m and 1200m levels (= 20 my; cf. ch. 6 and 9).

Site 3: Mkabene valley, south of Richmond in the hinterland of Durban, tributary to the Mkomazi. Altitude at mouth: 600m ab. s-l, in upper reaches: 1100m (= 35 my; cf. ch. 5 and 6). Ecca sandstones and Dwyka shales, with dolerite dykes.

Site 4: Valley of a Thousand Hills, near Durban (cf. Fig. 15. 4). TMS above granite-gneiss, TMS with thickness of 250m. Several possible time-spans: (a) Removal within the Maastrichtian (= 7 my); (b) during the Tertiary (= 65 my); (c) lowering of the granites on the interfluve between Maastrichtian and Lower Miocene (= 40 my); (d) lowering between Maastrichtian and Upper Miocene (= 58 my).

General remarks for calculating the time-span for sites 5 to 9:

Different calculations are necessary because the age for the forming of the 300m level cannot be fixed. Case (a) is always based on the assumption that the level was formed in the Lower Miocene, case (b) on an Upper Miocene age.

Site 5: Pongolo river, west of the fault-line in the west of Pongola; removal of Ecca sandstones and quartzites; thickness 200m (cf. Fig. 15. 2; Fig. 5. 1) (= 7 my).
Site 6: Pongolo river, east of the fault-line of Pongola (cf. Fig. 15. 2; Fig. 5. 7). Removal of Ecca series. (a) 40 my. (b) 58 my.
Site 7: Pongolo river area, basin of Paulpietersburg, west of the Pypklipberg fault-line; basin-bottom in 1100m (= intermediary level of the 1000m thalassoplain); escarpment in 1500m; Ecca sandstones and quartzites, dolerite sills; time-span: 20 my.
Site 8: Oribi Gorge, southern Natal; TMS on crystalline basement; altitude of TMS top: 600m; valley bottom in 200m, incised in the basement (= Plio-Pleistocene level); time: approx. 60 my.
Site 9: Mzumbe valley. Tributary of the Mzimkulu, near Port Shepstone (southern Natal); valley slopes: between the 300 and 200m levels (= base level of mouth); granite-gneiss; time: case A aprox. 20 my, case B 5 my.
Site 10: Vungu valley, near Margate (southern Natal: cf. Fig. 14. 3); rocks: TMS above crystalline basement; slopes of valley: within TMS; TMS on top in 300m; valley bottom in 200m; case A aprox. 20 my, case B 5 my.

With the data just given for the ten sites Table 15. 1 can be constructed. Equating the highest number in this table as 100 (i.e. site 4 with 0.03571) and correlating the other numbers to this value one is able to obtain a sequence of removal intensities.

On first sight certainly rather astonishing, the fact emerges that the shorter the period in question is the higher

Table 15.1: Rates of erosion and removal
Erosions- und Abtragungsraten

Sites	Rocks	Age	Million years	Difference between levels in m	Rates (mm/a)	Comparative intensities
1 A	quartzites	Upper Cretac.	22	400	0.01818	51
1 B	quartzites	Maastrichtian	7	200	0.02857	80
2	Ecca/Dwyka	Lower Cretac.	20	400	0.02000	56
	shales	Lower Cretac.	35	500	0.01429	40
3	" + sandst.					
4 A	quartzites	Maastrichtian	7	250	0.03571	100
		Upper Cretac.	22	"	0.01136	32
B	"	Tertiary	65	"	0.00385	11
A	granite	older Tertiary	40	300	0.00750	21
B	"	older and middle Tertiary	58	"	0.00517	14
5 A	sandstone, quartzites	Maastrichtian	7	200	0.02857	80
B	"	Upper Cretac.	22	"	0.0909	25
6 A	"	older Tertiary	40	"	0.00500	14
B	"	older and middle Tertiary	58	"	0.00345	10
7	quartzites, dolerites	Lower Cretac.	20	400	0.02000	56
8	quartzites	Tertiary	65	"	0.00667	18
9 A	granite-gneiss	younger Tert.	20	100	0.00500	14
B	"	youngest Tert.	5	"	0.02000	56
10A	quartzites	younger Tert.	20	"	0.00500	14
B	"	youngest Tert.	5	"	0.02000	56

the removal intensities are. The meaning of this result seems to be that removal is at its highest when the combined attack of the denudational forces first sets in. The attack seems to lose in impact after about 5 to 7 my. The example of the highest rate seems to corroborate this assumption (site 4: Maastrichtian). The high rate of removal during this particular period may be understood if one considers the fact that the sea level - serving as a base level - seems to have been a very constant one during this time, remaining at rather the same height throughout this period. This circumstance must have led to a continuous lowering of all neighbouring landforms down to this level.

The finding is in accordance with the rates AHNERT (1970) obtained for large mid-latitude drainage basins. Using quantitative methods, AHNERT came to the conclusion that a relief is reduced to 10% of its initial state within 11 my.

Two conclusions seem to be open:

(1) Denudational rates obtained on the basis of very long periods seem to be rather meaningless.
(2) The effect of a 'traditional forming on' process (German: 'traditionale Weiterbildung') cannot be very substantial.

AHNERT had a look also into the question whether the mean annual precipitation values of drainage basins had any effect at all. From his calculations AHNERT came to deny such effects.

In spite of this result the rates gained from the ten sites in southern Africa seem to suggest that there may be some relations to general and overall climatic and weathering conditions. The highest rates seem to have been reached in the Upper Cretaceous - a period that is generally considered to have been one of the warmest and most humid ages in the history of the earth (VALETON 1983).

In spite of the first conclusion above one might tentatively try and use all rates for an averaging procedure and consider time, landform and rock type on the purpose to see whether some correlations do show up.

A procedure for gaining some average values for landforms can run along the following lines: the particular rates for particular landforms are added and then divided by the number of cases. If one does so Table 15.2 is gained.

What has been said in part b) of this chapter seems to be fully corroborated by this table: all landforms with a larger groundwater table area are more favourable to denudation. The more linear the removing process is - as with fluvial erosion along valleys - the weaker the rate is. Such a weaker rate may be due to the scarce events of high water and floods, because it is only then that active erosion will take place (see ch. 16).

Table 15. 2: Average intensity rates for periods
Durchschnittliche Intensitätsraten für die Abtragung nach Perioden

Period	Intensity rate
Maastrichtian	87
Upper Cretaceous (without Maastrichtian)	36
Upper Cretaceous (including Maastrichtian)	61
Lower Cretaceous	51
Tertiary	15
Tertiary (old to middle)	15
Tertiary (younger, including the Pliocene)	35
Tertiary (younger, excluding the Pliocene)	14
Pliocene	56

Table 15. 3: Landforms and intensity rates (averages)
Durchschnittliche Intensitätsraten für die Abtragung nach Oberflächenformen

Landforms	Intensity rate	Sites
thalassoplains, basins of large removal	45	1,2,5,6
peneplains	56	7
interfluves	36	4
valleys	33	3,8,9,10

In comparing the effects of the rock material on erosion and removal rates it seems advisable to bracket out the Cretaceous examples because it can be assumed that the chemism of the different rocks did not play any decisive role under Cretaceous climatic conditions. Therefore only the rates from the Tertiary are here considered (Table 15. 4).

Table 15. 4: Rock materials and intensity rates in the Tertiary (averages only)
Durchschnittliche Intensitätsraten der Abtragung nach Gesteinen

Rock material	Intensity rates	Sites
quartzites	11	4b,6,8,10
granite-gneiss	16	4c,9

From Table 15. 4 it may be gathered that rock resistancies do play some part, but only a rather minor one.

Altogether however - in spite of the huge masses that certainly have been transported away throughout these million years: cf. Fig. 15. 3 - the general lowering of landforms seems to be very small indeed in an overall way. This result may be looked upon as especially surprising if one reminds oneself of the rates given by OLLIER (1981). OLLIER then estimated the denudational rate of lowlands to be in the neighbourhood of what he calls 50 Bubnoff (a Bubnoff meaning a rate of 1mm p.a.). The rates given in Table 15. 1 are only hundredths of a Bubnoff.

A general result of the argument here may be that the time-spans through which landforms will become fully formed must be regarded as rather long ones, longer, in any case, than has been assumed so far. BREMER (1981), for instance, thought that 1 my were sufficient for the reshaping of a plain and that a period from 1000 to 100000 years were enough for what she termed 'Einschwingzeiten' in German, i.e. the length of periods in which a new frame for new planatory forming processes could be set.

A last comparison with another author and area may be added. For Hawai, WIRTHMANN (1981, 1983) came to the conclusion that denudation there achieved remarkably high results amounting to 30cm within 800 y. - equivalent to ca. 40cm in

1000 y. or to nearly half of a Bubnoff (1 mm/a). Such amounts are obviously due only to fully tropical conditions with high precipitation as on the weathersides of the islands.

16. CLIMATIC GEOMORPHOLOGY OF SOUTHERN AFRICA

a) Introductory remarks

In the preceding chapters many references were made to climatic conditions, either recent or subrecent, or even for paleo-situations. Examples were: laterites, silcretes, pediments, serir-like covers of quartz fragments (cf. Fig. 7. 3 and 7. 2). It therefore seems to be time now - at long last - to deal with climate and its possible significances for the geomorphological processes in southern Africa in a more systematic way. Many questions may have arisen so far, but one of them will be given highest priority, the one whether climatic conditions have changed during the geologic ages and if so, in which way. Another top question is concerned with the regional climatic régimes, both in the present and in the past.

It seems best to start with conditions in the present and then go on to the past, comparing past and present. Reflections on the (un-) favourableness of climatic conditions for weathering and soils will follow. What all the findings will mean for denudation and erosion and the specific landforms deriving therefrom (again both in the present and in the past) will form a concluding part.

b) Climatic regions and some of the characteristics

The climate of southern Africa may generally be called a subtropical one - with variations of course according to height (hypso-metrical change), latitude (planetary or zonal change), continental situation (change from the west to the east), a more exterior or interior situation (change from the periphery to the core). In so outlining these four categories of change, fundamental for all physical aspects on earth, the author is indebted to LAUTENSACH (1950) and his rules for the global change of physical aspects (German: 'Formenwandel').

South African geographers propose climatic provinces as shown in Fig. 16. 1. From the figure it follows that the whole of southern Africa should be looked on as being fundamentally subtropical. Observations of plants, soils and

Fig.16.1: Climatic regions of southern Africa according to South African geographers (taken from Jutta's Magister Atlas)
Klimazonen Südafrikas nach Südafrikanischen Geographen (aus: Jutta's Magister Atlas)

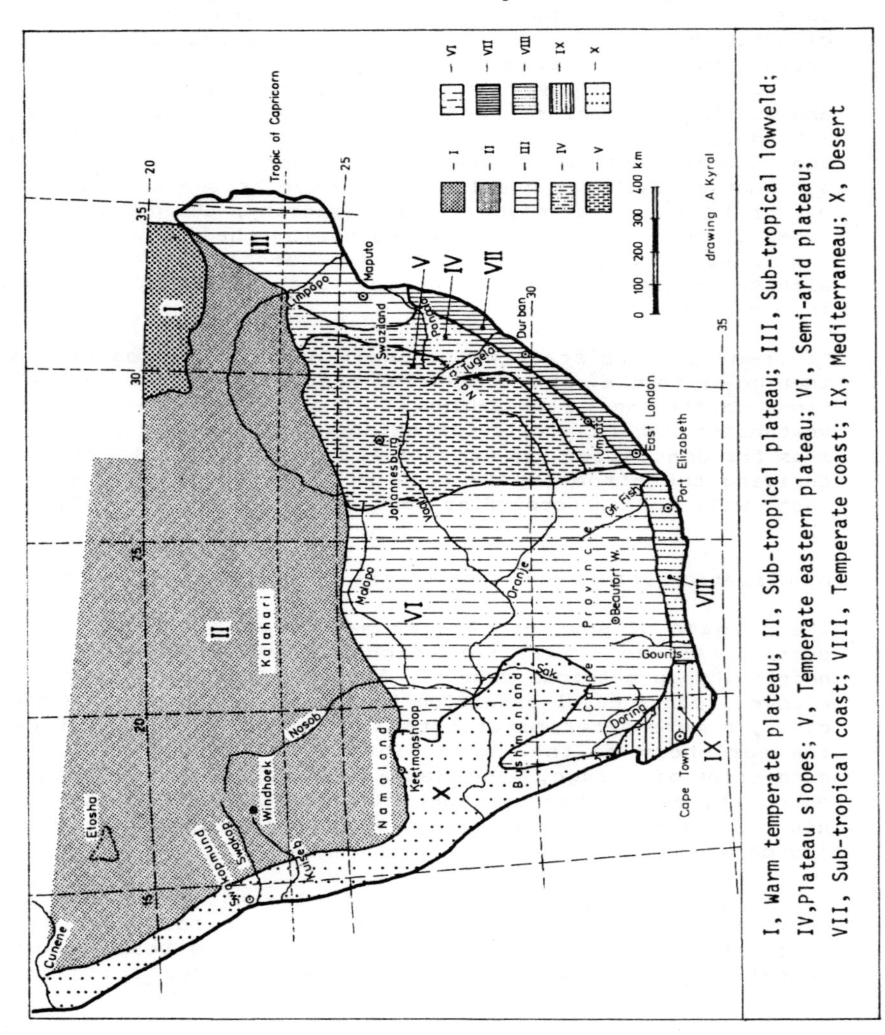

Fig. 16. 2: Climatic regions of southern Africa after KOEPPEN-GEIGER
Klimazonen in Südafrika nach KOEPPEN-GEIGER

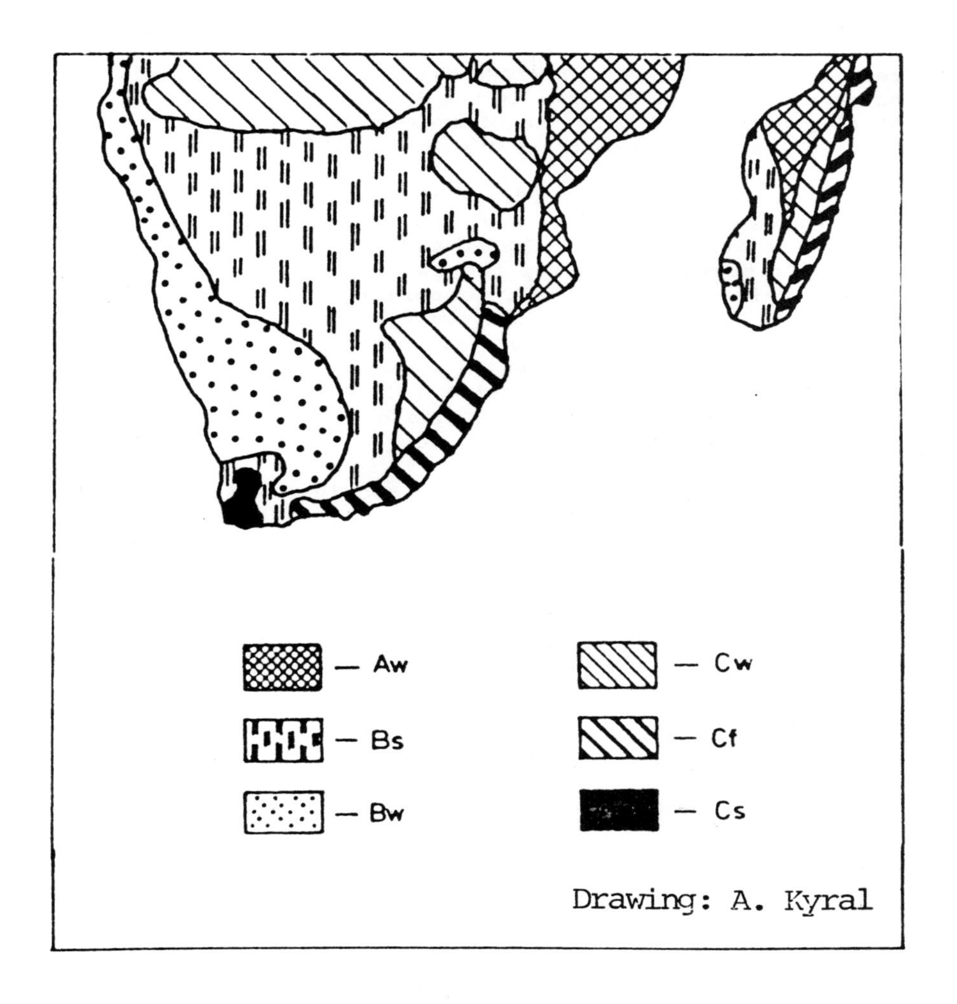

landforms seem to suggest however (at least to the present author) that at least two regions shown in Fig. 16. 1 must be regarded as marginally tropical. These two regions are the so-called subtropical plateau and the so-called subtropical lowveld. What in Fig. 16. 1 is termed 'temperate coast' must really be considered to be subtropical as the monthly averages for the winter season show (cf. Table 16. 1). The author's opinion about the two northern regions is corroborated by KOEPPEN's classification, here given as Fig. 16. 2 in the simplified version by GEIGER (1961). With respect to the KOEPPEN - GEIGER chart it must be pointed out however that their Cf-area in the African southeast cannot at all be compared with the Cf-area in Europe, as again the monthly averages demonstrate (Table 16. 1).

How far all such classifications differ, if generalised on a continental or even global scale, is shown by a comparison with a map drawn by FRANKENBERG and LAUER (1981). (See Fig. 16. 3.) On drawing this map the two authors made use of a hygrothermic classification. Comparing their map now with that of KOEPPEN-GEIGER one notices that a large part of the latter's Cf-area is incorporated into the tropics. Such an incorporation is however open to several doubts, being dealt with later on. What the FRANKENBERG-LAUER map however does achieve is giving a very good idea of how large the arid areas in southern Africa really are.

One variable for defining an area as tropical is the freedom of frost. Though this is certainly the case along the southern Indian Ocean coast the present writer would like to point out that this coastal area shows distinct temperature seasons (cf. Table 16. 1) and should therefore be regarded as subtropical. Moreover, the recent and sub-recent morphodynamic processes point into the same direction. Since this latter aspect is held to be the most conclusive it will be dealt with separately in ch. 16 c.

The hygric situation of a region seems to be most important for its morphodynamics. Therefore a closer look should be had at it.

Fig. 16. 4 (from WELLINGTON 1955) represents the mean monthly rainfall of selected stations. From the east to the west a sequence from the humid to arid regions (over semi-

Table 16.1: Selected climatic data from selected climatological stations in southern Africa
Ausgewählte Klimadaten ausgewählter Klimastationen im südlichen Afrila

Station	Altitude	Latitude	Aver. temp. of warmest (W) and coldest (C) months		Aver. rainfall in mm per season		Annual average of temperature	Days with frost (aver.)
			W	C	W	C		
Maputo	64	25.58	25.4	28.2	604	165	22.2	-
Nelspruit	300	25.25	23.0	15.0	593	166	19.0	-
Paulpietersburg	1100	27.25	20.8	12.3	771	159	12.3	3
Durban	5	29.50	24.7	17.9	695	308	21.4	-
East London	127	27.50	21.2	14.8	634	226	18.3	-
Kokstad	1300	29.30	19.1	9.2	600	128	9.2	34
P. Elizabeth	58	33.58	21.1	14.2	265	367	17.6	-
George	100	34.00	20.0	13.5	190	240	16.8	-
Oudtshoorn	335	33.35	23.7	11.7	157	97	17.9	1)
Cape Town	17	33.54	21.5	12.6	112	394	17.0	1)
P. Nolloth	10	29.15	16.0	12.0	10	40	13.9	-
Upington	809	28.26	27.8	10.6	178	26	19.7	1)
Keetmannshoop	1066	26.34	26.9	13.4	135	12	20.7	1)
Windhoek	1728	22.34	23.4	13.2	361	9	19.1	1)
Swakopmund	12	22.41	18.1	12.1	7	1	15.1	-
Victoria West	1200	31.25	23.0	1.0	180	70	15.8	1)

1) Infrequent days of frost have been recorded through the cooler months. The Table was compiled from data in (1) Handbuch ausgewählter Klimastationen der Erde, Trier 1979, (2) Jutta's Magister Atlas, London 1981, and (3) Climate of South Africa, Pretoria 1951.

humid and semi-arid ones) with typical seasonal variations can well be noticed.

Temperatures seem to be warm enough everywhere in order to uphold more intense chemic weathering. Compared with these the changes of the hygric situations must be regarded as being the more important ones in a geomorphic respect.

For such geomorphic purposes it seems to be sufficient to line out five climatical regions, i.e. two subtropical regions and three marginally tropical ones. These regions are the following ones:

Subtropical:	I	the humid and semi-humid east and south
	II	the semi-arid and arid west and northwest
Tropical:	I	the semi-arid highlands
	II	the semi-humid highlands
	III	the semi-humid lowlands.

The Roman numbers used in this classification are the ones used in Fig. 16. 4, based on the chart in WELLINGTON 1955. A climatic regionalisation as given in Fig. 16. 4 takes into consideration the hypsometric change resulting for instance in the leeward rainfall shadows and the continental changes and their consequences. The leeward (or 'föhn') effect makes itself felt especially in the rainfall shadows of the diverse Cape ranges. It may be pointed out here that these föhn-effects as well as the periphery-core changes must have been even more pronounced under Gondwanaland conditions than they are today, since southern Africa was then clinched in by southern America - with the Falkland Islands block reaching as high up the present-day Indian ocean coast as Durban.

On the whole, southern Africa - with its sequence of subtropical and tropical climates and their variations from humid to arid and from summer to winter rainfall - cf. Table 16. 1 - presents a favourable frame for studies in climatic geomorphology and the variations of geomorphic forms and processes within different climatic provinces.

As can be judged from the distribution of duri-crusts and the slope-forming processes, including pediment-forming (cf. Fig. 7. 3), climatic conditions must in the past have been roughly similar to those of today. The 'past' referred to here certainly goes as far back as the Cretaceous, because

Fig. 16. 3: Tropical and subtropical regions in southern Africa after FRANKENBERG and LAUER (1980)
Tropische und subtropische Zonen im südlichen Afrika nach FRANKENBERG und LAUER (1980)

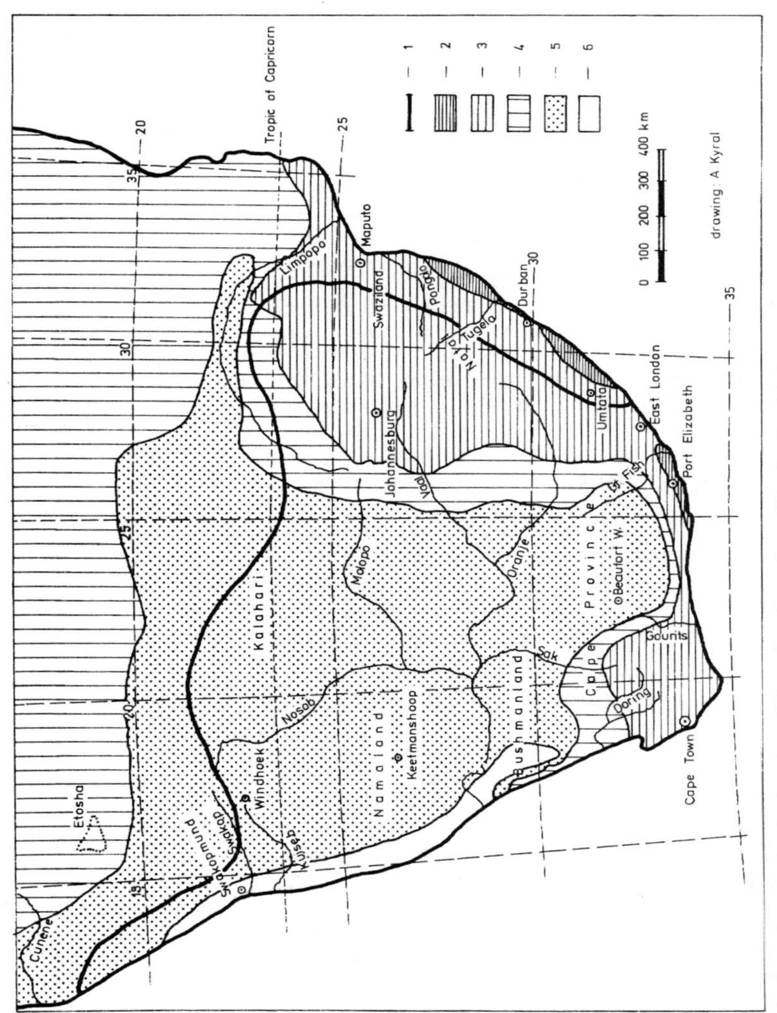

1: boundary between tropics (N) and subtropics (S); 2: humid (9-12 mths.); 3: semi-humid(5-9mths
4: semi-humid (3-5 mths.); 5: rather arid (1-2 mths.); 6: fully arid.

then a similar distribution of provinces can be found as today: warm and humid in the east, warm and semi-arid in the south, arid in the west. It should be stressed however that such similarity is certainly only a rough one, but, within certain boundaries, the similarity and thus stability of the broader climatic conditions are very striking features. This broad climatic stability together with the tectonic stability of southern Africa (this combination is important) forms a framework which alone seems to be able to explain the existence of so vast planations and pedimentations in the whole area.

It may be objected that in the Cretaceous southern Africa lay in a global position further down south than today (i.e. 48 degrees in the Cretaceous vs. 34 degrees today with respect to the Cape of Good Hope). Fig. 16. 5 shows the positions according to SIEVERS (1982). Such an objection may be right indeed; nevertheless one has to consider the fact that the Cretaceous was one of the most humid and warmest periods in the history of the earth, as has been pointed out above already, - and not only with regard to particular areas only, but for the whole globe which is shown from the fact that no polar ice caps existed then (cf. VALETON 1983). It is this global warmth and humidity that for example also explains why laterites could be formed in such a southern latitude.

In spite of such global conditions there can in the later Cretaceous be even observed a continental change which made itself felt in the distribution of the laterites in the east and the silcretes in the west, due to a change in humidity and warmth in the direction given (cf. ch. 7). The same can be noticed about the prevalence of pediplanation in the west. As VALETON (1983) showed the forming of duricrusts may well have gone on into the lower Tertiary, even on a world wide scale. According to VALETON climatic conditions began to change rather drastically in the Oligocene. Thus the climatic zonation of today can only have come into existence since then. Compared with the Cretaceous (as VALETON, 1983, equally points out) this climatic zonation must be regarded to be an extreme one, since the formation of latosols has since then been restricted to the present tropics.

At the same time when global conditions changed in this drastic way Africa however, from plate tectonic reasons,

Fig. 16. 4: Climatic regions in southern Africa (partly based on the rainfall regions after WELLINGTON 1955)

Klimazonen im südlichen Afrika unter Berücksichtigung der Niederschlagsregionen nach WELLINGTON (1955)

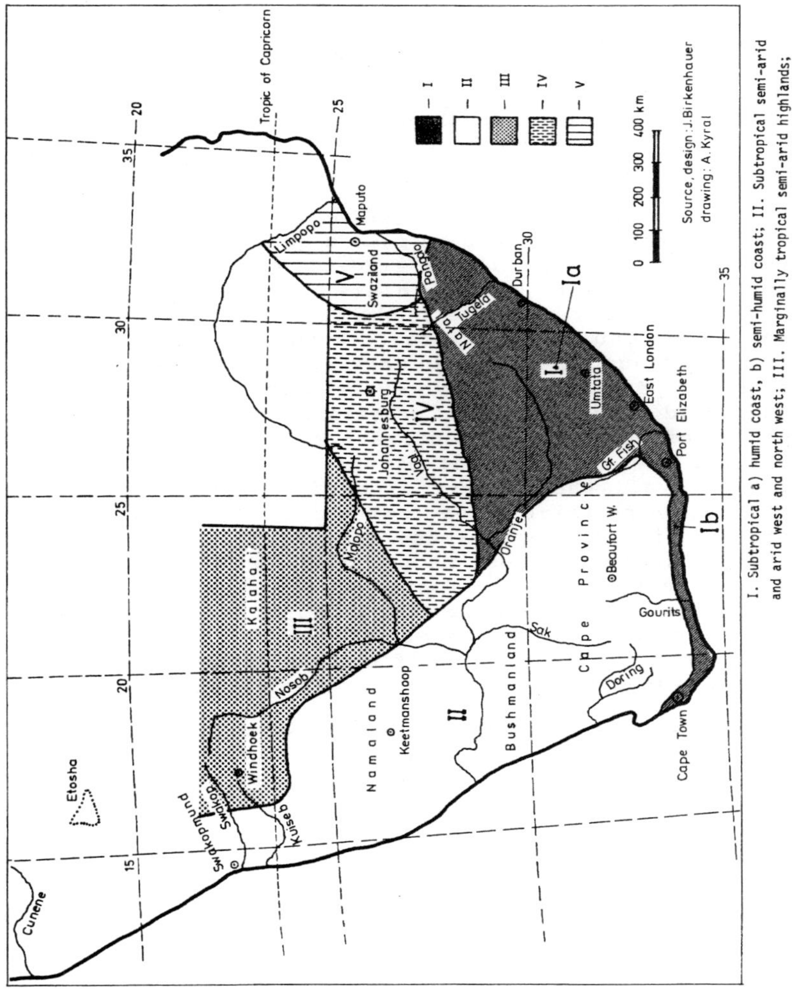

moved northward, eventually reaching its present global position - perhaps as early as the Oligocene or the Miocene. This northward move naturally means that the global climatic change was compensated by the drift into the warmer zones of today. This drift must then be taken as the only explanation why roughly similar climatic conditions could be upheld over all the long time-span and why the formation of latosols could go on in southern Africa until the on-set of the Ice Age (cf. MAUD, 1968; BUTZER and HELMGREN, 1972).

Though DOUGLAS's admonitions (1980) against a uniformitarian view of climatic geomorphology must certainly be taken care of they can however not be applied to the special situation of southern Africa.

Within the same frame the question of how long the Namib desert has existed should be put, a question dealt with by several authors. From all the evidence given in the previous chapters the present writer thinks himself to be justified to accord a rather high age to the desert - or to marginally desert-like conditions at least in the Namib area, conditions which must have lasted here since the Lower Cretaceous or may perhaps even have existed in the late Jurassic. Rather arid conditions must certainly have prevailed in the interior of such a huge continent as Gondwanaland was at that age. The vast spatial distribution of pediplanatary forms over all the western side of present-day southern Africa which certainly stem from the Cretaceous are another indication for the prevalence of rather arid conditions. These in themselves must be taken as a consequence of the two categories of change mentioned above: the continental one and the periphery-core one. It might be added that the existence of a rather extended interior Gondwanaland basin deduced from the discussion of the evolution ot the Orange drainage system fits very well into the present argument.

Equally, according to OLLIER (1978) the absence of subsoil weathering is in itself a sure indication of a great longetivy of the desert climate. According to the same author this desert climate may well have gone back far into the Mesozoic. According to WARD (1983, 175) "the stratigraphic record reveals a dominance of arid to semi-arid conditions throughout the history of the Namib, which dates back to the Cretaceous". WARD referred to evidence brought forward by KOCH, an entomologist, who ascribed the evolution of parti-

Fig. 16.5: Global positions of Africa between the Jurassic and the present (after SIEVERS 1982)
Lage Afrikas vom Jura bis zur Gegenwart (nach SIEVERS 1982)

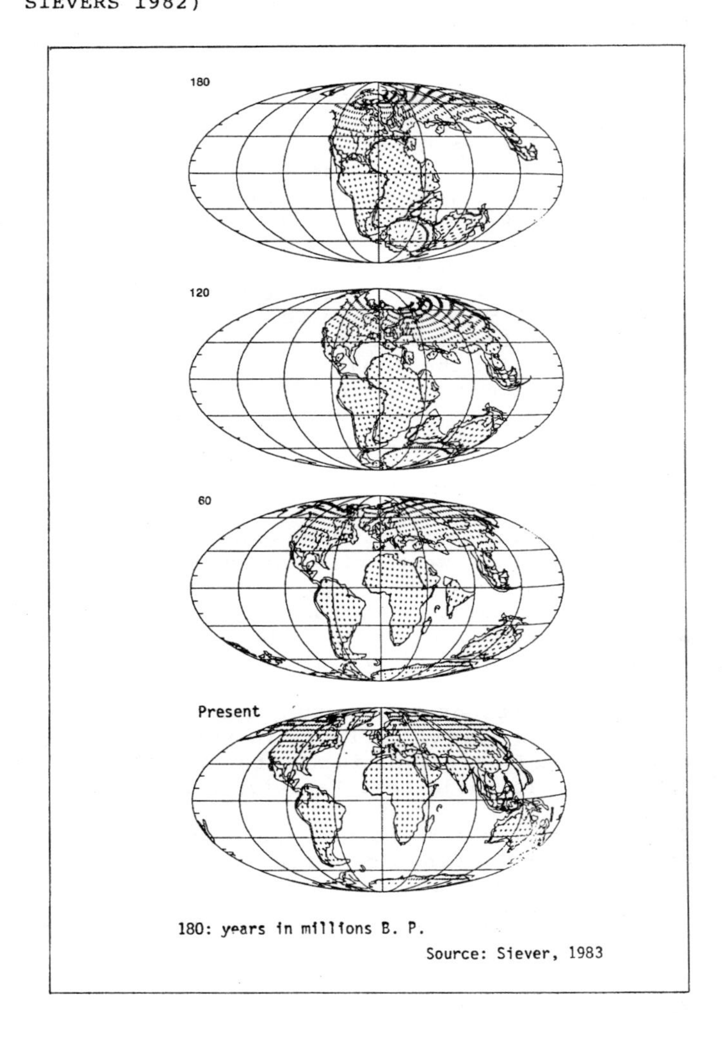

cular desert beetles to a long and undisturbed existence of arid desert conditions. It is because of these findings by KOCH that the Namib has been called 'the oldest desert on earth'. Though evidence for the beginning of the Benguela current system (to which the super-arid conditions are usually attributed) can only be found for as late as the Miocene (SIESSER 1980), an arid or at least a pronounced semi-arid climate must have existed before this time. Aeolian sandstones stemming from as far back as the Etendeka Formation (130 my BP) give support to this view as well as the fossil dunes within the Tsondab Formation. These dunes have an age of at least older Tertiary, as WARD admits. It might be mentioned here again that in the opinion of the present author the Tsondab Formation must be regarded to be of a Lower Cretaceous age since the sandstones rest in the neighbourhood of the 800m and 1000m thalassoplains (cf. ch. 6 d). WARD (1983, 182) came to the conclusion that the Namib must have existed already in Upper Cretaceous times in stating that "...the Namib tract...has not experienced climates significantly more humid than semi-arid for any length of time during the last 80 my".

c) Chemical weathering

Warm temperatures are necessary for the up-keeping of intense chemical weathering. As we have seen in the preceding part and can gather from the numbers given in Table 16.1 such warm temperatures present themselves everywhere and in a sufficient way in present-day southern Africa - even in winter and even in higher altitudes.

Similar conditions must have prevailed in the past as can be seen from the fossil latisols in altitudes as high up as the hinterland of Durban or in altitudes of 1200 to 1400m in the upper Pongolo basin near Paulpietersburg.

Yet, as we have seen before, warm temperatures alone do not mean much for intense chemical weathering if they are not accompanied by a sufficient humidity - here meaning especially a sufficient and constant soil moisture. As several other authors have shown soil moisture is, together with warmth, the most decisive parameter for chemical weathering (see for example FRÄNZLE, 1965, or BREMER, 1981). Regarding southern Africa today, soil moisture is sufficient in its eastern parts to favour chemical weathering - either throughout the year or at least through very long periods of the year.

Under certain topogenic circumstances (to be dealt with later) sufficient soil moisture prevails along the whole eastern front of the Great Escarpment and in the front of the Cape ranges to the south. In such topogenic circumstances a monthly rainfall average of only three inches even seems to be enough to guarantee an upholding of the weathering process once a certain amount of humid saturation has been gathered within these circumstances.

Such a monthly rainfall rate is the case along the eastern and southern coast for all year round or at least three quarters of the year as the charts show presented by WELLINGTON (1955, 241 - 243).

The favourable topogenic features referred to above can be found everywhere on the flat lower slope sections where, as a consequence of slope-wash, fine material has accumulated, forming bags as it were on these slope sections. It is these bags in which soil moisture is preserved over a long time even in periods of drought (as the ones Natal suffered between 1978 and 1983) and it is especially in these bags where the processes of hydration and even hydrolysis are launched and upheld and from where these processes can penetrate into the rock below.

Typical for the subtropical conditions of Natal is however that chemical weathering within and beneath these bags does not lead on to a complete decomposition of rocks and crystals (German: 'Totalzersatz') as in the full tropics, but only to a partial disintegration (German: 'Partialzersatz'). In the case of the tropics every rock component is decomposed totally (German: 'Komponentenzersatz'), in the subtropics of southern Africa only the texture of the rock itself disintegrates or disaggregates (German: 'Gefügezersatz'), whereas the components remain intact, even the micas and felspars. The former rock structure as such, complete with its folds and fissures, cracks and layers, is also preserved. Weathering thus results in granular disintegration for which angular rock fragments and angular particles are typical.

Naturally, granular disintegration depends on the kind of rock being weathered. It is especially the crystalline rocks, the intrusives and the granular sandstones or quartzites which are liable to this kind of disintegration. With shales, schists, graywackes and similar rocks disintegration is a little different, since these rocks fall into brittle or friable pieces the size of pennies or halfpennies. With weathering continuing, grit or sand or clay is left over in the end only, with the quartz fragments, brittle and angular, remaining as the only larger pieces.

This kind of disaggregational weathering seems to be typical for all regions in southern Africa that are intermediary between the temperate zone on the one hand in the south and the tropical one on the other hand in the north. As this weathering is a type of its own it may well be termed the intermediary type.

How quickly and completely disintegration takes place is a question of humidity. In the humid and semi-humid areas of southern Africa it is everywhere rather quick and complete (Fig. 16. 4) and as a consequence rocks in these areas are deeply weathered; this is especially the case with all granular rock types. No fundamental difference can be observed between the Mediterranean or Natalian sub-areas of southern Africa.

In contrast, weathering in the semi-arid or arid areas is not deep at all. Under a raw and skeletal 'soil', very shallow and coarse-textured, with a thickness of 5cm or of 10cm at the utmost, the fresh rocks appear, completely unweathered. The coarse texture of the 'soil' is due to the brittle nature of the fragments, even with shales and similar rocks. Over a longer time all fragments disintegrate into sand.

This arid type of rock weathering is to be found in all areas with pedimentation, pediplanation and silcretes (cf. Fig. 7. 3).

Therefore, climatic conditions that - together with these weathering processes attached to it - lead towards the forming of pediments might appropiately be called the pediment-forming climate. (Cf. Fig. 16. 4).

What has been described in the preceding paragraphs may be summed up with the following conclusion, namely that climatic influences seem to be the overriding ones, whereas lithogenic attributes are only subordinate (i.e. parent rock material does not become dominant). Such a statement should however not be taken to mean that lithological composition has no bearing at all. That lithological differences do play a certain part should sufficiently have become clear from the description of the inselberg fringes in crystalline rocks and of the ramps in the shaly rocks. Another example where the parent material makes itself felt is the weathering of the dolerites. Because of its importance dolerite weathering is dealt with separately in ch. 16 e.

The formation of inselbergs seems to be dependent on the fissures and cracks by which especially the granites, gneisses and intrusives are pervaded. These fissures and cracks are important because they allow the water to circulate along them, with the result that the rocks are dissolved here by hydration and granular disintegration is triggered off along these water tracks.

Though inselberg formation is certainly dominant within these arid and semi-arid climates this does not mean that inselbergs do not appear in other climatic regions. Rather, they also occur in the warm humid areas of the Natalian and Transvaal side - but only upon granites and granite-gneisses. They begin to appear in southern Natal (tentatively as it were) and become the more numerous the farther one comes into the northeast. In the lowveld of the eastern Transvaal inselbergs are formed on a rather broad scale. Yet even here their formation depends on the rock type (granites and the like). On the whole, in these more humid regions inselbergs are however not as dominant as in the more arid ones. (Bornhardts, as particular forms of inselbergs, also occur in both regions. They are dependent on lithology since they only appear on smaller or larger domes of granite proper.)

The formation of mesas, tafelbergs and spitskoppies can also be observed in both zones. In general, the process responsible for their formation and indeed for the whole 'chain' of forms (i.e. from mesas over tafelbergs to spitskoppies: cf. ch. 14) is, after a first dissection, the same everywhere: a consequence of the successive reduction and removal of the protecting hard cover. Nevertheless there are some differences between the humid and arid regions: (a) the process is slowed up in the arid ones; (b) the meaning of 'rock resistancy' changes in the same direction, with the greywackes (or mudstones) and schists being able to form cuestas, benchlands, mesas and so on (as has been mentioned above but is here repeated for systematic reasons).

Concluding this part, the present writer would like to point out once more his conviction that the two main climatogenetic provinces (together with their general weathering conditions and related landforms) do date back into the longer geological past, probably as far back as the late Jurassic. Three reasons can be given for such a conviction: (1) the ocurrence of fossil duri-crusts (both laterites and silcretes), some of them stemming from the Lower Cretaceous; (2) the pediplanatory landforms so strikingly extended over vast areas, together with the fossil sand dunes; (3) the long time any down-wearing must have taken. Therefore recent climatic geomorphology must be considered to have strong roots in the conditions of the long past.

d) The distribution of clay and heavy minerals

The distribution of the clay and heavy minerals is supposed to shed some more light on climatogenetic processen and their zonal appearance. Therefore pediment samples were gathered from the larger river basins along the Great Escarpment.

The samples were taken on an average spacing of 100km, starting with the Crocodile river in the eastern Transvaal and finishing with the Gamtos river in the eastern Cape Province. Usually, the samples were gathered from somewhere near the mouth of the rivers into the ocean. The reason for this was the assumption that what is carried by the rivers in their lower thalwegs must be the resultant of all weathering processes over the whole catchment area both recently and sub-recently. Samples of sand and silt were therefore regarded as probably being helpful about finding out about the possibly different weathering intensities over the overall distance between the Crocodile and the Gamtos.

(The samples were analysed by Dr. Salger from the Bavarian Geological Service. I am very thankful indeed to Dr. Salger. Dr. Salger's analyses enabled me to produce Table 16. 2.)

The results from the samples for the minerals concerned are presented in Table 16. 2, with an ordering of the samples from the northeast to the southwest.
What can be observed from the samples can be generalised as follows.

(1) Clays. Illite content is rather low in the NE (15 to 40%), but from northern Natal on, illite begins to become dominant (58 to 90%). Typically, the kaolinite contents are reversed and are rather low in the Transkei and the eastern Cape Province (6 to 8%). Gibbsite, a typcial representative of tropical weathering, has similarly completely vanished from the samples of the Transkei and the Cape.
(2) Iron oxides. Goethite is more present in the samples of the Transkei and the Cape.
(3) Heavy minerals. The less resistant ones (i.e. granate, epidote, augites, hornblende) increase steadily from the NE towards the SW. (Similarly, by the way, granate and glauconite are typical heavy minerals in the Lower Cretaceous of the Cape: SHONE 1978).

All these facts point into the same direction: More intense weathering conditions make themselves felt recently (and sub-recently) in the northeast (eastern Transkei, Swaziland) - in good relation to the warm and humid climate there (cf. Table 16. 1).

Inspite of this more intense weathering one is however not justified in calling this 'morphodynamic climate' a 'tropical' one - another proof why it is not suitable to follow FRANKENBERG's and LAUER's classification (cf. ch. 16 b). Natal is again in a middling position and the Cape and the Transkei show the least degree of intensity.

The samples represent three intensity areas within the subtropics. These are:

(1) the highly intense subtropics (easternTransvaal, northern Swaziland);

(2) the average or middling intense subtropics (Natal);
(3) the weakly intense subtropics (Transkei, Cape).

With respect to the western Cape Province the analyses in Table 16. 2 can be supplemented by the findings of BIRCH (1978). BIRCH analysed the clay mineral content of recent deposits on the Atlantic shelf, their illite and kaolinite contents being very similar to those in the eastern Cape (Illite: 56 to 90%; kaolinite: 5 to 20%).

e) Dolerite weathering

The soils derived from dolerite weathering present two different colours and each colour in different situations: either brown or red. These different situations are bound to distinct topogenic features.

(It may be mentioned here, for completeness' sake, that a certain variety of the granite-gneiss in Natal also tends towards the forming of red soils. Since the ocurrence of such pockets is very scarce indeed they are not dealt with here. Additionally, it should be pointed out that above a certain altitude, i.e. that of 2100 to 2400m, even on dolerite rocks, only brown soils occur whatever the specific topogenic situations are.)

Dolerites possess a particular lithology which makes them prone to red weathering even under the recent (or subrecent) climatic conditions of Natal. This is because they are suitably rich in augites, olivines and biotites - all of them minerals that are easily dissolved into haematites (VAN DER MERWE 1962).

Before going on with explaining why this red weathering only occurs in specific topogenic situations it may be stated that any red weathering on other rocks must have ceased at least with the end of the Tertiary - or even before. A cutting near Botha's Hill (in the hinterland of Durban in Natal) helps towards demonstrating this opinion. The cutting, in an altitude of 900m, lies in granite-gneiss on the flanks of the Valley of a Thousand Hills. The TMS, formerly on top of the granite, was worn off so that the granite became completely bared. The cutting within the granite shows a fossil red soil below, immediately developed

Table 16.2: Mineral contents of samples from selected rivers in eastern South Africa

Gehalte (Tone, Eisen, Schwermineralien) von ausgewählten Flußsedimenten im östlichen Südafrika

Numbers of sample	1	2	3	4	5	6	7	8	9	10	11	12	13	14
Rivers / mineral contents in %	Cro-co-dile	Usut-fu	Pon-go-lo	Mfo-lo-si	Tu-ge-la	Mge-ni	Mzim-ku-lu	Mzim-buvu	Tina	Ba-shee	Gr. Kei	Keis-kam-mer	Busch-manns	Gam-toos
(1) Clay[1)]														
vermiculite	25	9	-	-	-	-	14	-	27	24	-	4	10	15
montmorillonite	-	6	40	32	6	20	5	-	-	-	34	-	-	-
illite	40	15	39	58	83	60	59	89	67	70	60	90	83	77
kaolinite	35	70	21	10	11	20	22	11	6	6	6	6	7	8
gibbsite[2)]	.	.	-	-	.	.	.	-	-	-	-	-	-	-
(2) Iron[3)]	G	G	G	G	-	G	G	-	-	-	-	-	-	-
(3) Heavy[5)] minerals														
zircone	16	84	24	14	30	-[4)]	9	39	33	35	14	36	27	63
turmaline	2	1	-	.	1	-	1	.	-	.	-	-	-	-
rutile	.	-	5	4	3	-	2	15	21	7	3	2	13	7
granate	4	3	19	37	12	-	26	29	33	53	65	51	59	30
staurolithe	-	-	-	-	-	-	.	.	.	-	-	-	-	-
apatite	1	-	-	.	1	-	2	-	-	-	-	-	-	-
epidote/zoisite	27	3	8	1	-	-	-	-	3	3	3	6	1	-
amphiboles (horn-blende)	40	8	27	44	15	-	2	-	10	2	15	5	.	-
pyroxenes (augites)	10	1	13	.	38	-	58	17	-[6)]	-	-	-	-	-
(4) Opaque (ilmenite, magnetite)	90	90	86	67	76	-	48	61	50	81	75	85	61	70

1) relative content, without quartz and iron oxides; percentages of fraction (smaller than 0,002 mm) 2) dots: only in traces, less than 1% 3) G = goethite; other iron oxides except ilmenite and magnetite not present 4) heavy minerals in this sample not evaluable 5) Fraction 0,1 - 0,25, except opaque minerals. 6) Pyroxenes were not evaluable for samples 9 - 14.

on the granite, and made fossil by the fact that it is capped by brown material, derived from the same granite. The brown cap contains a Pleistocene pebble line (cf. Fig. 16. 6). The soil profile became preserved in a shallow depression within the general slope, not yet reached by the erosion gullies. At least during the Quaternary recent morphodynamics seem to have prevailed.

Now towards the specific topogenetic situations. Dolerites on steeper slopes always (and without any exception) weather into a brown soil consisting of sandy particles. The rock structure is also preserved on these steeper slopes, with the rock remaining firm and hard and keeping its fresh dark-bluish-grey colour. The same can be said about the dolerite boulders which everywhere occur on these steeper slopes.

In contrast to this dolerites present quite a different picture if weathering occurs on the flat lower sections of the slopes.(There is no exception to this in all Natal.) On these lower sections the dolerites show an enormous granular disintegration, resulting, on the one hand, in a rather thick cover of grit, and, on the other hand, in a complete through-and-through disintegration of the whole rock - a disintegration by which its former structure becomes fully disaggregated, sometimes with a thickness of 10m.

Again without any exception, the colour of this cover is not the dull brown of the steeper slopes but a vivid red. The gritty particles in the cover are friable, but usually not pliable (the latter being the case only under condition of complete discomposition, as, for instance, in the north of Swaziland). If such a dolerite cover is washed down over other rocks, the cover keeps its red colour, but the other rocks - even the TMS (which was liable to red weathering in the Upper Cretaceous) - retain their own specific colours (either fresh or weathered) (Cf. Foto 49). Wherever the granularised dolerite cover comes to an end (i.e. with the steepening of the slope) the red colour immediately disappears. Similarly wherever a slope of some steepness is kept free from any dolerite cover by slope wash so that no granular particles are able to accumulate the colour of the rock beneath and of the particles on top remains brown.

All these observations lead to the conclusion that under present climatic conditions dolerites can become rubefied

Fig.16.6: Soil section near Botha's Hill (Durban, central Natal)

Bodenprofil bei Botha's Hill (Durban, Zentral-Natal)

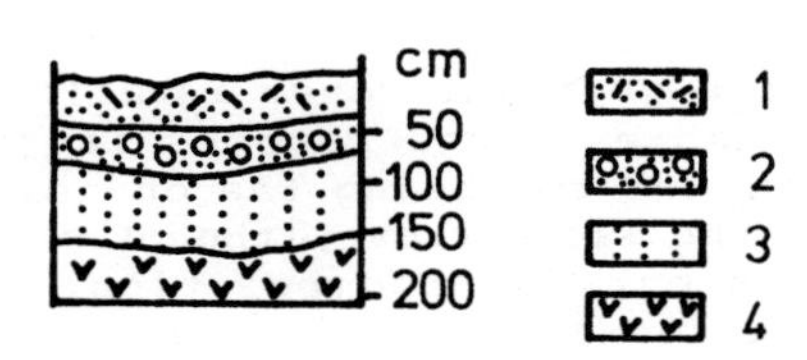

1 = middle brown, with angular components

2 = dark-brown with large rounded components

3 = red with gritty granite particles

4 = granite rock

Source: J. Birkenhauer
drawing: E. Glaser

only in the favourable topogenetic situation of the lower flat slope sections.

From these observations it is moreover obvious that it is the accumulated cover only that makes rubefaction possible. Rubefaction is obviously dependent on the particular geochemical environment that here furthers the exfoliation of biotites and all processes of hydration. The reason for this is that the granular cover is able to act as an efficient sponge. Such a sponge becomes larger and larger from two processes: the one being the on-going accumulation on the top from the slope above, the other being the forming of an on-going front of decomposition, 'biting' into the rock below. The formation of a sponge is therefore restrengthened on all sides. The importance of an ever-enlarging sponge lies in the fact that the more moisture can be preserved within it, even over long dry periods, the larger and thicker it becomes. Therefore even during long dry periods the rubefying process can still go on. On the other hand, wherever moisture is either not preserved - as on the steeper slopes - or is insufficient because of a change in climatic humidity - as towards the west - the red weathering of the dolerites comes to an abrupt end (even where the topogenic situation would be favourable as such). Other reasons for the stopping of any red dolerite weathering are (1) steep cracks in the rocks that guarantee a quick drainage of water and (2) a drop in the temperatures necessary for the upholding of the geochemical process - as is the case in the higher altitudes (see above) or towards the southwest.

Altogether, dolerite weathering is a good example to demonstrate how intricate and interdependent the processes are which are involved: be they lithogenic, climatogenic or topogenic.

What has been described above could be observed over the whole research area because of numerous fresh road cuttings allowing a detailed and close study of what might be called the 'dolerite environment'.

A comparison of all places has led to the conclusion that, under present conditions (and presumably through a considerable period of the Tertiary), no rubefaction in the sense of a zonal phenomenon can have taken place. Wherever red dole-

Fig.16.7: Paleo-soil on TMS near Umbumbulu (Central Natal)
Paläo-Boden auf TMS bei Umbumbulu (Zentral-Natal)

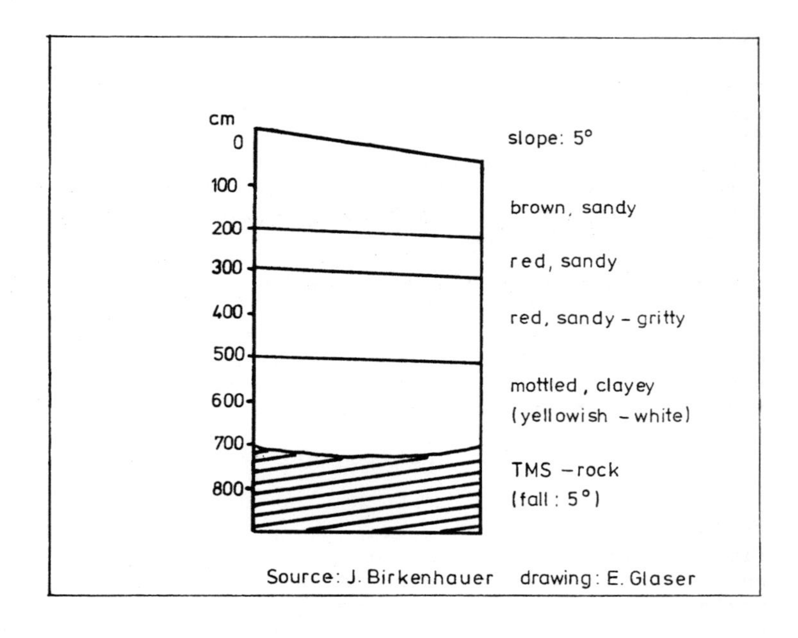

rite soils appear in Natal they must be regarded as extrazonal soils dependent on the specific topogenic situation.

f) Morphodynamic consequences

The results of rock weathering within the two main and broader regions of southern Africa (i.e. the more humid and the more arid one) must be considered to be fundamentally the same with regard to the main and decisive morphodynamic circumstances - with these two regions only differing in the length of time needed to achieve the same degree of weathering and of wearing down.

The general result for both broader areas is that only particles of sand if not silt and clay are washed into the rivers. The consequence of this is that rivers do not possess any sufficient weapons of erosion (at least not under average conditions). The consequence of this again is that linear erosion must naturally be a rather slow process - a consequence arrived at from the calculations done in some of the previous chapters that now can be put into an explanatory frame.

It is only during high waters when the whole river bed is flooded that a downward and deepening erosion can take place and really does take place. For it is only then that large boulders are carried by the waters. It is especially these boulders that - in thumping and bumping over the river bed - chisel away from the rocks below other lumps and even boulders, rather often being of a length of 1 to 2 metres.

Such chiselling away can naturally only happen where boulders can be disposed of. All observations have shown that this is the case only within the more humid regions, for it is here that slopes are kept steep enough for loosened boulders to roll down and to be eventually carried into the river bed. Where however the flattish pediments prevail as in the more arid areas the coarsest particles to reach the streams are only of the size of sand. Not much linear erosion can naturally happen there. This fact may perhaps serve as an explanation for the longevity of the land-forms in these regions.

The floodedness of the river beds thus seems to be the necessary condition for at least some amount of linear erosion. Such floodedness is certainly dependent on the amount of precipitation. The rainfall record of Natal for instance (PITMAN et al. 1981) shows that sufficient rainfalls resulting in floodedness is very scarce. Suitable conditions happen only once or twice during a century, - and this it the utmost even, as the records show. Such a fact naturally means that - inspite of the boulders - there is not much linear erosion going on even in the more humid climates.

The consequence is that it really must take millions of years to excavate deeper valleys (see ch. 15).

Though the high waters (after a sufficient rainfall event) develop terrific speed and power they can erosionally be active only as long as there are boulders to be made use of. Some observations to this respect could be made along the Oribi gorge in southern Natal. In the centre of the gorge (where the road towards the Oribi Flats crosses) huge table-like sheet boulders of TMS quartzites and of angular granite-gneisses (often with lengths of 2m) show, from their rounded features, that they must have been transported downstream at one time. In being transported downstream they obviously bumped against the rocky river bed (here in the crystalline basement) thereby tearing loose other rocks of equal size from the basement. Following the river downstream to a point 3km below the mouth of the gorge no more boulders are to be seen anywhere. What one notices is sand or, at the utmost, some small well-rounded pebbles. Thus, within a rather short distance, all the boulders must have become ground and pulverised.

Though conditions in the interior of the Cape Province and in South West Africa are arid enough high waters are able to follow from the occasional heavy (or sometimes even very heavy) downpours resulting from a thunderstorm. The surface seal in these regions helps towards a rapid accumulation over the pediplained flats so that all water sheetwashes into the shallow valleys or finds a quick run-off through the gullies. Rapidly the valleys are flooded, the waters tearing away masses of sand from the pediments and trees from the valley banks. Yet these floodings do altogether not result in any much linear erosion. This is why even larger rivers seem to have great difficulties in cutting back their valleys upstream from one level to the next. Falls and narrow canyons must be the outcomes - the Augrabies Falls of the Orange and the Fish river canyon are nothing but the more outstanding examples.

Consequently, knick-points in the river thalwegs will take a comparably long time to proceed upstream and to become graded down to a lesser gradient. This circumstance naturally is a contradiction to KING's assumption that scarp retreat follows from the retreat of knick-points. Moreover, the described circumstance is the reason why the knick-points (they themselves stemming of course from the downward crossing from one level to the next below) have remained intact so well and over such lengths of time. Therefore, their existence can really be taken as another proof for the existence of the levels (Fig. 12. 1).

The Oribi gorge can furthermore be used as an example for calculating the amount of linear erosion a river may achieve within a certain time. The river flowing through the gorge is the Mzilkuwana. Where it enters the gorge the TMS quartzites form flat plateaus in an altitude of 500m. Where the Mzilkuwana leaves the gorge the TMS plateaus stand at a height of 450 to 430m. Ostensibly, the valley was superimposed across the horst-like plateaus as a consequence of the Upper Cretaceous transgressions. When regression set in the river naturally was forced to cut through the quartzitic plateaus, this being the origin of the gorge. Today, the rocky river bed lies in an altitude of 300m where the river enters the gorge and at 200m where it leaves. The figures given mean that 200m of incision must have been executed during the last 65 my. The rate of incision can thus be calculated of being a little more that 3m per 1 my. If one now assumes that high waters occurred at the same rate as today (throughout this long span of time) one can go on to say that the rate of erosional deepening per every high water event cannot amount to more than 1 to 2mm (per event).

Calculations done for the inner tropics (BREMER, 1978, in New Guinea; STEIN, 1981, in Sri Lanka) have shown that the denudational rates there amount to 30 to 40m per my. The difference is indeed rather large and may serve as a measure for the erosional and denudational efficiency of the morphodynamic system in the inner tropics and even in the more humid subtropics.

g) Soils

One of the main characteristics of the subtropical morphodynamic system seems to be partial weathering. As we have seen above, the lithological rock types do play a certain part within the weathering processes. This fact naturally then means that the particular geochemistry of the rocks with its specific genetic features is not wiped out completely, as it seems to happen in the full tropics. A consequence is that the soils in subtropical southern Africa will be connected in some way with the lithological traits of the parent rocks.

Such a theoretical assumption is indeed carried by all observations, especially on the more humid eastern flank of the subcontinent.

In a certain contrast to this, the lithological connections in the arid and semi-arid areas are somewhat slurred over. Independent of the parent rock, yellowish-brownish soils prevail in the colder areas and yellowish-reddish soils in the warmer (and semi-arid) areas.

In humid Natal it is however possible to connect each particular soil colour with a particular parent rock. An example was given to a fuller extent above for the dolerites. The crystalline basement almost always shows yellowish-brownish colours, the Dwyka shales and tillite soils are always very yellowish with a slight brownish tint, the TMS and Ecca quartzite soils appear in a brownish grey and the Dwyka sandstones in a darker yellowish-brownish colour, whereas the Beaufort soils bear a light brown colour.

From this description it follows that the soils do pertain to the theoretical rule lined out above.The soils can therefore be regarded as truly subtropical ones.

In dealing with the soil colours and their relation to the parent rocks the author is in full agreement with KING (1972). It should be pointed out here that KING, in referring to 'soils', does not use the term in its strict pedological sense, but only in the broader geological sense, where the term means 'loose, weathered material'. RENK (1977), in his painstaking studies of the soils in Swaziland, used the term however in the strict pedological way. It is not very surprising then that he did not agree with KING. For RENK soils never develop immediately on top of any parent rock, but only on materials washed down from them. When discussing the formation of the red dolerite soils we have referred to exactly this situation.

In order to classify the southern African soils, pedologists in South Africa have partly developed a system of their own. Two notable classifications, together with maps, have been published for the whole subcontinent; the one is by VAN DER MERWE (1962), the other by the Department of Agricultural Technical Services in Pretoria (1973). The maps are here reproduced as Fig. 16. 8 (VAN DER MERWE) and Fig. 16. 9 (Department), condensed, generalised and in a smaller scale.

Though the terms used are differing, both maps show comparable features: for instance (1) the rather monotonous arid and semi-arid west and the various and even intricately manifold humid and semi-humid south and east with nearly the same boundary between these two larger subcontinental regions, (2) the extensive areas of weakly developed soils and lithosoils, (3) the large areas with latosols in the northeast and the differences between the lowveld and the highveld. What varies in both maps (sometimes to a rather great extent) is however the size of the areas.

In the opinion of the present writer, both maps falsely include the red dolerite soils within the latosols or lateritic soils. As has been shown above these soils should rather be treated as a-zonal ones, because such an inclusion suggests the extension of the marginal tropical soils of a savanna climate type into wide parts of Natal. The southern boundary of regular zonal red soils (common to both the northern and northeastern provinces in the maps) runs along a line demarked by the towns of Pary, Standerton and Mbabane, a line that roughly corresponds to the latitude of 27 degrees south. In the south of this boundary there occur yellowish brown (or brownish yellow) soils, in accordance with the lithological properties of the parent rock material (as was pointed out above). It is exactly these varying

Fig. 16.8: Soil groups of South Africa after VAN DER MERVE (taken from WELLINGTON 1955)
Böden in Südafrika nach VAN DER MERWE

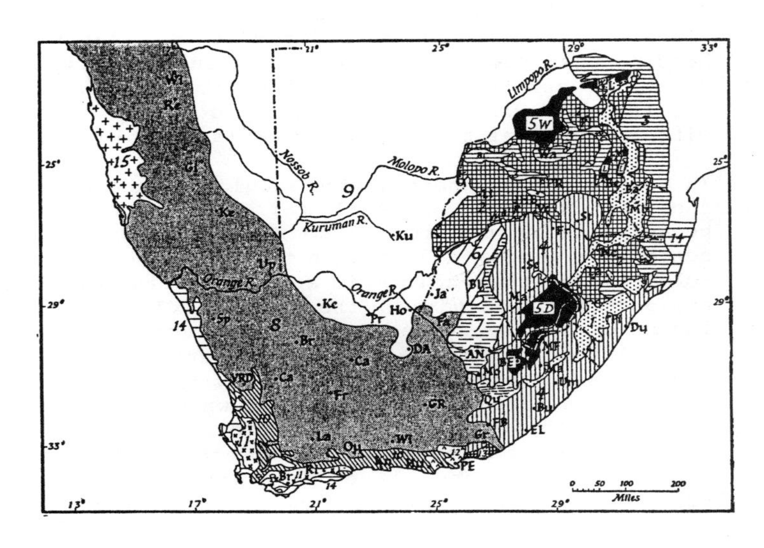

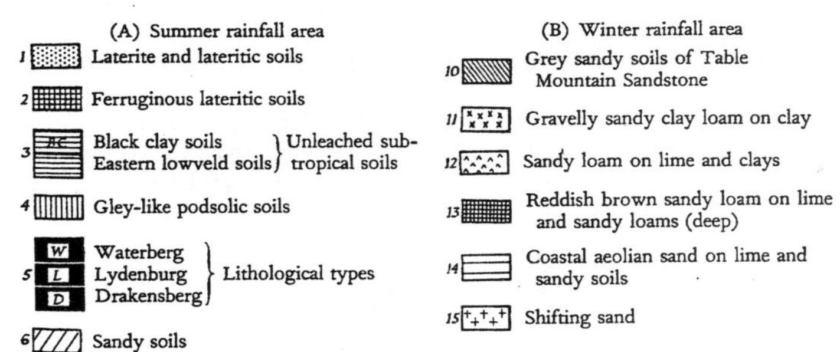

properties that are responsible for the intricate pattern the maps show in the east.

What is called the 'red-yellow-grey latosol plinthic catena' (Department), must be considered to be a far too general heading that really blurs over all the differences that do exist between the marginally 'tropical' soils in the north and the yellowish brown 'subtropical' soils in the south with goethite as their iron oxide (see Table 16. 1).

In passing, some short notice should be made with respect to the very generalised soil map in KUNTZE et al. (1981), because its result is too coarse even admitting that the scale is a very small one. In this map cambisols and luvisols fill up the whole arid and semi-arid west and podsols and podsoluvisols take up the whole middle and east with the inclusion of the red soils. Ochrid luvisols and rendzinas are allotted to Natal. These assignments do really make not much sense, neither in comparison with the two other soil maps nor in comparison with one's own observations.
On the whole, with some reservations (indicated in Fig. 16. 9), the map stemming from the Department seems to be suited best to try and generalise the manifold and intricate pattern into some broader and thus more handy regions.

For such generalisation the following reasons may be given. Three regions in the north (viz. I, II, III) form a super-region where red or yellowish red soils are predominant:

- from the red sands in the arid or semi-arid west (I)
- to the red loamy soils of the highland plateau (II)
- and the partly more gritty yellowish red soils in the coastal lowveld of the semi-humid east (III).

This super-region is altogether distinguished by the overall occurrence of red soils. It might therefore be called the marginally tropical area of a marginal savanna climate.

The four regions in the south (viz. IV, V, VI, VII) form the sub-tropical super-region, distinguished as a whole by the wide occurrence of weakly developed soils, both in the humid east (IV, V, VI), together here with solonetzic soils, and in the arid and semi-arid west (VI), including the coarse desert soils.

Fig. 16. 9: Simplified soil regions of southern Africa (based on the soil map of the Department of Agricultural Technical Services, Pretoria, 1973)
Bodenprovinzen im südlichen Afrika (vereinfacht, nach offizieller Bodenkarte)

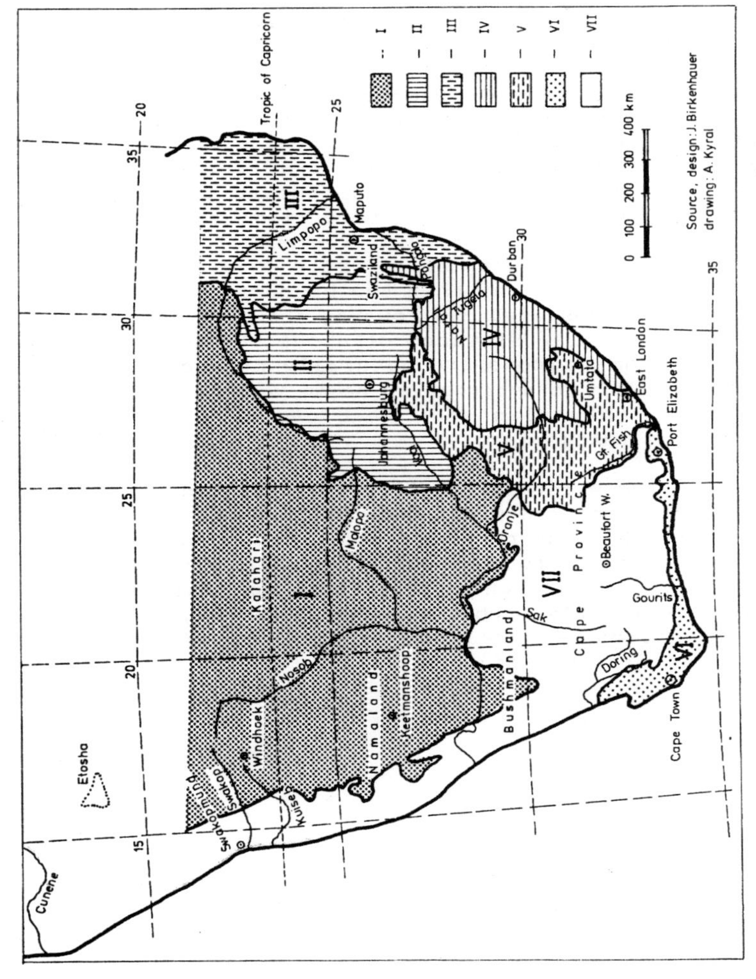

I. red sandy soils; II. red loamy soils of the highlandpl.; III. yellowish red soils of the Low Veld; IV. weakly developed soils, yellow-brown; V. with solonetzic soils; VI. with pod soils; VII. coarse desert soils.

The regions outlined here are shown in Fig. 16. 9.

One might go on trying to find out (certainly in a tentative way) whether there are some similarities between these recent (or sub-recent) soil-regions and the palaeo-regions. This may be assumed from what has been said about the earlier climatic conditions in the previous chapters. The similarities seem to consist of the following features.

The area of the Upper Cretaceous laterites can be compared with the present one of the latosols. The solonetzic area may be compared with the silcrete region. Doing so one sees that a shift seems to have occurred between the paleo-regions and the present ones amounting to a distance of roughly 500km. Continuing the comparison with a view (a) at the present eastern boundary of region VII and (b) with the former western boundary of silcrete development one again seems to be able to say that the boundary of the more arid zone of today has advanced towards the east from the paleo-zone for a similar distance of about 500 to 600km. Though all these comparisons are naturally mere conjectures the changes as such do not seem to be so very highly improbable. Altogether, they also show that overall drastic changes cannot have taken place.

Referring these findings to the main quest of the treatise, the evolution of the Great Escarpment, one can say, in conclusion of this chapter, that the Great Escarpment seems to have been subject to comparable conditions of climatic geomorphology (a) throughout a rather long timespan and (b) within the roughly stable super-regions (i.e. arid or semi-arid conditions in the west and the humid or semi-humid conditions in the east).

III: CONCLUDING PART

CHAPTER 17: **SUMMING-UP**

1. THE GREAT ESCARPMENT

a) Nature of the escarpment

The Great Escarpment should not be defined as a steepened continental margin or swell; neither it is just one giant cuesta; it is rather a chain of cuestas, and indeed a gigantic one, of varying rocks and heights between the Limpopo basin in the northeast and the Khomas Highland in the northwest (ch. 13). In this way it runs over an overall length of 2400km. As such it is probably the longest of its kind in the world and therefore a true form of macro-geomorphology (ch. 1). Although it runs roughly parallel to the continental margin, it should really not be considered to be a macro-form that is (directly or indirectly) related to the forming of continents as such or to plate-tectonics (cf. JESSEN 1948). The model designed by OLLIER (1985) seems to be oversimplified: the escarpment is certainly not the tectonically up-lifted rise or swell with just one basin in the centre caused by the marginal upswellings (ch. 1 a, 3) and subsiding when Gondwanaland broke up or a little prior to such upbreaking.

It should be born in mind that its structure and course is indeed controlled by three geological basins far older than the upbreaking. These basins are: (1) the Orange basin characterised by the so-called Karoo super group; (2) the Nosob-Molopo- (or Kalahari-) basin characterised by the Nama group; (3) the Olifants basin characterised by the Transvaal super group (cf. Fig. 13. 1; Table 1. 1; ch. 2; ch. 13).

It really is the strike of the cuesta bearing rocks of these three basins that fundamentally and originally determines the course and the appearance of the Great Escarpment.

Because of these geological and geomorphological features the Khomas Highland should really not be regarded as part of the Great Escarpment. It rather is an excentric horst region, only accidentally lying in the strike of the escarpment. The highland should better be regarded as a singular very ancient feature inherited from the Gondwana super-continent.

Similarly, the Swakop-Kuiseb Gap (German: 'Randstufenlücke') should equally not be regarded as a gap within the escarpment. It is, rather, nothing but an inheritance from an even older Pre-cambrian configuration.

Both features (i.e. the Khomas Highland and the Swakop-Kuiseb Gap) effectively separate the cuesta chain of southern Africa from what appears to be another cuesta chain of its own in north-western South West Africa and Angola. This separation has a width of roughly 500km. After a gap of such width one can surely not go on talking about the self-same escarpment. Moreover, the cuesta chain in the northwest of this gap does owe its structural control to another pre-rift tectonic basin and therefore has also nothing to do with any later rifting.

All this goes to say that the Great Escarpment cannot be regarded as the one super-unit of all Africa south of the Congo, with a super-length of 4800km. It is moreover virtually not continuous from northern Namibia right down to the western Cape, with the Orange river providing the only major break in the west, as OLLIER and MARKER (1985, 39) think. As has been pointed out the Swakop-Kuiseb Gap is really more than a mere break or interruption.

b) Age, sinuosity and position of the escarpment - age of the major drainage systems

The Great Escarpment - in the sense of a cuesta chain - can certainly not be taken as being a rather young feature of a merely Mio-Pliocene age (i.e. 5 to 9 my), as KING, KAYSER and OLLIER make us believe (ch. 3 and 4). It is certainly as old as the Lower Cretaceous (i.e. Valanginian = 130 my) - or perhaps even older. This statement can be made with some assurance, because certain landforms reach back to the very foot of the escarpment, landforms that can be dated, revealing that they are as old as the escarpment (Fig. 1. 1).

These landforms in particular are the P 2 peneplain (altitude: 1200 to 1700m) and the T 1 and T 2 thalassoplains, all of them of a Lower Cretaceous age (with the P 2 probably being of a Barrèmian age and the T 1 and T 2 of an Albian age).

These ages can be determined because these major landforms everywhere cut across the following four features:

(1) the Jurassic and Lower Cretaceous faults (everywhere)
(2) the Jurassic and Lower Cretaceous dolerites (everywhere)
(3) the Lower Cretaceous deposits (southern Cape),
(4) the Lower Cretaceous Etjo beds (central South West Africa).
(See chapters 2, 6 d, 9 a, 9 b, 9 c.)

From the reasons given the author feels sure that the ages of the planations are indubitable. Since these very old planations do reach back right to the foot of the escarpment the author feels equally sure when stating that the Great escarpment must always have lain (more or less) where it is situated today (cf. ch. 5 c, 9 e). It must moreover have always lain in a rather elevated position. This statement is corroborated by MARTIN et al. (1980, 192): According to them the Great Escarpment "has been in a relatively elevated position since the Jurassic".

As such, the Great Escarpment is surely one of the oldest macro-landforms preserved on earth.

One is even entitled to say that the Great Escarpment has surely not continuously receded into the interior - rather it is the forelands that successively have beome enlarged - both from uplift and from the regressions of the sea (cf.ch.5c, 13). And neither did the Great Escarpment develop according to the Piedmont treppen theory of Walther PENCK, as adopted by BORN, JAEGER, KAYSER (cf. ch. 3). (As a side-thought: this theory should altogether be abandoned.)

The sinuosity of the Great Escarpment along its course is nothing but the consequence of the fact that it has been deeply invaded by large embayments and inter-montane basins. Basins and embayments were both fundamentally shaped and reshaped by the processes of pedimentation and thalasso-planation, especially in the Lower Cretaceous and continued since then. It is these two processes that - in combination

- have helped to cut back the embayments against the cuestas (i.e. the socalled 'embayment retreat': cf. ch. 5 c, 9 a).

It is in this way that the Great Escarpment has become a landform of a truly and fundamentally erosional nature, structurally controlled. The series of cuestas are just not merely accidental features, as ROGERS and KAYSER thought, but essential ones (GRIESBACH, A. PENCK: cf. ch. 3). Research seems - in the course of a hundred years - to have come full circle.

Since pediplanational processes and thalassoplanation made use of and went into pre-existing drainage basins and valleys, following these upstream (as the surfaces show that run around basins and embayments), these basins must be considered to be even older than the Lower Cretaceous (i.e. more than 120 my). Some drainage basins may even be of an Upper Jurassic age, as the Orange system (with the Nosob, Molopo and Olifants-Doring-Tankwa sub-systems), the system of the Great Fish and that of the Transvaal Olifants (cf. ch. 4, 5 e, 12). These drainage basins seem to be some of the oldest on earth that have continuously functioned.

The working hypothesis formulated by OLLIER and MARKER (1985) - i.e. that the Great Escarpment owes its origin to the breakup of Gondwanaland - has surely to be modified. Erosion and denudation were at work all along the escarpment prior to the breakup. The sea, from tectono-eustatic fluctuations, would have penetrated anyway, whatever the particular coastline, by just following up the pre-existing continental drainage embayments.

c) Cuestas and climate

The cuestas are developed on differently layered rocks (see ch. 1 c, 13; Tables 1. 1, 1. 2; Fig. 1.). The rocks on which the cuestas are usually developed are the following ones: quartzites, sandstones, dolomites, layered dolerites. In the arid west even shales and mudstones (greywacke) can bear cuestas. From this we see that humidity is an important climatic factor even for the development of cuestas. The more humid a climate generally is, the more cuestas will develop only on the more resistant rocks. The transition between the more arid and the more humid areas takes place in the southern Cape Province. This zone of transition must have been situated here for rather a long time, since broadly similar climatic conditions must have reigned in southern Africa since at least the Lower Cretaceous. Such an assertion is made possible from the comparison of several circumstances: the undercutting of the same rock types by the Lower Cretaceous planations (Fig. 1. 2, 9. 2, 11. 2), the fossil provinces of the laterites (east) and silcretes (west), the cover of the serir-like quartz fragments in the west, pediments and pedimentation (west) (ch. 8, 10, 16 g; Fig. 7. 3).

Fig 17. 1 shows a comparison of the position and the course of the Great Escarpment as outlined by ROGERS (1928), OLLIER and MARKER (1985) and the present writer. For ROGERS 'steepness' is the decisive argument, though applied somewhat arbitrarily in some occasions (i.e. east of the Pietersburg highlands, northwest of Beaufort West). OLLIER and MARKER do not name their criterion, but it does seem to be the run of the highest crest (comparison of their line with a physical map), though they do not adhere to this criterion everywhere (exceptions being, without giving a reason, Swaziland, the area northeast of Port Elizabeth, the area west of Beaufort West). For the present author the one criterion to be used is that of the cuestas: the line should be drawn where the first true cuestas or their main outliers appear (in coming from the sea).

d) Scarp retreat, pediplanation, cuesta development

Scarp retreat as a whole and on a larger scale (KING, OLLIER and MARKER) must be refuted, since all slopes in southern Africa retreat parallel to each other only on a very minor scale. After a rapid initial phase slope development seems to come to a standstill or near-standstill. Not more than 300 to 500m of slope retreat can really be proved, and not more than 500 to 1000m of cuesta retreat (if one uses the planations as guiding lines: see ch. 14 d, 15 c). It is only in the embayments that a wider backwearing can be observed, with an average of 5 to 10km, sometimes even up to 25km. Compared to KING's assumption and in view to the length of time involved (at least 120 my), even such an amount must be regarded as a rather small one. The varying amount depends upon the rock type (deeper backwearing in shales and the like). (Cf. ch. 14 d, 15 a).

Wherever slopes in the humid region occur they are always of the same nature, independent of rocks, independent of the age of the planation above which they occur. From this fact it must be concluded that similar forming processes must have prevailed in the east since at least the Upper Cretaceous (ch. 14 b). The lower slope segments have not at all merged together with parallel forms and therefore do not form continuous pediments - in spite of the time-span at disposal (ch. 14 b, 14 d). In humid subtropical southern Africa pedimentation - as an efficient process of backwearing - can moreover be ruled out with certainty, both for the present and the geological past. It is only in the semi-arid and arid subtropical regions that pediments are truly striking gemorphic features. Here - and only here, in the past as well as in the present - pedimentation is to be observed as one of the most significant forming processes (ch. 10). Dry conditions must have prevailed in roughly the same regions - again both in the past and today -,these being the semi-arid basins in the rain-shadow of the Cape ranges (lee or föhn effect) and in the arid west and northwest. Such an assertion can be well proved from the rather wide intermediary Cretaceous pedimentation levels in the altitudes of 700, 900 and 1100m (ch. 7 d, 8, 10) as well as from the pedimentation level above the P 2 peneplain in 1500 to 1750m (ch. 9 a).

It was necessary to deal explicitly with slope forming processes in a special chapter (cf. ch. 14) because of the theory of pedimentation so central to KING's argumentation. Yet - apart from pedimentation in the more arid regions - the one process that still goes on to some extent (and must have gone on so in the past) is the one by which the cuestas have been formed, especially under the more humid conditions of the east, but also (though retarded) in the more arid regions of the west and north-west. The 'jumping-up' and 'amphitheatre' effects must be judged to be the rather efficient processes in cuesta forming, themselves dependent on the local watertables and spring horizons beneath any cracky rock (cf. ch. 14 c).

Fig. 17.1: Margins of the Great Escarpment in southern Africa according to different authors
Verlauf der Großen Randstufe im südlichen Afrika, nach verschiedenen Autoren

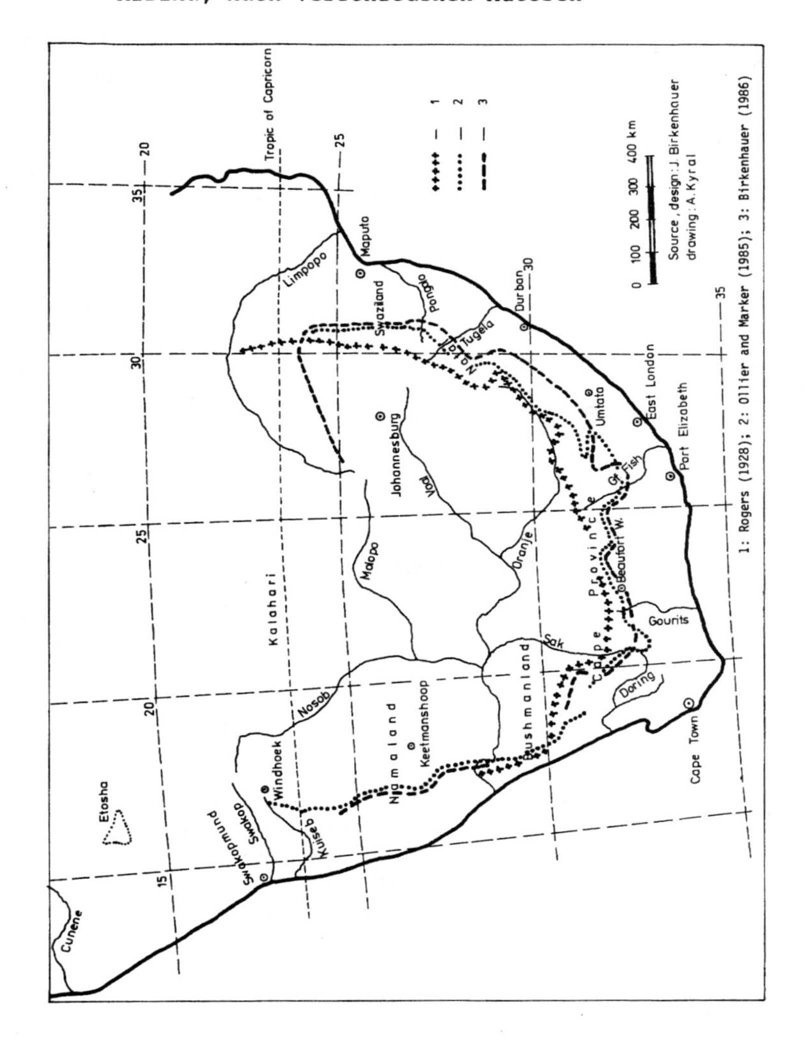

2. THE FORELANDS

a) Thalassoplains: nature, number, meaning

The horizontally level planations stretching as such over hundreds and even thousands of kilometres all around the southern African coastline, with several of them lying one above the other, are indeed very striking features of the forelands. These horizontally level plains must have been formed within the neighbourhood of a sea level as a consequence of tectono-eustatic transgressions. The different 'storeys' of these planations are usually separated from each other by distinct scarps as especially the lower ones (T 3 to T 8) or by intermediary levels of pedimentation as especially the higher ones (T 1 to T 3) (see ch. 14 b, 8).

Altogether eight 'storeys' of thalassoplains can be observed in southern Africa, presented in Table 17. 1.

The planations, in their horizontal nature, do not show any signs of tilting, warping (neither up nor down), faulting and so on. No faulting and so on has taken place since the Valanginian in the Lower Cretaceous. Rather these planations

Table 17. 1: Ages and altitudes of the thalassoplains in southern Africa

Thalasso-plains	Ages	Altitudes (average; in m ab. s-l)
T 8	Pleistocene	100
T 7	Plio-Pleistocene	200
T 6	Lower Miocene	300
T 5	Upper Cretaceous	400
T 4	Upper Cretaceous	500
T 3	Upper Cretaceous	600
T 2	Lower Cretaceous	800
T 1	Lower Cretaceous	1000

(Cf. ch. 5, 6, 9 d; Fig. 9. 2, 11. 2, 11. 1; Table 9. 1; for the averages of the levels: see ch. 11.)

do cut across all faults, folds, intrusives, rocks as old as the Valanginian (135 to 130 my ago) or even somewhat older ones, as the Etjo beds for instance.

Southern Africa cannot have suffered from any tectonic instabilities since then; it must have remained as a rather stable block with no uplifts at all since that last faulting age in the Valanginian. Such a view is - the author is afraid to say - in complete contrariness to what all other authors have so far believed (i.e. KING, OLLIER and MARKER, KAYSER, TANKARD et al.).

All tectonic activities and uplifts must have occurred either before the Lower Cretaceous or lastly within the Lower Cretaceous. (Cf. ch. 4, 5, 6, 7 g, 9 c, 11; Fig. 11.1.)

Starting from these grounds it is possible to construct a model of how the Great Escarpment together with its forelands, must have developed. This model is represented in Fig. 17. 2.

In spite of this model the 'oblique plain' (German: 'Schiefe Ebene') of the Namib desert, especially in the front of the Swakop-Kuiseb Gap, seems on first sight to be an argument against the existence of thalassoplains here (cf. JAEGER 1930, ABEL 1953 and 1959, KAYSER 1949; cf. ch. 3). Careful observation in this particular region has however shown that even this 'oblique plain' is stepped (ch. 6). The steps have however become veiled over by fine gravel and sand being blown across all features. In this they are very similar to the other 'low gradient embayments' (German: 'Flachböschungsbuchten'). (Cf. ch. 5.)

Since the thalassoplains always occur in the same altitudes the 'rule of series relation' (German: 'Relation der Serie') can be applied (cf. ch. 6 c). The rule means that the height intervals between one level and the next are everywhere the same; by this a definite altitudinal relation is established between the levels, grouping them into a real series.

Because of this it is possible to identify the levels from their position within the series and within these relational heights. The rule does not only apply to the planations but also to the 'terraces' that swing back and to within the drainage basins; it can moreover be applied to the 'valley bottom widenings' and the corresponding 'valley bottom closes' (German: 'Talbodenschlüsse') above as well as to the regular knickpoints along the thalwegs of the rivers appearing always in the areas of the valley bottom closes, i.e. where the scarps are crossed by the rivers between the level higher up and the one lower down (ch. 12). In this way the knickpoints themselves are connected with the thalassoplains.

b) Thalassoplanation

The great extent of the thalassoplains shows that the process (or processes rather) leading to their shaping must have indeed been an efficient one. Thalassoplains are etch-plains in the broadest sense of the term. Nevertheless, their forming process (or processes) must certainly have been different from the processes usually attributed to the formation of peneplains or etch-plains. The processes usually brought into connection with their shaping are the following two ones: (1) the one proposed by DAVIS, (2) the climatogenic one, fostered especially by BÜDEL and LOUIS and in broad use in Germany throughout the last decades.

In order to distinguish between etch-plains of different origins, it was thought advisable to introduce a new name for the particular kind of etch-plains as described by the present author. Since they must have originated in the neighbourhood of a then sea level (ch. 5 d, 5 e, 5 g) the term thalassoplain was thought to be appropiate. Consequently, the process (or processes) leading up to the formation of such plains should be called thalassoplanation.

The following combination of six circumstances and processes seems to be sufficient for an explanation as well as for defining the difference to other planational forms (cf. 5 d, 5 g, 6 d, 7 g, 7 a, 7 b):

(1) active advance of the sea against the land;
(2) development of flood plains and shallow coasts with a
    - slackening of water flow near the base level newly established by the advance of the sea,

Fig. 17. 2: Author's model of the development of the Great Escarpment and its forelands
Modell der Entwicklung der Großen Randstufe und ihres Vorlandes nach Auffassung des Autors

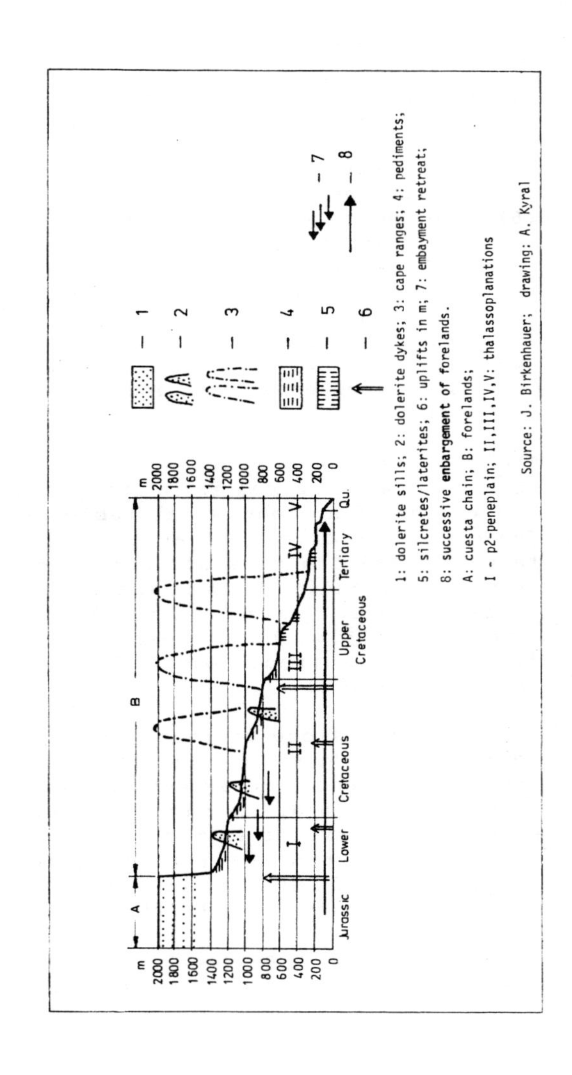

- damming up of ground water, especially in
  - o interior basins
  - o embayments;

(3) a high ground water table, presenting a favourable condition for
- silicification,
- laterisation,
- kaolinisation;

(4) extensive braidening and meandering of rivers near the shallow coast (forming of lagoons, spread of sands and gravels, formation of conglomerates);

(5) down-levelling of deeply weathered tracts of rock towards the base level;

(6) development of 'landscapes of wide removal' (German: 'Ausräumungslandschaften') (cf. ch. 15 a) along and over a blocked-up water table serving as a 'reservoir surface' (German: 'Staufläche') (cf. ch. 15 b), to which the following processes are related:
- cuesta effect,
- wearing down effect (as an 'expanded cuesta effect'),
- basin and embayment widening,
- subsequential effect.

The measure and the extent of denudation and removal of rocks over wider areas is in itself a function of the depths and widths of a groundwater table dammed both up and back. The older the origin of a level is (and the higher up it therefore is) the more all processes (going on along this level and spreading from it into the interior) have had time to take place. Equivalently, the pediments on the oldest levels are the widest: those in 1100 to 1200m (developed from the T 1 surface) and in 1550 to 1750 (developed from the P 2 surface).

Compared with slope retreat the efficiency of these planatory processes must be considered to be rather great. Nevertheless the speed and extent of actual denudation proved to be rather slow and, one might say, even disappointing if one expresses the extent in the measure introduced for denudational processes: the Bubnoff (1mm p.a.). Denudation in southern Africa seems to be far less: in the neighbourhood of around a hundredth Bubnoff.

Processes fixed to an established surface (i.e. still serving as a base level) tend to go on through the ages, these processes are: pediplanation and removal, all of them forms 'tradited on'; (German: 'traditionale Weiterbildung'). (Cf. ch. 8, 9 a.)

All this goes to say that thalassoplains (together with the planatory processes related to them) should be regarded a class of macro-geomorphic phenomena of its own - though under the heading of etch-plains.

SUESS (1885) and HASSINGER (1905) were the first to think seriously of the eustatic sea level rises and the landforms these rises must have brought with them. These two authors should really be reintegrated into the tradition of research. Their findings should be furthermore be rounded off now by the conviction (uttered here for the first time) that thalassoplanation must be considered to be one of the most decisive land forming processes.

c) The question of uplift and rejuvenation

The geologic times in which uplifts occurred are, at least in southern Africa, limited and these uplifts are moreover of a rather old age: Upper Jurassic to Lower Cretaceous. These tectonical facts can be gathered from the altitudinal relation of the higher thalassoplains (T 1 and T 2) in respect to the two peneplains (P 1 and P 2) (cf. ch. 6, 9 c; Fig. 11. 1).

Uplift amounts altogether to 1800 to 2000m. By the middle of the Cretaceous southern Africa had gained the altitudinal position in which it has stood ever since. All other landform variations depended on the tectono-eustatic sea level rises and fluctuations. There has never been just one period of rejuvenation only - and a rather young one at that - as believed by KING and OLLIER and MARKER. With fluctuations in between because of marine trans- and regressions rejuvenation has certainly lasted now for about at least 120 my. (see ch. 4, 6).

This goes to say that, within the same time-span, there cannot have been any 'cycles' of erosion in the sense of KING and DAVIS. Whether the Lower Cretaceous and Jurassic planations should be regarded as erosion cycles is a question of what aspect one likes to stress.

d) Climate and soils

The overriding and fundamental climatic circumstances of southern Africa apparently depend on the LAUTENSACHian category of continental change (ch. 16 b). The climatic circumstances must be regarded as principally alike and stable throughout the development of the Great Escarpment and its forelands since the Jurassic. The fundamental climatic circumstance meant here is the difference between the arid or semi-arid west and centre and the humid or semi-humid east. Equivalently, one finds the (arid) province of pediplantion and the (humid) province of slope forming. There are indications that, throughout this long time, at least semi-arid conditions must have prevailed for instance in the region of the present Namib desert.

Changes in climatic conditions apparently followed along the LAUTENSACHian category of planetay change. The tropical conditions for which the Natal laterites bear witness must have gradually shrunk to their present zonation in the time since the Upper Cretaceous. (As the silcretes show there must have been differences between the east and the west as well, both in the Lower and in the Upper Cretaceous, differences within the category of continental change which today do not seem to be as marked as in the Cretaceous.)

The shift from the tropical conditions to the subtropical ones of today was reduced because of Africa's plate-tectonical drift to the north. Altogether the climatic conditions (here judged by the one variable of the red soils) have receded to the northeast by only 500km since the Upper Cretaceous (cf. ch. 16 g), so that the northern parts of southern Africa have at least marginally remained within tropical conditions. Such an assumption can be made from a comparison of today's soil provinces with the occurrence of the fossil laterites in the east and again from having a look at the distribution of clay and heavy minerals (see Fig. 16. 10, Fig. 16. 5, Table 16. 2).

Regarding soils and weathering processes the main distinctions between the tropics and subtropics are the following ones: for the tropics a complete discomposition (German: 'Totalzersatz' or 'Komponentenzersatz') is typical, for the subtropics a partial disaggregation and disintegration only (German: 'Partialzersatz'). (Cf. ch. 16 c, 16 g.) This fact explains why lithosols are rather common in the subtropical parts of southern Africa.

Within the broader soil (or climatic) provinces topogenic and lithogenic modalities make themselves felt. A first instance is the rubefaction of the dolerite soils in the warm and humid east only in the favourable topogenic situation of the lower slope sections where a higher moisture content can be preserved. Rubefaction must here considered to be an a-zonal phenomenon, to be taken care of in delineat-ing the broader climatogenic soil regions of southern Africa (Fig. 16. 9 - in comparison with the official soil map: see ch. 16 e, which must be revised). True zonal latisols begin to appear today only in the north of Natal. A second instance is the variety of soils in the east where the chemical and lithogenic nature of the rocks can make itself felt (which certainly would not be the case in the full tropics, at least not to such an extent of varieties). Under arid and semi-arid conditions the lithogenic nature of the rocks is somewhat blurred over as Fig. 16. 9 shows.

A third instance are the different types of partial disintegration. Two classes of rock material can be made out. The first group consists of the intrusives, granites, gneisses and similar crystalline rocks. The disintegration of all these is granular. The second group consists of shales, mudstones, schists. Their disintegration can be classed as brittle or friable even, with the former under more arid conditions and the latter under more humid conditions. The different weathering behaviour explains the fact why in areas built up of type 1 rocks inselberg fringes are common in and around the embayments, whereas pediments are 'peeled off', as one might say, in the areas built up by the second group. The rocks of the first group can therefore be called 'inselberg-affined', those of the second as 'pediment affined', with the forming 'ramp belts' (German: 'Rahmenhöhen'). (Cf. ch. 8).

Subtropical weathering does not leave over weapons for an efficient erosion. Therefore knick-points are so well preserved and falls and canyons must be regarded as typical (see ch. 16 f).

CHAPTER 17: GERMAN VERSION: ZUSAMMENFASSUNG - ÜBERBLICK ÜBER DIE ERGEBNISSE

1. DIE GROSSE RANDSTUFE

a) Artung

Die Große Randstufe sollte nicht als ein versteilter Kontinentalrand oder eine versteilte Randschwelle bezeichnet werden; sie ist zwar auch keine einheitliche Schichtstufe im Sinne von BLUME (1970); doch ist sie als eine riesengroße Schichtungsstufe eine besondere makro-geomorphologische Form auf der Erde. Unter Schichtungsstufe ist eine Kette von Schichtstufen dann zu verstehen, wenn jede einzelne Schichtstufe über einem je anderen Schichtglied ausgebildet ist und somit im Verlauf der gesamten Schichtungsstufe die Einzelstufen von Schicht zu Schicht gewissermaßen hinüberspringen und dabei auch ihre Höhen ändern. In diesem Sinne beschrieb bereits A. PENCK (1908) Teile des Schichtstufensystems in Natal. Im Sinne einer Schichtungsstufe erstreckt sich die Große Randstufe vom Limpopo-Becken im Nordosten bis zum Khomas Hochland im Nordwesten (Kap. 13). Auf dieser Strecke verläuft sie über eine Länge von 2400km. Sie ist damit eine der längsten Schichtungsstufen der Erde, wenn nicht gar die längste. Sie ist damit wirklich eine makro-geomorphologische Form (Kap. 1).

Zwar verläuft die Große Randstufe mehr oder weniger parallel zum Kontinentalrand; doch sollte man sie deswegen nicht als eine Makro-Form betrachten, die direkt (oder auch indirekt) mit der Formung und Gestaltung des Kontinents als solchem bzw. mit plattentektonischen Vorgängen in irgendeiner Weise etwas zu tun hätte (vgl. JESSEN 1948, OLLIER 1985). Das von OLLIER vorgelegte Modell erscheint als zu sehr vereinfacht; es geht nämlich nicht an, die Bildung der Großen Randstufe in den Zusammenhang mit einem zentralen tektonischen Becken zu bringen, das, sich absenkend, gleichzeitig mit dem Auseinanderbrechen des Gondwanakontinents zur Ausbildung kam (Kap. 1a, 3).

Tatsächlich werden nämlich die Struktur der Randstufe wie aber auch ihr Verlauf durch drei deutlich ältere geologische Becken beeinflußt, und zwar auf ganz entscheidende Weise. Diese drei Becken sind:

1. das Oranje Becken mit den dort umlaufenden Karu-Schichten,
2. das Nosob-Molopo-Becken (auch Kalahari-Becken genannt) mit den im Streichen dieses Beckens verlaufenden Schichten besonders der Nama-Formation,
3. das des Olifants-Flusses in Transvaal mit den umlaufenden Schichten der Transvaal-Formation.
   (Vgl. Fig. 13.1; Tab. 1.1 und 1.2; Kap 2 und 13.)

Wie aus der Beschreibung zum Ausdruck kommt, ist es das Streichen der Schichtgesteine dieser drei Becken (jedes von ihnen deutlich älter als das Auseinanderbrechen des alten Super-Kontinentes), die den Verlauf der Großen Randstufe als Schichtungsstufe im einzelnen bestimmen.

Aus den genannten Gründen stellt auch das Khomas Hochland nicht einen Teil der Großen Randstufe dar. Das Khomas Hochland (zusammen mit dem Naukluft-Gebirge) ist in Wirklichkeit nichts als eine exzentrisch liegende, horstartige Region, die durch nichts als einen zufälligen geologischen Werdegang genau im Streichen der Schichtungsstufe liegt. Daher muß das Khomas Hochland als ein singulärer, eigener Grundzug der Reliefgestaltung gewertet werden, ein Grundzug, der - ebenso wie jene drei genannten Becken - aus dem Gondwanakontinent vererbt worden ist. Gleicherweise sollte auch die sog. 'Randstufenlücke' zwischen dem Swakop und dem Ugab im zentralen Südwestafrika nicht als Lücke der Randstufe als solcher aufgefaßt werden, sondern durchaus ebenfalls als ein Grundzug, der aus der präkambrischen Gestaltung Gondwanalands vererbt worden ist.

Khomas-Hochland und 'Randstufenlücke' zusammen trennen die Schichtungsstufe des südlichen Afrika effektiv auf einer Länge von 500km von der weiter nördlich liegenden Schichtungsstufe des nord-westlichen Südwestafrikas und des süd-westlichen Angola. M.a.W.: Man sollte sich nach einer so breiten Unterbrechung ernstlich davor hüten, noch von der selben Großen Randstufe zu sprechen. Zusätzlich zur großen Lücke kommt noch ein weiteres Moment hinzu: diese nordwestliche Schichtungsstufe ist ebenfalls an ein eigenes tektonisches Becken gebunden, das genau so wie die anderen schon genannten längst vor dem Auseinanderbrechen des Gondwanakontinentes angelegt war.

Berücksichtigt man die genannten Umstände, geht es keinesfalls mehr an, die Große Randstufe als eine Super-Einheit des Reliefs im gesamten südlichen Afrika südlich des Kongobeckens zu bezeichnen, sich über eine Super-Länge von 4800km erstreckend. Die Große Randstufe ist eben nicht, wie OLLIER and MARKER (1985) meinen, "virtually continuous" zwischen dem nordwestlichen Namibia und der südwestlichen Kapprovinz. Innerhalb dieser Erstreckung betrachten die beiden Autoren nur das sich nach W hin öffnende Becken des Oranje als eine größere Unterbrechung. Während der Vf. jedoch dieser Unterbrechung keine allzu große Bedeutung beimißt (wegen der auch in diesen hineinziehenden Schichtungsstufen), bewertet er hingegen die Lücke im mittleren Südwestafrika als eine sehr wesentliche und tektonisch begrenzende Erscheinung.

b) Alter, Verlauf und Lage sowie die größeren Flußeinzugsgebiete

Die Große Randstufe - stets im Sinne einer Schichtungsstufe - ist ganz und gar nicht ein recht junger Zug in der Gestaltung des südlichen Afrika mit einem nur mio-pliozänen Alter (5 bis max. 9 Mill. Jahre), wie KING, KAYSER und OLLIER glauben (Kap. 3 und 4). Sie ist vielmehr mit ziemlicher Sicherheit auf Unterkreide zu datieren (nach dem Valanginien, d.h. 130 - 120 Mill. J. vor heute). Diese Aussage beruht darauf, daß gut datierbare Landformen bis ganz an den Fuß der Großen Randstufe zurückreichen (Fig. 1.1).

Bei diesen Landformen handelt es sich besonders um die P 2 - Rumpffläche, in einer Höhe von 1200m ansetzend, und die T 1 und T 2 - Thalassoplains, die beide ebenfalls ein Unterkreidealter besitzen. P 2 hat möglicherweise ein Barrème-Alter, die beiden T 1 und T 2 wahrscheinlich ein Alb-Alter.

Das Alter der Verebnungen kann aufgrund folgender Umstände bestimmt werden. Sie schneiden nämlich über folgende ältere Schichten und Strukturen hinweg:

- Bruchlinien aus Jura und Unterkreide: überall im südlichen Afrika,
- Dolerite aus Jura und Unterkreide: überall,
- Sedimente der Unterkreide (endend mit dem Valanginien): in der südlichen Kapprovinz,
- Schichten der Etjo-Formation aus der frühen Unterkreide bzw. dem obersten Jura: im mittleren Südwestafrika. (Vgl. Kap. 2, 6d, 9a, 9b, 9c.)

Die genannten Verebnungen (P 2, T 1, T 2) mit ihren sehr wahrscheinlichen Altersdatierungen müssen als die entscheidenden Argumente für die Datierung der Großen Randstufe deswegen angesehen werden, weil sie, wie gesagt, unmittelbar bis an den Fuß der Randstufe heranreichen und damit zeigen, daß durch eine sehr lange Zeit hindurch (eine Zeit, deren Länge vermutlich bisher als recht unwahrscheinlich anzusehen gewesen wäre) die Große Randstufe dort verlaufen ist, wo sie im großen ganzen auch heute noch verläuft - und, daß die Große Randstufe dann durch diese ganze lange Zeit hindurch bereits als Schichtungsstufe bestanden haben muß (Kap. 5e, 9e). Zudem muß sie sich auch schon damals in einer relativ hohen Position der Höhe nach befunden haben. Unabhängig hiervon schreiben MARTIN et al. (1982, 192), daß sich die Randstufe immer schon seit dem Jura in einer relativ hohen Lage befunden hat.

Mit diesem hohen Alter ist die Große Randstufe sicher eine der ältesten Makro-Formen, die auf der Erde bis heute intakt erhalten geblieben sind. Betrachtet man die Lage und den Verlauf der Großen Randstufe zu Ende der Unterkreide, wird man feststellen müssen, daß es nicht die Randstufe war, die kontinuierlich von einer präsumptiven Bildungslage an der Küste ins Innere zurückgewichen und zurückgewandert ist, wie besonders KING immer wieder ausgeführt hat, sondern daß es vielmehr die Vorländer gewesen sind, die sich sukzessive vergrößerten, sei es durch Hebungen, sei es durch das Zurückweichen des Meeres (Kap. 5e, 13).

Ferner entwickelte sich die Große Randstufe nicht nach dem Modell der Rumpfstufentheorie W. PENCKs, wie diese von BORN, JAEGER und KAYSER auf das südliche Afrika zu übertragen versucht wurde. (Man täte vermutlich gut daran, die W. PENCKsche Rumpftreppentheorie ganz aus der Geomorphologie zu verabschieden.)

Der sehr gebuchtete Verlauf der Randstufe ist eine Folge der von außen in sie eingreifenden Abtragungsbuchten sowie der intramontanen Becken. Diese Becken und Einbuchtungen bzw. auch regelrechte Stufenrandbuchten wurden grundsätzlich unter dem Einfluß der Thalassoplanation geformt und ebenso grundsätzlich in den arideren Teilen durch Pediplanation überformt - und zwar seit der Unterkreide. Thalassoplanation und Pediplanation sind verantwortlich dafür, daß die Buchten sich auf Kosten der Schichtungsstufe in diese hinein zurückverlegten. Man könnte daher von einem 'Einbuchtungsrückzug' der Stufe sprechen, im Gegensatz zu einem Zurückwandern der Stufe parallel zu sich selbst (vgl. Kap. 5c, 9a).

Auf diese Weise wurde die Randstufe in den sie bildenden Schichten zu einer wirklich erosiven Reliefform ausgestaltet.

Die Serie der aufeinanderfolgenden, aneinandergereihten Schichtstufen gehört nicht zu den bloß akzidentiellen Eigenschaften der Stufe, wie ROGERS and KAYSER meinten, sondern diese ist vielmehr als eine fundamentale Eigenschaft anzusehen, die gesamte Stufe über 2400km bestimmend.

Mit dieser Aussage schließt sich der Bogen der Forschung; er kommt nach 100 Jahren zur Auffassung von GRIESBACH und A. PENCK wieder zurück (Kap. 3).

Insofern Thalassoplanation und Pedimentation in die Gebiete der noch heute bestehenden, aber auch schon damals vorweg bestehenden Flußeinzüge eindrangen und diese benutzten (wie die in den Flußeinzügen terrassenartig umlaufenden Unterkreideniveaus ausweisen), müssen diese Flußeinzüge mindestens ebenso alt wie Unterkreide sein, vermutlich sind sie aber durchaus noch älter - also jurassisch. Eine solche Altersannahme ist mit großer Sicherheit für den Oranje-Einzug (mit Nosob, Molopo, Fischfluß und dem Olifants-Doring -Tankwa-System), den des Großen Fischflusses und den des Systems des Transvaaler Olifants zu machen (vgl. Kap. 4, 5e, 12) - womit ein bisher für ziemlich unmöglich gehaltenes hohes Alter auch für Flußeinzüge zu konstatieren wäre. Nach Auffassung des Vf. kann kein Zweifel daran bestehen, daß die genannten Einzugsgebiete einige der ältesten auf der Erde darstellen, die seit 140 oder 150 Mill. Jahren bis heute ihre Funktion in der selben Abflußrichtung beibehalten haben.

Nach allem muß somit die Arbeitshypothese von OLLIER and MARKER (1985), derzufolge die Große Randstufe ihren Ursprung dem Auseinanderbrechen des Gondwanakontinents verdankt, erheblich modifiziert werden. Erosion und Denudation waren entlang der ganzen Stufe längst vor dem Auseinanderbrechen an der Arbeit.

Was die Thalassoplanationen aufgrund von marinen Ingressionen angeht, so würden diese Ingressionen auf jeden Fall den vorweg bestehenden Entwässerungsbahnen nach oben aufwärts gefolgt sein, ohne Rücksicht darauf, wie die Küstenlinie vor dem oder bei dem Auseinanderbrechen verlief.

c) Schichtungsstufenbildung und Klima

Wie schon angedeutet, sind die Schichtstufen auf verschiedenartigen geschichteten Gesteinen entstanden (Kap. 1c, 13; Tab. 1.1, 1.2). Bei den geschichteten Gesteinen kann es sich um Quarzite und Dolomite verschiedenen Alters handeln, aber auch um die endjurassischen Dolerite (die unterkretazischen Dolerite tragen, da ungeschichtet, keine Schichtstufen). Ferner können Stufen auch auf weicheren Gesteinen wie Sandsteinen oder gar Grauwacken und Schiefern ausgebildet sein. Ob nun eine Schichtungsstufe über den widerständigeren Gesteinen oder den weniger widerständigen entwickelt ist, hängt von einem fundamentalen klimatischen Faktor ab: dem der Humidität. Grundsätzlich gilt nämlich, daß Stufenbildung an die widerständigen Gesteine allein dann gebunden ist, wenn es sich um ein humides Klima handelt; umgekehrt gilt genau so grundsätzlich, daß die weniger widerständigen Gesteine als Stufenbildner erst in den arideren Bereichen fungieren.

Der Übergang zwischen mehr humid und mehr arid (und damit zu der Palette weiterer Stufenbildner) vollzieht sich in der östlichen und südlichen Kapprovinz (in den intramontanen Becken). Einerseits vollzieht sich dieser Übergang heute noch dort, andererseits zeigen die bis an die weniger widerständigen Stufenbildner zurückreichenden unterkretazischen Verebnungen, daß der Übergang in ähnlicher Weise bereits in der Unterkreide bestanden hat. Damit muß diese Übergangszone schon seit sehr langer Zeit im groben wie heute vorfindlich vorliegen. M. a. W.: Seit der Unterkreide ist die Verteilung der humiden und ariden Regionen derjenigen entsprechend, die heute noch besteht - zumindest im groben betrachtet (vgl. Fig. 1.2, 9.2, 11.2). Die Provinzen fossiler (oberkretazischer) Laterite im Osten und ebenso fossiler (oberkretazischer und tertiärer) Kieselhartkrusten (silcretes) im Westen und Süden, wie aber auch die auffällige, serirartige Decke der losen, nicht zementierten Quarzfragmente im Westen und Nordwesten weisen grundsätzlich (zeitlich als auch lagemäßig) in dieselbe Richtung, desgleichen auch das Vorkommen der Pedimente und das Einsetzen der Pedimentation als eines sehr bedeutsamen Denudationsvorganges. Fossile Pedimente, rezente und subrezente bzw. traditionale Pedimentation (bzw.

Pediplanation) und die Kieselhartkrusten beginnen (nach W hin) alle erst in dieser noch heute bestehenden Übergangsregion (vgl. Kap. 7a, 7b, 7d, 8, 10; Fig. 7.3).

Fig. 17.1 vergleicht Lage und Verlauf der Großen Randstufe bei ROGERS (1928), OLLIER und MARKER (1985) und beim Vf. Für ROGERS ist die Versteilung des Anstiegs von der Küste aus das entscheidende Element für die Festlegung des Verlaufs, allerdings nicht frei von gelegentlicher Willkür (z.B. im Osten der Pietersburger Hochlande im nordöstlichen Transvaal oder nordwestlich von Beaufort West in der südlichen Kapprovinz). OLLIER und MARKER nennen kein Kriterium, haben aber anscheinend den Verlauf des jeweils höchsten Kammes gewählt. Dieser aber ist von der Schichtungsstufe als solcher abhängig, weil sonst ganz und gar nicht festzulegen. Doch beide Autoren halten sich ebenfalls nicht durchgängig an dieses Kriterium, so z.B. in Swasiland, dann nordöstlich von Port Elizabeth, westlich von Beaufort West. Auch wird es mit dem Kriterium des höchsten Kammes überall dort schwierig, wo die Schichtungsstufe aussetzt und kristalline Gesteine beherrschend werden (wie auch schon ROGERS selbst und auch KAYSER bemerkten).

Für den Vf. bildet das entscheidende Kriterium das Auftreten der ersten echten Schichstufen (von außen nach innen gesehen) bzw. ihrer großen Ausliegerpakete.

d) Hangrückverlegung, Pediplanation und Stufenentwicklung

Die Vorstellung einer Hangrückverlegung über weitere Strecken, parallel zu sich selbst, wie sie KING, aber auch OLLIER und MARKER vertreten, muß mit Entschiedenheit zurückgewiesen werden. Alle Befunde zeigen, daß eine solche Rückverlegung nur in einer sehr untergeordneten Weise erfolgt. Die Hangentwicklung kommt nämlich nach einer ersten raschen Phase zum Stillstand oder doch nahezu zu einem solchen. Nicht mehr als 300 bis 500m an Hangrückverlegungen können nachgewiesen werden und nicht mehr als 500 bis 1000m an Stufenrückwandern. Solche Werte können annäherungsweise bestimmt werden, wenn man die Thalassoplains als Leit- und Zeitmarken verwendet (Kap. 14d, 15c).

Nur in den Stufenrandbuchten kann eine stärkere Rückverlegung beobachtet werden; doch dort handelt es sich nicht um eine Rückverlegung des Traufs, sondern um eine Ausdehnung der Bucht. Eine solche Buchtrückverlagerung (Englisch: 'embayment retreat') macht im Mittel 5 - 10km aus, gelegentlich 25km. Letzterer Wert wird in der Randbucht von Piet Retief erreicht, deren enormes Ausmaß bereits von KAYSER (1986) beschrieben worden ist. Dieses Ausmaß scheint eine Folge der jeweils auszuräumenden, mehr oder weniger widerstandsfähigen Gesteine zu sein. Zumindest in der Bucht von Piet Retief handelt es sich um weiche Ecca-Tonschiefer. (Vgl. Kap. 14d, 15a). Insgesamt aber können auch die oben genannten Werte für die Buchtrückverlagerung als nicht besonders exzeptionell angesehen werden - erst recht nicht, wenn man die Länge der Zeit berücksichtigt, die zur Verfügung gestanden hat (jeweils seit der Unterkreide).

Wo immer nun Hänge auf der humiden Ostseite Südafrikas auftreten - und zwar, in welcher Höhe auch immer und über welchem Gestein auch immer - besitzen sie grundsätzlich das gleiche Hangprofil. Für dieses ist stets ein unterer sehr flach auslaufender Hangabschnitt sehr charakteristisch. Von den Beobachtungen im Zusammenhang mit den thalassogenen Niveaus ausgehend ist es sogar möglich festzustellen, daß sehr ähnliche Bedingungen für die Hangformung mindestens seit der Oberkreide ausgebildet gewesen sind (Kap. 14b).

Trotz dieses doch recht langen Zeitraumes haben sich die unteren flachen Hangabschnitte nicht zu Flachformen vereinigt, die sowohl flächen- als auch linienhaft geschlossen aufträten (Kap. 14b, 14d). Es haben sich somit nirgendwo Pedimente herausgebildet. Somit kann die Pediplanation in den humiden Bereichen seit der Oberkreide mit großer Sicherheit ausgeschlossen werden. Es ist aber gerade hier, in diesen humiden Bereichen, daß KING seine Pediplanationstheorie nicht nur entwickelt, sondern auch mit großer Entschiedenheit (und anscheinend auch mit Überzeugungskraft) vertreten hat.

Erst in den semiariden und ariden Bereichen (von der oben genannten Übergangszone ab nach W und NW) werden Pedimente und die damit verbundene Pedimentation einerseits zu geomorphologisch wirklich flächenhaft weiten, außerordentlich auffälligen Formen, andererseits aber auch zu recht wirksamen Vorgängen, die durchaus zu Pediplanation führen können bzw. in weiten Bereichen auch dazu geführt haben. Nur in diesen arideren Bereichen - und wirklich nur in diesen - kann der Pedimentationsprozeß als der bedeutsamste Formungsprozeß konstatiert werden - sowohl rezent als auch in der langen geologischen Vergangenheit (Kap. 10). Wie schon ausgeführt, ergibt sich im Zusammenhang mit den Verebnungsniveaus, daß aridere Bedingungen in ähnlicher Weise wie heute schon seit langem wirken müssen. Diese Ähnlichkeit bezieht sich auch auf die Tatsache, daß Aridität (bzw. Semi-Aridität) in der südlichen Kapprovinz aus der Leelage des Binnenlandes hinter den Kapketten und damit einem wirksamen Föhneffekt heute wie in der Vergangenheit hervorgerufen wird, wie ferner darauf, daß die kontinentalen Verhältnisse im W und NW schon seit langem herrschen müssen.

Daß ähnliche Abtragungsverhältnisse in diesen Bereichen in der geologischen Vergangenheit genau so wie heute abgelaufen sein müssen, zeigt sich sehr schön an den sog. intermediären Niveaus der Hauptverebnungsbereiche. Es sind dies die Stock werke in 700, 900 und 1100m sowie in 1500 - 1750m, letzteres oberhalb der P 2 - Rumpffläche gelegen (vgl. Kap. 9a).

Im Rahmen dieser Arbeit erwies es sich als notwendig, sich mit den Hangbildungsprozessen besonders zu befassen, da diese ja für das Theorie-Gebäude KINGs von entscheidender Bedeutung sind (Kap. 14). Derjenige Hangbildungsvorgang, der sowohl in den humiden als auch in den ariden Gebieten auftritt, ist jener der Schichtstufenbildung und -abtragung. Ein Unterschied besteht nur insofern, als dieser Vorgang in den arideren Gebieten wesentlich langsamer abläuft als in den humiden. Für den Abtragungsprozeß an den Stufen ist es in beiden Bereichen möglich, die selben 'Anlasser' und 'Betreiber' zu identifizieren. Diese sind die lokalen Grundwasserträger unter den durchlässigen Stufenbildnern mit den sich unter diesen ausbildenden Naß- und Quellhorizonten. Von jeder einzelnen Quelle (oder auch jeder bloßen Naßstelle) aus innerhalb eines solchen Horizontes springen Unterschneidungen und Erosionsrinnen nach oben (sog. 'jumping-up-effect'), fächern über der je neuen weicheren Zwischenschicht weiter auseinander, wobei sich dann regelrechte Abtragungsamphitheater bilden (sog. 'amphitheatre effect') (vgl. Kap. 14c).

2. DIE VORLÄNDER

a) Die Thalassoplains: Artung, Anzahl und Bedeutung

Eine Anzahl von Verebnungen mit absoluter Höhenkontanz über Hunderte, ja Tausende von km hinweg ist um die ganze Küste des südlichen Afrika herum außerordentlich verbreitet und daher sind diese Verebnungen recht auffällige morphologische Erscheinungen. Wie deren absolute Horizontalität und Höhenkonstanz über solch weite Entfernungen schon eo ipso ausweist, müssen sich diese Verebnungen in der Nachbarschaft eines Meeresspiegels gebildet haben, und zwar im Zusammenhang mit unterschiedlich hoch reichenden marinen Transgressionen. Diese Transgressionen sind zu verstehen als verursacht durch tekto-eustatische Meeresspiegelanstiege.

In der Regel werden die einzelnen Horizontalniveaus durch deutliche Stufen voneinander getrennt. Dies gilt besonders für die unteren Niveaus (T 3 - T 8), während die oberen Niveaus, besonders im arideren W und NW, (es handelt sich um die T 1 - T 3 - Niveaus) durch intermediäre pedimentäre Verebnungen und Inselberglandschaften gekennzeichnet werden (vgl. Kap. 14b, 8).

Wie die Bezeichnungen zeigen, können insgesamt acht solcher distinkter Verebnungen ausgemacht werden. Diese werden aufgrund ihrer Entstehung in der Nachbarschaft eines Meeresspiegels als 'Thalassoplains' bezeichnet (ein Begriff, der in den vorigen Abschnitten bereits vorgreifend gelegentlich verwendet werden mußte). Tabelle 17. 1 gibt eine Übersicht über Alter und Höhenlage dieser Thalassoplains.

Tabelle 17.1: Alter und Höhen der Thalassoplains im südlichen Afrika

Thalassoplains	Alter	Höhen in m üb. dem Meeressp. (Durchschnitt)
T 8	Pleistozän	100
T 7	Plio-Pleistozän	200
T 6	Altmiozän	300
T 5	Oberkreide	400
T 4	Oberkreide	500
T 3	Oberkreide	600
T 2	Unterkreide	800
T 1	Unterkreide	1000

(Vgl. Kap. 5, 6, 9d; Fig. 9.2, 11.1, 11.2; Tab. 9.1. Für die Höhenangaben: vgl. Kap. 11)

Diese Verebnungen weisen keinerlei Anzeichen von Verbiegung, Verstellung und dgl. auf; vielmehr ziehen sie über Falten und Verwerfungen sowie Bruchlinien und unterschiedlichste

Gesteine gleichermaßen hinweg. Das Hinwegschneiden betrifft alle Gesteine und Strukturen, die gleichalt wie das Valanginien (135 - 130 Mill. J.) oder älter sind. Da sie selber von keinen Verstellungen (s.o.) betroffen sind, kann es seit dem Valanginien keinerlei tektonische Bewegungen mehr im südlichen Afrika gegeben haben. Das südliche Afrika muß seitdem als tektonisch völlig stabil betrachtet werden. Mit dieser Auffassung steht der Vf. allerdings ziemlich allein auf sehr weiter Flur; denn alle bisherigen Autoritäten, wie KING, OLLIER und MARKER, KAYSER, auch TANKARD et al., sind entschieden anderer Meinung. Alle tektonischen Beanspruchungen, die das südliche Afrika betroffen haben (Verwerfungen, Verstellungen, Hebungen), haben sich trotz dieser anderen Auffassungen jedoch nur bis einschließlich Unterkreide abgespielt (Kap. 4, 5, 6, 7g, 9c, 11; Fig. 11.1).

Von den bisher im Voranstehenden mitgeteilten Überlegungen aus ist es - aufgrund einer Vielzahl entsprechender Beobachtungen im Gelände - möglich, ein Modell zu entwerfen, wie nach Überzeugung des Vf. sich die Große Randstufe samt ihren Vorländern entwickelt haben muß (vgl. Fig. 17.2).

Einen Sonderfall stellt die 'Schiefe Ebene' der Namib dar, insbesondere dort, wo sie sich aus der sog. Randstufenlücke heraus zur Küste absenkt. Als solche scheint diese Ebene ein Argument gegen die Existenz hier durchlaufender Thalassoplains zu sein (cf. JAEGER 1930, ABEL 1953, 1959, KAYSER 1949; vgl. Kap. 3). Wie die Beobachtungen im Gelände jedoch gezeigt haben, ist auch diese sog. 'Schiefe Ebene' durchaus gestuft (Kap. 6), wenn auch die früher einmal angelegten Stufen durch Feinkies und Sand verhüllt worden sind - vergleichbar den sog. Flachböschungsbuchten, von denen drei in Kap. 5 näher beschrieben wurden. Die 'Schiefe Ebene' wie auch die Flachböschungsbuchten sind ein Ergebnis überprägender arider Morphodynamik.

Da die Thalassoplains immer in derselben Höhenlage und mit demselben Abstand voneinander verlaufen, kann bei ihnen auch die sog. 'Relation der Serie' (vgl. Kap. 6c), wie sie erstmals durch VON KLEBELSBERG für die Alpen dargestellt wurde, angewendet werden. Diese Relation stellt ein regelhaftes Moment der Festlegung von Höhenniveaus dar und kann im südlichen Afrika überall angewendet werden, gleichgültig wohin man auch immer kommt. Diese Regel kann nämlich dazu benutzt werden, im voraus für unbekanntes Gelände das Erscheinen von Verflachungen in bestimmten Höhenlagen vorherzusagen, um dann dort beim Hinkommen diese Vorhersage als Tatsache zu überprüfen. An keiner einzigen Stelle ergab sich eine Abweichung von der Vorhersage.

Die Regel gilt nicht nur für breitere Verebnungen, sondern auch für die sog. 'Terrassen' in den Talbuchten und in den Talschaften; sie gilt ferner a) für die sog. 'Talbodenschlüsse' (d.h. für jene Geländeabschnitte in den Talschaften, wo die 'Terrassen' sich im Halbkreis oder Halboval umlaufend von beiden Talseiten her zusammenschließen) und b) für die von dort nach talaufwärts zum nächsten Talbodenschluß hin auftretenden Gefällsknicke im Tallängsprofil. Von Felsschwellen (also zufälligen Momenten) abgesehen, treten Gefällsknicke als völlig regelhafte Erscheinung immer in der Höhenlage der Stufe zwischen zwei Verflachungen im Sinne von Hauptniveaus auf (vgl. Tab. 17.1 und Kap. 12).

b) Flächenbildung

Die große Ausdehnung der Thalassoplains weist darauf hin, daß der Vorgang (bzw. die Vorgänge), der zu ihrer Entstehung führte, recht wirksam gewesen sein muß. Thalassoplains sind 'etch-plains' und damit Rumpfflächen im weitesten Sinne. Jedoch muß der Vorgang ihres Entstehens (bzw. die Vorgänge) ein anderer gewesen sein als jener (bzw. jene), der üblicherweise in der deutschen, klimagenetisch orientierten Geomorphologie herangezogen wird oder wie er nach DAVIS das Entstehen von Rumpfflächen bewirken soll.

Um zwischen 'etch-plains' verschiedener Entstehungsweise unterscheiden zu können, erschien es ratsam, für die südafrikanischen 'etch-plains' einen besonderen Namen einzuführen. Hierfür bot sich die Tatsache an, daß die größeren und durchlaufenden Verflachungen sämtlich in der Nachbarschaft eines Meeresniveaus entstanden sind (vgl. Kap. 5d, 5e, 5g). Daher wurde für diese Art von Rumpfflächen der Name Thalassoplain gewählt.

Eine Thalassoplain verdankt ihr Entstehen allerdings nicht (etwa) der Arbeit der Brandungserosion in der Nachbarschaft dieses Meeresspiegels, sondern einer Kombination von miteinander verbundenen Einflüssen. Diese Kombination führt zu einem regelrechten Ursachengefüge für den Entstehungsprozeß (vgl. Kap. 5d, 5g, 6d, 7g, 7a, 7b), und zwar handelt es sich im einzelnen um folgende Kombination:

1. aktives Vorrücken des Meeres gegen das Land aufgrund tekto-eustatischer Meeresspiegelanstiege,
2. Entwicklung von Überschwemmungsebenen und seichten Küsten wegen
   - o starker Gefällsverminderung in der Nähe des Meeresspiegels als der Erosionsbasis,
   - o des mit der Nähe des Meeresspiegels verbundenen Grundwasserstaus, der besonders hineinwirkt in
     - die Becken des Hinterlandes und
     - sonstigen Einbuchtungen,
3. ein wegen der Nähe des Meeresspiegels ständig hohes Grundwasserniveau, das unter entsprechenden Klimabedingungen hervorragende Voraussetzungen schafft für
   - o Silifizierung,
   - o Laterisierung,
   - o Kaolinisierung unter den Hartkrusten,
4. ein häufiges Verlagern und Mäandrieren der Flüsse in der Nähe der seichten Küste (mit Bildung von Lagunen, Sandflächen, Konglomeraten),
5. Einebnung der vorweg tief verwitterten Gesteine in das Niveau der Küstenebene hinein,
6. Entwicklung von weiten Ausräumungslandschaften im Staubereich über einer Staufläche (vgl. Kap. 15a, 15b), die mit sich bringen
   - o einen sog. 'Schichtstufeneffekt',
   - o einen Ausraumeffekt (als 'erweiterten Schichtstufeneffekt')
   - o eine Ausweitung aller Becken und Buchten,
   - o einen sog. 'Subsequenzeffekt'.

Das Ausmaß der Denudation und der Ausräumung von Gesteinen über weitere Flächen hin ist offensichtlich eine Funktion der Tiefe und der Ausdehnung eines auf- und (nach gerinneaufwärts) zurückgestauten Grundwasserspiegels.

Je älter eine Verebnung ist (d. h. je höher sie liegt), umso mehr hatten die auf diese Verebnung eingestellten Vorgänge Zeit, von dieser Verebnung aus Platz zu greifen und die Verebnung zu verbreitern. Dies gilt besonders für die von den Verebnungen gegen die Bergländer vor sich gehenden Pediplanationen. Entsprechend sind die Pedimente über den ältesten Niveaus die ausgedehntesten, sind dort die Bergländer am stärksten aufgelöst worden. Diese Feststellung gilt somit besonders für die Pedimente in 1100m Höhe (von der T 1 in

1000m Höhe aus) und für jene in 1550 - 1750m (von der P 2 - Rumpffläche aus).

Vergleicht man die auf den Thalassoplains initiierten Vorgänge mit jenen, die sich bei der Rückverlegung von Hängen abspielen, so muß man feststellen, daß die thalassoplanen Vorgänge eindeutig die wirksameren gewesen sein müssen. Trotz dieser größeren Wirksamkeit läuft jedoch auch hier der Abtragungsvorgang recht langsam ab. Wählt man als Maß der Abtragung 1 Bubnoff (1B = 1mm Abtrag pro Jahr), so betragen die rekonstruierbaren Werte nur Hundertstel B. Diese Feststellung schließt beileibe nicht aus, daß das über Millionen Jahre hinweg akkumulierte Volumen des Gesteinsabtrags durchaus astronomische Summen erreicht (vgl. Kap. 15).

Prozesse, die ursprünglich einmal auf eine Thalassoplain (bzw. auch der P 2 - Rumpffläche) als Erosionsbasis eingestellt waren, tendieren offensichtlich dazu weiterzuverlaufen. Die wichtigsten weiterlaufenden Vorgänge sind die der Pedimentbildung und der Ab- und Ausräumung von Landschaften (vgl. Kap. 5, Kap. 9a). Es handelt sich somit dabei um eine Form der traditionalen Weiterbildung (wie sie von BREMER erstmals beschrieben worden ist). Damit diese Weiterbildung vor sich gehen kann, scheint eine wichtige Voraussetzung zu sein, daß sich die allgemeinen klimatischen Verhältnisse nicht grundlegend ändern.

Insgesamt sollten Thalassoplains und die sich auf ihnen abspielenden Verebnungsvorgänge als eine eigene Klasse von Rumpfflächen und somit auch von makrogeomorphologischen Formen aufgefaßt werden. Steht die Bildung dieser Plains im Zusammenhang mit tekto-eustatischen Meeresspiegelanstiegen, so ist zu folgern, daß Thalassoplains weltweit verbreitet sein müssen. (Appendix C.)

SUESS (1885) und HASSINGER (1905) waren die ersten, die in Geologie und Geomorphologie auf tekto-eustatische Meeresspiegelschwankungen verwiesen haben und entsprechende Reliefformen nachzuweisen suchten. Die mit dieser Studie vorgelegten Befunde können in diese - lang unterbrochene - Forschungstradition eingeordnet werden, wobei in dieser Studie die Thalassoplanation erstmals als ein außerordentlich bedeut- und wirksamer Vorgang dargestellt wird.

c) Hebungen und Reliefverjüngungen

Die Zeitalter, während denen im südlichen Afrika Hebungen abliefen, sind offensichtlich begrenzt gewesen und sie liegen ebenso offensichtlich weit zurück (soweit sie für das heutige Relief noch von Einfluß sind): es handelt sich um oberen Jura und untere Kreide. Daß dieser Tatbestand vorhanden ist, kann aus dem Zueinander der beiden oberen Thalassoplains (T 1 und T 2) zu den beiden Rumpfflächen (P 2 in 1200 bis 1750m Höhe, P 1 in 1800 bis 2000m Höhe) erschlossen und nachgewiesen werden (vgl. Kap. 6, 9c; Fig. 11. 1). Insgesamt beträgt die Hebung in dem genannten Zeitalter 1800 bis 2000m. Sie erfolgte in mehreren Phasen (Fig. 1.1) und war mit der Mittelkreide abgeschlossen.

Nach Abschluß der Hebungen hatte das südliche Afrika seine heutige Höhenlage erreicht, d.h. spätestens seit der mittleren Kreide.

Alle weiteren Variationen der Erosionsbasis erfolgten allein im Zusammenhang mit den Trans- und Regressionen des Meeres. Diese Vorgänge hielten durch die Oberkreide und das Tertiär hindurch bis ins Pleistozän an. Daher geht es nicht an, von einer einzigen und dazu noch sehr jungen Phase der Reliefverjüngung ('rejuvenation') zu sprechen, wie dies KING und OLLIER getan haben. Vielmehr hat es mit dem ständigen Wechsel von Trans- und Regression in den letzten 130 Mill. J. entsprechende Verjüngungsphasen gegeben. Daher ist es kein Wunder, daß besonders die humiden Bereiche, wie Natal, in jedem Betrachter spontan den Eindruck der 'Jugendlichkeit' des Reliefs dort hervorgerufen haben.

Ob nun die jurassischen (P 1) und unterkretazischen Verebnungen (P 2, T 1, T 2) als jeweils eigene 'Erosionszyklen' gewertet werden sollen, wie es der Tradition nach DAVIS entspräche, ist nach Auffassung des Vf. eine Frage, deren Beantwortung davon abhängt, welchen Aspekt man betonen möchte. Zwar erstrecken sich die Verebnungen von P 2 und T 1 häufig über Hunderte von Kilometern und bestimmen damit die südafrikanische Landschaft flächenhaft weit - womit sie seit dem Auseinanderbrechen des Gondwanakontinentes wahrlich die eigentlichen Afrika-Niveaus sind -, so haben diese weiten Verebnungen dennoch nicht zur Ausbildung eines Endrumpfes im Sinne von DAVIS geführt. Insofern ist auch kein voller Erosionszyklus abgelaufen.

d) Klima und Böden

Die bestimmenden und fundamentalen Klimaumstände des südlichen Afrika können sehr gut mit der Kategorie des kontinentalen Wandels nach LAUTENSACH (Kap. 16b) zusammengebracht werden, z.T. akzentuiert durch die Kategorie des Peripherie-Kern-Wandels. Diese Klimaumstände müssen während der ganzen Zeit der Entwicklung der Großen Randstufe und ihrer Vorländer als stets prinzipiell ähnlich betrachtet werden: nämlich der aride bzw. semiaride Westen und Süden (nördlich der Kapketten) und der humide bzw. semihumide Osten. Dieser fundamentale Unterschied hat durch den sehr langen Zeitraum hinweg bestanden und seinen gleichbleibenden Ausdruck darin gefunden, daß im arideren Bereich stets Pedimentationsvorgänge abliefen, während im humideren Bereich die Hangformung ausschlaggebender blieb, somit sich eine 'Pediment-Provinz' von einer 'Hang-Provinz' deutlich sowohl in der Vergangenheit wie aber auch in der Gegenwart unterscheiden läßt.

(Daß die arideren Bedingungen im W und NW von langer Dauer gewesen sind, läßt sich z.B. am Alter der Namib als eines bereits in der ausgehenden Jurazeit ariden bzw. semiariden Gebiets belegen.)

Klimatische Wandlungen über diesen langen Zeitraum hinweg erfolgten im Sinne eines planetarischen Wandels. Die fossilen Laterite der Oberkreide (bzw. Unterkreide) in Natal weisen daraufhin, daß dort in den genannten Zeiten ein tropisches bzw. randtropisches Klima geherrscht haben muß. Dieses ist seitdem zunehmend auf die äquatornäheren Bereiche hin eingeengt worden. Allerdings müssen auch in der Kreide Klimaunterschiede zwischen Ost und West bestanden haben, worauf die in der Ostkappprovinz einsetzenden Silcret-Bildungen verweisen (vgl. Kap. 7). Die Unterschiede müssen sogar etwas deutlicher ausgeprägt gewesen sein als heute (vgl. Tab. 16.1).

Der Wechsel von tropischen (bzw. randtropischen) zu den heute überwiegend subtropischen Bedingungen erfolgte mit der allgemeinen globalen Klimaänderung seit der Oberkreide, reduziert jedoch durch den Umstand, daß Afrika gleichzeitig nach N driftete. Vergleicht man nämlich die Lage der fossilen, oberkretazischen Latosolbildungen mit der rezenten Lage der zonalen Roterden (also ausschließlich der Roterden auf Doleriten), so kommt man zum Ergebnis, daß die Klimaregionen sich insgesamt um den relativ geringen Betrag von nur 500km nach NO verschoben haben können. Ferner liegt das nordöstliche Südafrika ja auch heute noch unter randtropischen Bedingungen (vgl. Kap. 16g) - sowohl klimatisch als auch edaphisch (Fig. 16.9). Von NO nach SW läßt sich der Wandel von noch randtropisch zu subtropisch - über verschiedene Intensitätsbereiche - gut auch an den Ton-, Schwer- und Eisenmineralen ablesen (Tab. 16.2).

Als typisch für tropische und randtropische Bedingungen kann der Totalzersatz der Gesteine bis zur völligen Zersetzung ihrer einzelnen Komponenten betrachtet werden, als typisch für subtropische Bedingungen der Partialzersatz, bei dem zwar das Gesteinsgefüge aufgelockert wird, die Komponenten aber erhalten bleiben (vgl. Kap. 16c, 16g). Dieser Unterschied erklärt z.B., warum Lithosole in den subtropischen Bereichen des südlichen Afrika recht weit verbreitet sind.

Innerhalb der jeweiligen Klima- bzw. Bodenprovinz machen sich Modalitäten bemerkbar, die von der Lithologie der Gesteine, wie aber auch von der jeweiligen Reliefsituation abhängen. Als ein erstes wichtiges Beispiel dafür ist die Rubifizierung der Doleritverwitterungsdecken im warmen und feuchten Osten (Bereich Natal) zu nennen. Diese Rubifizierung tritt allein auf den flachen Unterhängen ein, d.h. unter lokalen Bedingungen, in denen eine dauernde oder zumindest doch längere Durchfeuchtung gegeben ist. Wegen der Bindung der Rubifizierung an solche rein lokalen Gegebenheiten, zumal sie unter heutigen klimatischen Umständen allein auf die Dolerite beschränkt ist, muß sie völlig als a-zonale Erscheinung aufgefaßt werden.

In allen anderen Situationen verwittern alle Gesteine, auch die Dolerite, im subtropischen Bereich mit Braunfärbung. Die Braunfärbung wird allerdings je nach der Lithologie und dem damit vorgegebenen je eigenen Chemismus der Gesteine bei der Verwitterung ins Gelbliche oder Ockerfarbene abgewandelt. Dies ist ein weiteres Beispiel dafür, daß unter den heutigen subtropischen Klimabedingungen sich auch im warm-feuchten Natal die Lithologie durchzusetzen vermag.

Ähnliches gilt zu einem größeren Teil auch für das nördlich anschließende Transvaal mit der Vielfalt seiner Bodenausprägungen. Allerdings treten hier auch unter heutigen Bedingungen die ersten roten zonalen Böden auf. Wir gelangen hier somit in randtropische Bereiche.

Die vorstehend mitgeteilten Befunde flossen in die Darstellung der Bodenprovinzen im südlichen Afrika ein (Fig. 6.9). Die in dieser Figur gegebene Darstellung ist gegenüber der offiziellen südafrikanischen Bodenkarte (vgl. Kap. 16e) insofern und entsprechend abgeändert worden. Wie die Karte zeigt, sind unter arideren Verhältnissen allerdings auch in den subtropischen Bereichen des südlichen Afrika die lithologischen Eigenschaften der Böden (vielfach nur Lithosole) nicht durchschlagend.

Als ein drittes Beispiel muß angeführt werden, daß der Partialzersatz ebenfalls in Entsprechung zu den lithologischen Eigenschaften erfolgt. Bei einer ersten Gruppe von Gesteinen, umfassend Intrusiva, Vulkanite, Granite, Gneise, auch körnige Quarzite und ebensolche Sandsteine, ist das Endprodukt des Partialzersatzes Grus. Bei einer zweiten Gruppe, umfassend Schiefer und Grauwacken und dgl., besteht das Endprodukt zunächst aus kleinen und kleinsten Plättchen und Scherbchen, die sich je nach Humidität lehmig zerdrücken lassen. Das verschiedene Verwitterungsverhalten führt in der ersten Gruppe dazu, daß in solchen Gesteinen Inselbergformen als Abtragungsformen dominieren (mit regelrechten Rändern um die Becken und Buchten herum), in der zweiten Gruppe dagegen Pedimente, die gewissermaßen von diesen Gesteinen abgeschält werden. Man könnte daher insgesamt die erste Gruppe als inselberg-affin, die zweite Gruppe als pediment-affin bezeichnen. Bei letzteren entstehen um Becken und Buchten herum sog. Rahmenrampen (Kap. 8).

Schließlich verbleiben (aufgrund des Partialzersatzes) der Erosion keine sehr wirksamen Waffen. Das erklärt, warum (auch in den Subtropen) Gefällsknicke und Wasserfälle gut über lange Zeit erhalten bleiben, Kanyons typisch sind (Kap. 16f).

APPENDICES/ANHANG

APPENDIX A

Dixey's views on the age and altitudes of peneplains in southern Africa

The writings and findings of DIXEY (1942 and 1944) seem to have become completely forgotten, perhaps because of the impact KING's ideas have had. DIXEY's description of the altitudes of certain surfaces is in some respects remarkably close to those of the present writer, especially for the higher surfaces - though there are differences on the ages.

DIXEY held the watersheds and river systems to be of Jurassic age. The planation in 2000m ab. s-l was equally regarded to be Jurassic and it was "on the western side of the continent ... fully as high as in the east" (131). Below it follows the peneplain in an altitude of 1800m, to which Dixey gave an age of Cretaceous/early Tertiary. The upper and the lower peneplain were regarded as two full cycles in the sense of DAVIS. A third cycle followed in the Miocene after a phase of uplift, in which another peneplain came into existence, lying in the same altitude as our P 2 peneplain (i.e. 1200 to 1400m ab. s-l). According to DIXEY this surface was distributed very widely over southern Africa (corresponding to our observations), with the watershed between the Congo and Zambezi basins in 1200m being a part of it. The uplift leading to this cycle "was remarkably uniform from one end of Africa to the other" (126), in this referring to "early Cretaceous marine sediments" in Abyssinia in an altitude of 1000 to 1200m (127). The next cycle was the Pliocene one. Surfaces belonging to it occur for instance around Grahamstown and Willowmore in the eastern and southern Cape or around Pietermaritzburg in Natal. The altitudes vary between 450 and 900m. These altitudes correspond more or less exactly to our T 2 to T 5 thalassoplains; but DIXEY regarded them as one peneplain that was thought to have suffered from a widespread gentle warping. All lower planations were regarded as Quaternary and to be nothing but "successive marine-cut terraces" (173). Apart from the dating the observation as such is very near to those of the present writer's, as is another of DIXEY's findings, viz. the "successive inter-montane valley plains" with a "step-like relation to each other" (173). As an example DIXEY cited the valley plains along the upper reaches of the Great Fish river.

APPENDIX B

Lower coastal terraces

Below the T 8 thalassoplain several more 'marine-cut' terraces can be observed. They were first painstakingly looked into by KRIGE (1926). All of them occur along the whole coast from the Transkei to the western Cape Province. They are - according to KRIGE - horizontally level in the same way as the 'terraces' higher up (a finding that must be considered to be a confirmation of our observations by an older and independent author).

KRIGE described terraces in the following altitudes above sea level l.:

60m, 45m, 18m, 6m, 4m.

MAUD described similar terraces in Natal - all held to be Quaternary. Some of the terraces appear in the same altitudes of those of KRIGE. The altitudes (heights in m ab. s-l.) are:

70, 45, 33, 18, 12, 8, 4.5, 2.4, 1.0, 0.3.

APPENDIX C

Evidence for coastal planations elsewhere

If the coastal planations of southern Africa were caused by eustatic sea level rises, both the planations and the rises must by definition occur on a world-wide scale, i.e. on other continental plates as well. (The defining element of 'eustatic' is 'of a world-wide nature': VAIL et al. 1977).

The evidence gathered for other regions partly depends on chance and partly on the author's work in the Rhenish Massif and its borderlands as well as along the southern borders of the Bohemian Massif in Bavaria and Austria.

With regard to Africa OBST and KAYSER (1949, 144, 147) pointed out that the 200m level runs along the Lebombo range far into the north. DU TOIT (1922) showed that the sea must here have invaded all over Mocambique and must have ingressed inland in this altitude over more than 160km.

BREMER (1981) described planations running around Ceylon. The levels described always occur in the same altitude ab. s-l, which is a remarkable fact in itself and only to be understood within the framework of the present treatise. The altitudes and ages are (according to BREMER, 1981, 12 ff, 36 ff, 44, 112, 113) the following ones (heights in m ab. s-l):

700	
500	Pre-Miocene
300	
200	Miocene
100	Plio-Pleistocene

BREMER also pointed out that each lower (and younger) surface ingresses into the higher (and older) one in form of embayments which is rather similar to southern Africa.

BIRKENHAUER (1973) described the well-known 'trough-planes' (German: 'Trogflächen') of the Rhenish Massif and its surroundings. These 'trough-plains' run along the valleys of major and minor rivers as 'terraces' - sometimes as wide as 10km or more on both sides of the valley (for instance along the Moselle trough); they equally surround embayments and tectonic basins and even run around the outsides of the Massif. From there they spread over to the adjacent regions without any interruption. The plains or 'terraces' are everywhere independent of rock formation, faults, folds and so on. They are not at all structurally controlled. BIRKEN-HAUER was the first to notice that the levels keep the same altitudes over hundreds of kilometres. He was able to connect the levels with datable marine and lacustrine deposits, proving thereby that the levels were related to marine transgression during the Tertiary. The surfaces are the following ones (altitudes in m ab. s-l.):

400	late Oligocene/early Miocene
360	a middle Oligocene sub-level
300	Upper Pliocene/earliest Pleistocene

BÜDEL (1957, 1977) described levels which, according to the cross sections given by him, obviously occur in the same altitudes everywhere. Their altitudes are:

500, 390 to 400, 340 to 350, 300.

The levels were found by him both along the valleys and the scarplands of southern Germany and their bordering mountain ranges. The surfaces could easily be connected by BIRKEN-HAUER by extending his 'Rhenish' surfaces continuously into parts of southern Germany. Remarkable are moreover the same altitudes.

After this 'coincidence' BIRKENHAUER began to sight the numerous papers and theses dealing with the morphogeny of the German Hill Belt and found that different authors in different regions were indeed agreed on at least one thing: a planation surface occurring in an altitude of 300m, with an Upper Pliocene/early Pleistocene age. Moreover, the same authors described two more Tertiary surfaces higher up, in similar altitudes and in the same relations towards each other as had been observed in the Rhenish Massif. All these similar observations cannot be mere coincidences and are best explained (and without any forcing) by all of these levels being connected to tectono-eustatic sea level fluctuations. (It should duly be pointed out that the dates given for the higher surfaces vary from author to author, because every author is forced to fix his dates according to the often very scarce or even missing sediments in his own local region).

From these comparisons the opinion has once more been justified that it is no use to call certain surfaces by the name of 'peneplain' if their one and only common denominator is the fact only that they are not structurally controlled. It should moreover be proved by careful observations in every case whether the development of a surface in the neighbourhood of the sea can be excluded.

The phenomenon of the horizontally level surfaces and 'terraces' in the German Hill Belt were originally believed by BIRKENHAUER to have developed in connection with transgressions caused by subsidence. In a project sponsored by the German Research Society (named 'Vertikalbewegungen und ihre Ursachen am Beispiel des rheinischen Schildes') it was proved that subsidence did not happen and that vertical uplift amounted to only 130 to 200m - and all of this during the Pleistocene only. The transgressions first proposed by BIRKENHAUER were thus proved to be of a tectono-eustatic origin. The ages given by BIRKENHAUER and the totally level nature of the surfaces over long distances were equally confirmed by the results of the project - and by independent workers (BIBUS 1984, SEMMEL 1984).

Other levels of a wide occurrence in higher altitudes (between 600 and 800m ab. s-l) had also been known before. On these levels Upper Cretaceous sediments had been found, stemming from marine transgressions (cf. PFEFFER 1984).

Similarly, Upper Cretaceous marine deposits occur on the southern border of the Bohemian Massif, together with horizontally flat levels in 600 and 700m ab. s-l. The Upper Cretaceous transgressions from which the sediments stem lasted for about 20 to 22 my. This time-span is in good accordance with the charts in VAIL et al. (1977) and HANCOCK and KAUFFMAN (1979). At the same time the advancing southeastern alpine nappes were covered by the sediments of the well-known Upper Cretaceous Gosau transgression.

The Alps and the Bohemian Massif were equally affected also by Tertiary transgressions, in each case connected with wide-spread planatory surfaces. The two outstanding transgressions were the ones in the Oligocene and in the Lower Miocene. The former led to the development of the well-known 'Augenstein-Landschaft', the latter to that of the 'Rax-Landschaft'. Both landscapes are prominent in the eastern Alps. Miocene transgressions went up as far as 500m ab. s-l

both in the Bohemian Massif and in the Swabian Jura. Pliocene surfaces are well preserved in altitudes of 400 and
460m ab. s-1.

APPENDIX D

Inter-montane basins

Inter-montane basins seem to have first been described by CREDNER (1931) from South East Asia. CREDNER attributed their development to the special weathering conditions of the humid tropics. BREMER (1975) took up this opinion and demonstrated the evolution of similar basins under humid tropical conditions with examples from Australia, South East Asia and Nigeria. According to her research the upkeeping of a sufficient moisture content on the soil surface is considered to be the most decisive factor. From mere patches with high moisture central clay flats begin to develop.

Denudation spreads quickly and laterally, independent of sea level. In the distal parts of the flats, with gradients of more than two degrees, surface moistening can less well be preserved and is intermittent only. Eventually a scarp (at first very low) will be formed between the flat and the slope. Gradually the scarp will become higher and steeper because the steeper parts are topogenically more resistant. Simultaneously, the central flat or plain is gradually switched lower. The process leading to this landform differentiation is called 'diverging denudation' by BREMER. She holds such divergence to be typical for the humid tropics and thinks it to be far more decisive than lithology.

Going on to inter-montane basins in arid and semi-arid regions BREMER thinks that these also owe their origin to a former humid tropical climate. From this she feels sure to generalise that inter-montane basins - wherever they occur - are witnesses for an at least initial humid tropical climate, under which 'diverging denudation' (German: 'divergierende Abtragung') can start with its work.

In contrast to this theory the many inter-montane basins along the Great Escarpment seem to tell a different story. One can rather certainly say that here they are quite independent of any humid tropical conditions, because many of them occur where arid or semi-arid subtropical conditions must have prevailed for a very long time. It has to be admitted that moisture is indeed a decisive circumstance (as has been shown in ch. 16e) - yet not as surface moisture but as an efficient groundwater table. For the sustenance of such a groundwater table two reasons can be given - one of them is lithogeny, the other the neighbourhood of the sea level (leading towards a damming up of the groundwater table). (Fotos 30, 46, 52, 53.)

Comparing his own observations with the elements of BREMER's theory the present writer feels sure to disagree with all the elements in her theory, as the following juxtaposition will make clear.

Elements for the development of inter-montane basins

	Bremer	Birkenhauer
moisture	soil surface	**sustained groundwater table**
lithogeny	not important	**important**
sea level	independent	**important**
climate	humid **tropics**	**tropics/subtropics even semi-arid**

APPENDIX E

Recent and subrecent peneplanation

As was shown (and perhaps conclusively) thalassoplanation must be regarded as a genuine planatory process of its own, distinct from other concepts of extensive planation - might they be founded on DAVIS or on recent German ideas of climatogenic planation (i.e. BÜDEL, LOUIS, BREMER).

The DAVISian concept seems to be the broader of the two because it 'works' even under temperate humid conditions. A peneplain may - according to this concept - develop partly by lateral planation starting from rivers and partly by a gradual and general lowering of slopes, being directed to a coast as the final base level of all denudation and erosion. The younger German concept sees all peneplains or 'Rumpfflächen' linked to the more or less humid tropics. Main characteristics are rubefaction, deep weathering and sheet flood denudation following deep weathering. The neighbourhood of the sea level is a prerequisite for BÜDEL, but not so for LOUIS and BREMER. After a first initiation in a favourable climate peneplanation may go on within a tradited frame even in marginally tropical climates and subtropical climates.

In the study area of the present writer only a very marginal region can be considered to represent conditions suitable for peneplanation according to the German concept. This region can be found in the eastern and northeastern parts of Transvaal, in Mocambique and in the north of Swaziland. Though the climatic conditions are marginally tropical ones only, peneplanation in the German sense does seem to go on here, having reached the lowveld north of Swaziland, with its extension growing towards the north and northeast. The peneplanation (and with it the plain) is terminated in the south and southwest of the area by an erosional scarp. This scarp is developed within the crystalline basement, in front of the sedimentary rocks of the Transvaal super group with their cuestas in the west. The plain is not structurally controlled. Upstream the lower Crocodile, Komati and Sabie rivers the plain ends with triangular embayments. The plain itself is characterised by valleys with very flat and wide bottoms (which are called in German 'Flachmuldentäler'by LOUIS or 'Spülmulden' by BÜDEL). The surface pattern can be compared to a very flat corrugated iron ('Wellblechmuster'). Towards the scarp and along the distal margins of the triangular embayments inselbergs - isolated or as fringes - occur. The inselbergs can serve as good examples in order to prove that the plain has been formed as a consequence both of granular disintegration and sheet wash.

Along the thalwegs and within the embayments there also occur the 'terraces' of the different thalassoplains. As long as these are situated within the crystalline basement, the 'terraces' are dissected by closely spaced gullies so that the 'terraces' are shaped into an even stronger pattern of corrugated iron. Sheet washing continuously lowers the bottom of the gullies and the interfluval ridges between them, sweeping away the particles of the crystalline rocks (which everywhere are deeply weathered). All transitory stages from a stronger relief within the embayments towards the flattish relief of the exterior can be observed.

KAYSER (1986) also described the 'ramps' ('Rampen') of interfluves that lead down from the scarp to the lowveld peneplain within the crystalline rocks as well as the flat broad gullies ('Spülmulden'). (Cf. Fig.7 in KAYSER 1986, especially in the northeastern corner of the chart.)

On the whole, conditions of a tropical climate must be considered to be marginal only. The reasons for this conclusion can be gathered from the following observations: complete decomposition - a characteristic of the more humid tropics - is scarce (cf. ch. 16); disaggregation - a subtropical characteristic - is abundant; kaolinite, though dominant, is by no means the only clay mineral; goethite is more dominant than haematite.

According to RENK (1977, 124 ff) sheet washing is made possible on the Swaziland luvisols from three reasons: the A - horizon is missing; the B - horizon is cemented by iron oxides; infiltration is scarce. These observations also point to marginally tropical conditions. The latitudes of Nelspruit (25.5 degrees south) and Mbabane (26.3 degrees south) seem to be the poleward limits of such marginally tropical morphodynamics. (Cf. ch. 16.)

APPENDIX F

Inselberg formation

Inselberg formation has formerly been regarded to be a constitutive par excellence of tropical morphogenetics, especially with regard to peneplanation (at least by the German school of geomorphologists referred to in Appendix E). For BREMER inselbergs are "impressive forms of the humid tropics" (1975, 28; the other most impressive features being for her the great planations). KAYSER (1986, for instance) connected them with crystalline and igneous rocks only. RUST (1970) believed the inselbergs of central South West Africa - in altitudes between 1000 and 1200m ab. s-1 - to be connected both with the granites, extant there, and a former tropical climate in which alterites could be formed. Such a climate had reigned there, according to him, during the Lower Tertiary. He assumed that such a climate led to conditions of deep weathering in the sense of BÜDEL's 'double planation surface' ('doppelte Einebnungsfläche'), which sheet floods planating the surface above (i.e. the upper 'surface'), following the deeper front of weathering below (the lower 'surface'). As we have however seen in chapters 6, 8, 9, 10 such a climate (and particularly not in the altitudes given) cannot have reigned in the region.

MEYER (1967), who studied inselbergs and peneplains in northern Transvaal, regarded the inselbergs there as forms that developed continuously throughout all these millions of years. He thought their development to be steered by deep disaggregation and by slope wash, the latter eventually merging into sheet floods. Except for the sheet flooding the present author fully agrees with MEYER.

As was shown in chapters 8, 9, 10 inselbergs do occur outside the humid tropics. They even seem to be more abundant in the semiarid and arid subtropics, where they - with certainty - are not forms only 'tradited on' from tropical (humid) conditions, but must be regarded as an independent and original offspring of subtropical arid circumstances.

Altogether, inselbergs should therefore be regarded as nothing but residual hills, both in the tropics and the subtropics. Their development is best on granites, gneisses and intrusives, but they can equally be residual forms of former tafelbergs, as the embayments of the Great Escarpment demonstrate.

Inselbergs are not at all connected with tropical planation processes only, but originate on every kind of planation, especially at the scarps between one surface and the other. As such, inselberg fringes are nothing but witnesses of newly established planatory fronts that ingress towards higher surfaces. It is at these fronts that the erosional and denudational work is done. (In German one might say: Inselberglandschaften entstehen in den Tropen und Subtropen bei jeder Art von Verebnung, die von unten her an einer 'Arbeitsfront' in ein höheres Niveau vordringt.)

As a consequence, one should either use the term inselberg without any climatogenic or lithogenetic connotation or just speak of 'residual hills' (in German: 'Restberge' oder 'Restberglandschaften').

(Fotos 44, 18, 22, 26 will be of good service for this appendix.)

APPENDIX G

Knickpoints and 'valley bottom closes'

BIRKENHAUER (1973) described similar phenomena from all valleys in the Rhenish Massif and its borderlands (both from the major as well as from the minor ones). The 'valley bottom closes' (German: 'Talbodenschüsse') correspond to the scarps between the major surfaces (here the 'trough-planes' oder 'Trogflächen) and those between the broad main terraces 'German: 'Hauptterrassen') of the older Pleistocene in exactly the same way as they do in southern Africa. (Cf. Appendix C.)

COOKS and DE VILLIERS (1979) tried to use knickpoints of the Mbilo river in the hinterland of Durban in order to prove erosion cycles in the sense of KING or DAVIS (cf. Appendix E). The two authors described three major knickpoints and, by using very ingenuous mathematical procedures, related them to the three major Cenozoic erosion cycles postulated by KING. For them, each lower knickpoint represents a new cycle and corresponds to a period of strong uplift.

This knickpoint model must however be strongly refuted - from several reasons. The first is that the thalweg of the Mbilo river cannot at all be regarded to be in any way representative because the valley is neither cut back towards and into the Great Escarpment. Secondly, it does not even reach the 600m level. It therefore must be considered to be one of the younger valleys. Thirdly, no uplifts can be proved during the Cenozoic. Fourthly, the knickpoints are a consequence of the tectono-eustatic levels.

APPENDIX H

Measurements of denudational processes

DU TOIT (1933) estimated that denudation and erosion in the drainage basin of the Orange and Vaal rivers would amount to an average of 0.1mm/a for the time-span between the Eocene and today. Such an estimation would however lead - for the roughly 30my in question - to an overall surface lowering of the whole basin of 3.8km - which is obviously completely impossible.

As was shown in ch. 15 the general lowering must have been far less, with overall rates of some hundredths of a mm per year. If this is the general case a Bubnoff (B) - as the proposed inter-national measure for denudation - should really not be related to the denudational amount of just one year but to that of a thousand years. 1B would then stand for the general surface lowering in mm for thousand years (1B = mm/1000a.).

With regard to the time-span in which peneplains come to be fully developed BÜDEL (1977) estimated 3 to 5my. BREMER (1978) spoke of similar intervals, but, from her studies in Ceylon, warned against the assumption of too short time-spans.

OLLIER (1981) reckoned the lowering effect of flat lowlands to be ca. 50mm/a. Multiplying this amount for 1my a general lowering of 50km would be achieved, surely a not very realistic idea.

APPENDIX I

The problem of cuesta retreat in other regions

The amount of cuesta retreat has also been a matter of interest in the scarplands of Germany. WAGNER (1923, 1958) always held that the Malm limestone cuesta of the Swabian Jura retreated actively, parallel to itself, over a distance of some 20km within 15 to 17my. This would mean an average retreat of 1km per 1my. GERMAN (1965) and BLEICH (1960) showed however that the cuesta as it stands today in its pronounced way above the lowlands of the Neckar basin is nothing but the result of periglacial processes during the Pleistocene. The cuesta must therefore be regarded as a rather young landform.

DÜRR (1970) studied the retreat of the cuesta-like scarps or walls in the Dolomites of the central Alps. He also showed that it is the periglacial processes that are responsible both for the forming of walls and their back-wearing. The parallel scarp retreat initiated by the falling of stones (German: 'Steinschlag') amounted to no more than just 1m within several thousand years.

DÜRR (1970, 104) came to the conclusion that the walls in the Dolomites cannot have receded more than 50m - within 5my.

In the Sani Pass area of the Natal Drakensberge similar steep walls can be observed, especially if Cave Sandstone is covered by thick sheets of dolerites. Beneath these towering walls - in altitudes of 2400m and more - one can find huge cones of fallen stones that are very similar to those described by DÜRR from the Dolomites, with the same cone angles as there (i.e. 34 degrees). By comparison it is therefore possible to say that the walls in the Drakensberge were shaped by similar periglacial processes, i.e where the altitudes were favourable.

SEMMEL (1965) reported a cuesta retreat of ca. 150m (at the utmost) under maximum conditions during the Würm stage. He concluded from his observations that layer removal was the decisive denudational process. SCHUPP (1962) - from similar work in the German Hill Belt - came to comparable results.

Going back to the amount of cuesta retreat postulated by WAGNER one feels justified in having severe doubts about it from all the other observations mentioned so far. And indeed OLBERT (1975) showed that, during the younger Tertiary, no cuesta in the proper sense could have been lying as far north as WAGNER thought. What really must have happened - since layers altogether amounting to a thickness of about 900m must have been removed during the younger Tertiary, a fact stated by WAGNER that certainly must be considered true - is a very efficient removal of all these layers. Such an efficient removal can well be envisaged along the lines outlined by the author in chapters 14 and 15.

APPENDIX K

Dongas

Dongas are gullies with depths of 0.3 to 6m and are forms of soil erosion. They dissect the disintegrated material washed down from the slope sections on to the lower ones. In most cases dissection does not reach into the rock below. Uphill, donga formation usually comes to an end where the cover of weathered material is not thick enough any more. This is the case usually where the cover thickness is less than 0.3m. On the flat lower slope sections the cover is sometimes completely denuded so that the rocks are bared all over the slope. Dongas are formed only in the humid and semi-humid east and southeast of southern Africa. They usually occur here on grasslands, especially in the native areas, and are a sure sign that the original sensitive ecosystem including the morphodynamic subsystem has effectively been disturbed over wide areas.

Dongas are recent forms, having come into appearance since about a hundred years ago. They owe their origin usually to the overstocking of the grasslands by too much cattle. Apart from that the necessary circumstances for the formation of dongas are the following ones:

- o a forceful run-off after strong rainfalls,
- o the gathering of precipitation in small depressions and hollows from where the run-off is 'channelled' in a linear way,
- o an accidental incision of the turf and soil as a consequence of this channelling,
- o followed by dissection both up- and down-donga,
- o the destruction of the formerly coherent turf by the cattle hooves so that the turf is prevented from regenerating.

The size of the area with the destroyed turf is very often no more than a mere patch of about 0.3 to 0.5 sq. m.
The kinetic energy necessary for the incision by the channelled run-off has been calculated by SCHIEBER (1983).

When the cover over the lower slope section dries, vertical cracks are formed within the weathered cover material. Along these cracks vertical walls are torn down by the run-off. Nearly everywhere vertical walls have come into existence. Immediately above their upper edges the turf remains intact.

Occasionally, earth pyramids are being cut out from the walls. Downstream, dongas usually begin to widen because of lateral erosion. It is this lateral erosion that easily tears away parts of the walls along the vertical cracks, gradually widening the donga channel and eventually leading to the complete uncovering of the rocks. (Lateral erosion is a consequence of the whirling and eddying of the water during the rapid run-off.)

Donga formation has so far affected about 10% of the surface of the region where it occurs. The morphodynamic system and balance is still kept over 90% of the area.

APPENDIX L

Slopes and climate

TOY (1977) investigated the relationship between climate and slope form in the USA, using five slope variables against nine climatic variables. His findings can be summed up as follows:

Slope and segments	Percentages of climate variables 'explaining' the slope segments
curvature of the convex slope segment	59
straight segment	43
slope of regression	
line (fitted through the entire profile)	37
slope length ratio	26

Comparing his results for different areas TOY came to the conclusion that slopes in arid regions tend to be shorter and steeper than slopes in humid areas.

From what the present writer has observed in southern Africa he is able to agree to this rule.

APPENDIX M

River meanders

KING always used meanders as a kind of cornerstone in his theory about scarp retreat. According to him meanders were inherited from a higher pediplanatory surface above that gradually became reduced by 'rejuvenation'. Meanders develop on the pediplains because their gradient is considered to be very low. Since the gradients of the present thalwegs are steeper the meanders are to KING a sure sign that they have been tradited on from the pediplain above and thereby been impressed into the present-day landscape (in spite of 're-juvenation' and steepening).

The meandering of rivers and streams is indeed a phenomenon to be seen everywhere in Natal and in the more humid regions of southern Africa as a whole. Yet they also occur in the arid regions - the Fish river being one of the outstanding examples. The Fish river canyon is moreover an example for the fact that meanders were indeed inherited from higher plains that can still be widely observed in the neighbour-hood of the canyon. Another example are the inherited mean-ders of the Oribi gorge in Natal. Yet these two examples are not at all inherited from a pediplain, but from a thalasso-plain.

Apart from that the present writer feels rather sure that meanders are not only formed in such landscapes of low gradients. Most of them are indeed formed in areas with higher gradients and occur everywhere when a river or even stream is forced to incise its bed. Lithology seems to play an important part in meander forming. This can be seen from the fact that meanders are most abundant where the river beds are incised in crystalline rocks and in shales. Out-standing examples all over southern Africa are the Fish river canyon, the Orange canyon below the Augrabies Falls, the canyons of the Kuiseb, Swakop and Khan.

APPENDIX N

Fluvial terraces

True fluvial terraces (i.e. terraces developed from river work and fluvial deposits and with gradients following that of the thalweg) seem to have been formed in southern Africa only during the Pleistocene and were preserved for example in the canyons of the rivers that traverse the Namib desert (WARD 1983, RUST and WIENEKE 1973).

If such terraces were formed in the more humid parts of southern Africa they must have been completely washed away since the Pleistocene. The present writer was not able to discover a single one of them.

If one has a look at the temperate zones, true fluvial terraces with gradients seem to make their appearance equally only through the Pleistocene. Many examples can be found in the Alps, the Alpine forelands, the German Hill Belt. Outstanding examples are the terraces of the Rhine (BIRKENHAUER 1970c). But it is only the terraces below the so-called 'main terraces' ('Hauptterrassen') - with a Günz or even pre-Günz age - that show gradients in compliance with that of the thalweg. All terraces in the areas mentioned were formed under the condition that more sedimentary matter was accumulated (much of this material stemming from periglacial processes) than could be transported away by the rivers. When climatic conditions changed the accumulated terraces were successively undercut. The terrace material was removed to such an extent in some cases that only remnants of them can be observed today. A similar removal must have happened to the terraces of the rivers crossing the Namib desert (RUST and WIENEKE 1973).

In southern Africa all sedimentary matter is today completely carried away, either to the sea and onto the shelf or to tectonic basins, with the basins serving as sedimentary 'traps'. In Natal , for instance, as MAUD (1975) pointed out, all the deeply incised river mouths became drowned because of a post-glacial eustatic sea level rise, with the mouths becoming completely filled up by large deposits of sands. The surface of these deposits reaches as high up as the present sea level. (It is from these places that the masses of sands are won for the building industry.) What

Maud described for the Natal rivers is equally true for all river mouths around the southern African coasts.

Occasionally fluvial terraces seem to occur upstream in Natal, independent of such mouth infill or of the coastal 'terraces'. Yet closer observation shows that these 'terraces' are accidental formations only of gliding slopes intersecting and merging with each other on the interior sides of meanders. When the gradients of the merged slopes alter the 'terraces' at once completely vanish. The slope profiles along the thalweg show continuously changing gradients. Because of this it is impossible to find any cross profiles through the valleys that show some regularity (whatever place one would choose for such a profile). The only rule one can find is that of total irregularity. The reason for this are the meanders themselves: irregular in size, radius, breadth, width, slope angles etc. The landscape resulting might be called 'turbulent' because of its indeed unbelievably quick change over short distances (especially in shaly and crystalline rocks of the humid and semi-humid areas). Straightened valley or river courses are extraordinary scarce (as a consequence of such 'turbulence'). The relief is such that really no true fluvial terraces can be formed, neither locally, nor continuously over longer stretches.

APPENDIX O

Tertiary climate in central Europe

In central Europe an intermediary climate (between temperate and tropical zones) similar to the one described in southern Africa seems to have prevailed in the Tertiary, i.e. especially between the Oligocene and Quaternary (BIRKENHAUER 1970, using as witnesses fossil plants, heavy minerals, iron oxides, clay minerals, pebbles and gravel). From these witnesses the climate could be reconstructed leading to the conviction that the red soils (or latisols even) formed on the basalts still during the Pliocene could not be regarded as proofs for a tropical or even only marginally tropical climate, still lingering on and allowing for the formation of peneplains, as BÜDEL (1977) and BREMER (1978) thought. These soils should however be regarded as a-zonal lithogenic soils - as which they are similar to the a-zonal red soils on the dolerites described in ch.16. If the Pliocene red soils in central Europe must be regarded as a-zonal another 'proof' for marginally tropical conditions must be done away with (cf. Appendix E).

APPENDIX P

A comparison with WIRTHMANN's views

Rather categorically WIRTHMANN (1981, p.176) states that the development of peneplains cannot be regarded as a problem at all where the rate of linear erosion is nearing zero. (He then connects this statement with barriers formed by resistant rocks in the river thalwegs, with these barriers also serving as local bases for erosion.) WIRTHMANN's statement must be even truer if erosion is nearing zero over wide distances in the neighbourhood of a sea level. In such a case another statement of his of an equally categorical nature will be relevant indeed: "Any kind of relief will be reduced that is not subject to erosion."

Regarding knick-points WIRTHMANN (1981, p.174) seems to be of the opinion that these are due only to barriers formed on resistant rocks, and that peneplains develop from these points and, moreover, are even preserved upstream of them. According to our observations, this statement should be extended especially to those which appear when thalwegs cross the scarps between one thalassoplain and the next farther down. From our observations it is definitely this type of knick-points that is not only most frequent (only sometimes additionally accentuated by rock resistances), but also occurs with greatest regularity.

Regarding the widening of valleys it is not at all possible (at least from our observations in southern Africa) to follow WIRTHMANN's statement (1981, p.183): "the widening of valleys is obviously bound to areas that are subject to stronger tectonic activity". Moreover, opposite to his statement, the older morphodynamics and the different peneplanations can be reconstructed rather well "on the South African highlands" - at least for the whole time-span of the Cretaceous and the Tertiary.

Finally WIRTHMANN (1985) is of the opinion that the coexistence of two or more planation surfaces at different levels is limited to the Gondwana continents. He goes on to state that it is not yet possible to solve all open points and contradictions in the development of such surfaces. The present writer is of the opinion, however, that sea level changes and the planatory processes necessarily combined with them provide a good key to an understanding, even outside Gondwanaland (c.f. Appendix C).Another problem not solved so far is, according to WIRTHMANN's (a) the conservation and (b) the still on-going development of these surfaces. For a solution, WIRTHMANN takes refuge in the resistancy of basement rocks. For him this is the main factor to prevent up-river incision. As we have seen, the main point however is the existence of knick-points be they related to rock resistances or to scarps between planatory levels.

APPENDIX Q

Connection with the Wilson- or Supercontinent-cycle

NANCE, WORSLEY and MOODY have recently presented their global model of continental developments based on the findings of GROVE, ANDERSON and WILSON.
Especially the latter developed the notion of cyclic continental developments, after him called the Wilson-cycle. It is especially this notion that has been adopted by the three authors first mentioned as one of the decisive backgrounds for their global model. Out of the many aspects involved in the model there is one that is particularly interesting in connection with our theory of thalassoplains - i.e. the explanation of tectono-eustatic sea-level changes. According to the authors these changes seem to occur rather regularly within specific periods of the Wilson-cycle and are caused by (a) the rifting process and the development of marine basins and (b) the thermic up-lift of the continents. Within the following table some findings by NANCE et al. are made use of and compared with the development of southern Africa since the Jurassic, i.e. since the beginning of the present Wilson-cycle.

Table Q.1: Connections with the Wilson-Cycle

a) According to Moody - Nance - Worsley (1988)

Stage	Cyclical event	Time-span	Tectonic consequence
1.	Accumulation of heat under super-continent		Updoming and uplift of continent (at least 400m), sea-level at its lowest
2.	Tephrogenesis (growth of rift valleys)		Volcanism
3.	Break-up of supercontinent		First forming of oceans
4.	Beginning of oceanic spreading	)	Forming of new and lighter crust and first sea-level rises
5.	Up-heating under new continents continued	) 80 my	Uplift; at least 400m
6.	Continued spreading and maximum dispersal of continents	)	Forming of larger marine areas with lighter crust, - resulting in highest sea-level rises
7.	First ageing of oceanic crust	)	Down-warping of ocean basins at their continental margins and gradual fall of sea-level
8.	a) Spreading still continued, but slower; less forming of new crust	) 80 my	Fall of sea-level continued, but because of (a) and (b) several (smaller) sea-level rises
	b) several subductions	)	

b) According to author

Stage	Events and landforms in southern Africa	Time scale in southern Africa in my
1.	Highest and oldest peneplain (P_1, in 2000m); uplift: ca. 1000m	Jurassic 200 - 150
2.	Dolerites; Etendeka-formation	Upper Jurassic/Lower Cretaceous 170 - 130
3.	Opening-up of southern Atlantic horst and graben-structures in Natal, Cape range basins, Algoabay	Lower Cretaceous 140 - 120
4.	Forming of P_2, T_1, T_2	Lower Cretaceous 140 - 120
5.	Several phases of uplifts: a) between T_2 and P_1 b) between T_2 and T_3 (together 400m)	Lower Cretaceous 140 - 100
6.	Forming of T_3 - T_5	Upper Cretaceous 100 - 65
7.	Fall of sea-level	Tertiary
8.	Forming of T_6, T_7...	

FOTOS

Preliminary remarks

The fotos are thematically grouped.
Their sequence is arranged as in the following Table. Similar themes in the fotos of other groups are referred to under the heading 'references'.

Vorbemerkungen zum Bildteil

Die Fotos sind nach inhaltlichen Gruppen zusammengefaßt, und zwar in der Reihenfolge wie in der folgenden Tabelle. Wenn vergleichbare Inhalte in Fotos anderer Gruppen angesprochen werden, so ist dies unter 'Hinweise' in der Tabelle vermerkt.

Groups	Fotos	References
I. Great Escarpment	1, 2	9, 11, 36, 37
II. Horizontal levels (Thalassoplains)		
1. 100 - 600m	3, 4, 5, 6, 7, 8, 9, 10, 11, 12, 13	26, 27, 31, 33
2. 800 - 1000m	14, 15, 16, 17, 18, 19, 20, 21, 22	1, 2, 9, 11, 28, 30, 31, 32
III. Peneplains	32, 33, 34, 35	16, 30, 43
IV. Breachings, gorges, poorts	23, 24, 25	31, 40
V. Low gradient embayments, 'oblique plain' etc.	30, 31	24, 25, 54
VI. Weathering, denudation etc.		
o Slope, cuesta	36, 37, 38, 39, 40, 41, 42	7, 21, 32, 55
o Pediplanation, pediments, shield-like pediments	43, 44	18, 25, 27, 30, 32, 33, 35, 38, 45, 46
o inselbergs	44	18, 22, 26
o disgranulation etc., valleys with flat bottoms	46	1, 33, 45
o triangular embayments		1, 2, 27
o Silcretes	47, 48	20, 21, 24, 31
o groundwater table, removal	52, 53	26, 31, 45
o fine gravel, soils	49, 50, 51	29
o calcretes		34
o incised levels	55	38
o cones of falling stones		38
o meander		38

Inhaltsgruppen	Nummer der Fotos	Hinweise
I. Große Randstufe	1, 2	9, 11, 36, 37
II. Höhenbeständige Verebnungen (Thalassoplains)		
1.100 - 600m	3, 4, 5, 6, 7, 8, 9, 10, 11, 12, 13	26, 27, 31, 33
2. 800 - 1000m	14, 15, 16, 17, 18, 19, 20, 21, 22	1, 2, 9, 11, 28, 30, 31, 32
III. Rumpfflächen	32, 33, 34, 35	16, 3o, 43
IV. Durchbrüche	23, 24, 25	31, 40
V. Flachböschungsbuchten, 'Schiefe Ebene', auslaufende Niveaus	30, 31	24, 25, 54
VI. Prozesse der Abtragung, Verwitterung u.dgl.		
o Hänge, Schichtstufen	36, 37, 38, 39, 40, 41, 42	7, 21, 32, 55
o Pediplanation, Pedimente, 'Schildpedimente'	42, 44	18, 25, 27, 30, 31, 33, 35, 38, 45, 46
o Inselberge	44	18, 22, 26
o Verwitterung, Vergrusung, Spülwannen, Flachmuldentäler	46	1, 33, 45
o Dreiecksbuchten		1, 2, 27
o Silcretes	47, 48	20, 21, 24, 31
o Grundwasser, Ausraum	52, 53	26, 31, 45
o Feinkies (Serir), Böden	49, 50, 51	29
o Kalkkruste		34
o Zerschneidung von Niveaus	54	
o Mäander	55	38
o Steinschlaghalden		38

Foto 1. The Great Escarpment near Aberdeen (southern Cape Province, near Graaf-Reinet). From S to N

Die Große Randstufe bei Aberdeen (südliche Kapprovinz, bei Graaf-Reinet). Von S nach N

The 800m-level, forming wide flats, ingresses into the Great Escarpment in a broad triangular embayment, together with smaller embayments. The escarpment itself stands at a height of 2000m. Where the slopes steepen, dolerite sheets take over. The 800m-level forms the base level for the pediplained level above (up to 900m; cf. Foto 16 for 1000m-level). Within the embayments, both levels merge into each other over flat 'ramps', especial-ly where shales form the base (here between softer Stormberg sandstones). Most inselbergs have been eroded away on the soft layers in the vast time-span since the Lower Cretaceous. They have been retained only on the dolerites next to the scarp itself. All rock material weathers into yellowish-brownish sands, sheet-washed into the 'valleys' with very flat bottoms.

Buchtartig springt das breit verebnete 800m-Niveau in die Randstufe hinein vor (Dreiecksbucht mit Seitenbuch-ten). Die Randstufe erreicht hier eine Höhe von über 2000m und ist bei den Steilanstiegen in Dolerit ausgebildet. Die 800m-Thalassoplain ist hier - ähnlich wie die 1000m-Fläche in Foto 16 - die Erosionsbasis für das bis auf 900m reichende Pediplanationsniveau. Beide Niveaus gehen in den breiten Buchten sanft ineinander über, besonders dann, wenn wie hier, die Basis aus weichen Schiefern zwischen weicheren Sandsteinen (hier Stormberg-Formation) besteht. Sehr flache Rampen vermitteln zwischen den beiden Niveaus. Inselberge sind auf den weichen Gesteinen und wegen der langen Zeit der Abtragung (seit Unterkreide) bis auf wenige Reste verschwunden; erst unmittelbar am Rand der Stufe stellen sie sich auf den härteren Doleriten wieder vermehrt ein. In den flachen Spülwannen wird das gelbbraune Material (Sand) abgeführt, das bei der semi-ariden Verwitterung allein übrigbleibt.

Foto 2. The Great Escarpment near Calvinia (northwestern Cape Province)

Die Große Randstufe bei Calvinia (nordwestliche Kapprovinz)

The scarp (or cuesta), in 1660m, looked at from the south, is bound to the dolerite sills (steep walls, ver-tical cracks) above softer Ecca shales and sandstones. The cuesta is broadly cut open by a triangular embayment in the 1000m level, stretching here from the W (left) towards the E (right), with the thalassoplain completely transsecting the cuesta and merging with the same level along the Sakrivier (cf. Foto 18). Thus a cuesta is form-ed in the back of the escarpment, fully equivalent to the one on its front. (A little further on to the N, not to be seen in the foto, the cuesta-bearing dolerite sills disappear and with them the escarpment as such.)

Man erkennt hier wiederum die Bindung der Stufe an die Doleritschichten (steile Wände, senkrecht geklüftet) über den weicheren Ecca-Schiefern und Sandsteinen. Die Große Randstufe erreicht hier noch eine Höhe von 1660m. Sie wird bei Calvinia von einer breiten Dreiecksbucht aufgeschlitzt. Die Bucht selbst ist im Niveau der 1000m-Thalassoplain ausgebildet. Diese zieht sich von W (links) nach E in den Schichtkörper hinein. Im O durchbricht die Plain den Schichtkörper völlig und vereinigt sich mit dem selben Niveau, das vom Sakrivier her zurückgreift (vgl. Foto 18). Die Große Randstufe besitzt hier somit eine Achterstufe. (Etwas weiter nördlich, nicht im Bild, ver-schwinden die Schichtdecken der Dolerite; mit ihnen ver-schwindet die Große Randstufe als zusammenhängendes Ge-bilde.)

Foto 3. Sunday river area near the river mouth into the Algoa Bay (E of Port Elizabeth, eastern Cape Province). View to the N and NW

Sonntagsfluß oberhalb der Mündung in die Algoa-Bucht (östlich Port Elizabeth, östliche Kapprovinz). Blick nach N und NW

The levels of the 'terraces' to be seen here are in 6m, 10 to 12m (youngest Pleistocene), 100m (older Pleisto-cene). The 'terraces' form wide levels along the valley and run along with an absolutely horizontal constancy of height. In the background: the 200m level.

Die Niveaus der 'Terrassen' sind hier in 6m und 10 - 12m (Jungpleistozän) sowie in 100m (Altpleistozän) ausgebildet. Die 'Terrassen' laufen flächenhaft weit und mit absoluter Höhenkonstanz durch. Im Hintergrund: das 200m-Niveau.

Foto 4. Kliprivier canyon in the Tsitsikama natural park (W of Port Elizabeth, southern Cape Province). From N to S

Kanon des Klip-Riviers im Tsitsikama Naturpark (westlich Port Elizabeth, südliche Kapprovinz) Blick nach S

The absolutely horizontal 'terrace' is the one in 200m (Plio-Pleistocene), forming a wide plain that completely cuts across the TMS quartzites folded vertically. The plain also cuts across the Lower Cretaceous sediments, similarly smooth and flat, and at the same altitude.

Man blickt hier über die absolut horizontale, flächenhaft weit ausgebildete 'Terrasse' in 200m (Plio-Pleistozän). Die Fläche schneidet die senkrecht stehenden gefalteten TMS-Quarzite glatt ab und zieht von dort genau so flach und glatt auf die Unterkreide-Sedimente in gleicher Höhe hinüber (nicht im Bild).

Foto 5. Near Umbumbulu (Kwazulu, W of Durban, Natal). View towards NE and E

Bei Umbumbulu (westlich Durban, Natal). Blick nach NO und O

The Upper Cretaceous level in 600m forms the horizon in the background, here cutting across the strongly dissected crystalline basement. Below the 600 m level, the one in 500m (equally Upper Cretaceous) ingresses, forming an embayment.

Das oberkretazische Niveau in 600m bildet den Horizont. Es schneidet über das stark zertalte kristalline Grund-gebirge hinweg. In dieses greift das tiefere Niveau in 500m (ebenfalls Oberkreide) buchtartig ein.

Foto 6. West of Ramsgate (Kwazulu, southern Natal). View to the N and NE

Bei Ramsgate (südliches Natal). Blick nach N bzw. NO

The cuesta on the left is formed on flat layers of TMS quartzites, topping the crystalline basement (to the right, strongly dissected). The top of the cuesta (also in the background) lies in the Upper Cretaceous 400m level. The hilly tops of the basement belong to the 300m 'terrace' (Miocene). The 200m level (Plio-Pleistocene) ingresses from the right, forming an embayment. The gene-ral characteristics of the slopes (both on the quartzites and on the crystalline) can well be observed: steep upper sections and flat, 'pediment'-like lower ones. The over-all length of the slope projecting from the TMS cuesta in the left is 500m.

In den flachlagernden TMS Quarziten (links) ist eine lokale Schichtstufe ausgebildet (auch im Hintergrund). Die Quarzite lagern dem Grundgebirge (rechts) auf. Das flächenhafte weite Niveau auf der Stufe liegt in 400m (oberkretazisch), während das Grundgebirge oben von der 300m-Verebnung (Miozän) überzogen wird, in die die 200m- 'Terrasse' (Plio-Pleistozän) buchtartig von rechts her eingreift. - Gut erkennbar sind die allgemeinen Merkmale der Hanggestaltung, sowohl in den Quarziten als auch im Grundgebirge: nämlich mit steilen Oberhängen und flach auslaufenden, 'pedimentartigen' Unterhängen. Die Länge des gesamten Hanges links von der Stufenkante ab nach rechts beträgt 500m.

Foto 7. The lower reaches of the Great Fish river near Hunts Hoek (E of Grahamstown, eastern Cape Province). View to the NE

Am Unterlauf des Großen Fischflusses (östlich Grahamstown, östliche Kapprovinz). Blick nach NO

The 'terraces' ingress widely upstream along the course of the valley (from right to left), all of them partly dissected by fluvial erosion. The lowest 'terrace' (or plain) to be seen is here formed by the one in 100m (ol-der Pleistocene); the next one higher up is the one in 200m (Plio-Pleistocene); between the two one observes a well-developed scarp. A similar scarp leads up to the 'terrace' in 300m

(Miocene; in the middle left and right, more to the back). The background and horizon is dominated by the wide Upper Cretaceous 'terraces' or plains in 400m and in 500m.

Buchtartig weit greifen im Talungsbereich die Verebnungen (heute stärker zerschnitten) nach oben aufwärts zurück. Die unterste Verflachung im Bild ist die 'Terrasse' in 100m (älteres Pleistozän); darüber steht - mit deutlicher Stufe dazwischen - das 200m-Niveau (Plio-Pleistozän) an. Wiederum mit einem steilen Anstieg geht es zum nächsthöheren Niveau in 300m (Miozän) hinauf (linker und rechter hinterer Mittelgrund). Darüber stellen sich im Hintergrund die oberkretazischen Niveaus in 400 und 500m ein.

Foto 8. Scarp of the Tsitsikama Range (eastern part of the Outeniekwasberge) in the neighbourhood of Port Elizabeth (eastern Cape). View to the N

Abfall der Tsitsikama-Kette (östlicher Ausläufer der Outeniekwasberge); Nähe von Port Elizabeth (südliche Kapprovinz). Blick nach N

The range is built up by the heavily folded TMS quartzites. The plain in front is the 200m 'terrace' (Plio-Pleistocene; cf. Foto 2). The scarp itself is stepped because of the 300m (Miocene), 400 and 500m 'terraces' (Upper Cretaceous) being carved into the quartzites, forming ledges. The level on top is the one in 600m (Upper Cretaceous). Levels and ledges, always in their altitudes, stretch along from here to Cape Town in the west over a distance of 560km.

Die Tsitsikama-Kette besteht aus gefalteten TMS-Quarziten. Unten vor der Kette erstreckt sich die 200m-'Terrasse' (Plio-Pleistozän; vgl. Foto 2). In die Quarzite eingearbeitet sind die terrassenartigen Simse der Verebnungen in 300m (Miozän), 400m und 500m (oberkretazisch). Das Flachniveau oben wird gebildet von der 600m-'Terrasse' (oberkretazisch). Die Verflachungen und Simse ziehen sich in gleicher Höhenlage von hier bis Kapstadt im W über 560km hin.

Foto 9. Bokkeveldberge in the northeast of Vanrhynsdorp (north-western Cape Province). View to the S

Bokkeveldberge nordöstlich Vanrhynsdorp (nordwestliche Kapprovinz). Blick nach S

The cuesta (on TMS quartzites, not folded, with a slight fall to the left = E) looks towards the Atlantic. It is an erosional feature above the widely removed Nama shales (bottom, to the right = W) with their surface in 200m (Plio-Pleistocene level). The flat and wide plains on top of the cuesta lie in 800m (Lower Cretaceous level). The 'terraces' between the plain at the bottom and the one on top form ledges on the slopes beneath the cuesta, i.e. especially the 'terraces' in 300 and 600m.

Der in Richtung Atlantik schauende Steilabfall der (flachlagernden, nach E = links flach einfallenden) TMS-Quarzite ist ein Erosionsrand über Namaschiefern, die breit im pliopleistozänen 200m-Niveau ausgeräumt wurden (rechts unten). Die Stufe wird oben abgeschnitten durch eine weite Verflachung in 800m (unterkretazisch). Die zwischenliegenden Niveaus sind weitgehend aufgezehrt worden, bilden aber verflachte Simse entlang der Stufe (insgesamt 600m hoch), besonders in 300 und in 600m.

Foto 10. Keurboom valley in the Outeniekwasberge, east of Knysna (southern Cape Province). View to E and SE

Die Gebirgsbucht der Keurboom-Talung in den Outeniewasbergen, östlich Knysna (südliche Kapprovinz). Blick nach O und SO, von Klein-Bavaria aus

The valley lies within a broad embayment, opening towards Plettenberg Bay. The embayment is characterised by 'terraces' which cut over from the softer Bokkeveld shales into the hard TMS quartzites strongly folded in the Tsitsikama range (background): The 'terraces' (or plains rather) are the Upper Cretaceous ones in 600, 500 and 400m on the broad TMS 'hogbacks', dissected by the Keurboom river in the centre. The 300m level (Miocene) ingresses into the smaller embayments between the 'hogbacks' and is itself dissected by the meandering river. The level is (near Uplands, outside the foto, on the right) terminated by the 200m level (Plio-Pleistocene) along a very long and well-formed scarp between the two levels. - The Bokkeveld shales form shallow synclines between the quartzite anticlines of the 'hogbacks'. The structurally not controlled 'terraces' evenly cut across both tectonic forms.

Die Verflachungen, die die Gebirgsbucht deutlich überprägen, ziehen sich in gleicher Höhe von den weichen Bokkeveld-Schiefern auf die harten TMS-Quarzite, die im Hintergrund, in der Tsitsikama-Kette, steil gefaltet sind und den Gebirgsrahmen bilden. Bei den Verflachungen handelt es sich um die oberkretazischen Thalassoplains in 600, 500 und 400m, die über die breiten Quarzitrücken in die Gebirgsbucht hineinziehen, zerschnitten vom Keurboom (mittlerer Vordergrund). Die 300m-Verflachung zieht in die Buchten zwischen die Quarzitrücken hinein, von den Mäandern des Keurboom unregelmäßig angeschnitten. Die 300m-'Terrasse' ist hier in der Tat terrassenartig ausgebildet, vergleichbar einer 'Trogfläche' im Rheinischen Schiefergebirge. Die 300m-Verebnung (Miozän) grenzt bei Uplands (rechts, außerhalb des Bildes) mit einer gut ausgebildeten und sich lang hinziehenden Erosionsstufe an die 200m-Verflachung (Plio-Pleistozän). - Die Bokkeveld-Schiefer bilden schmale Muldenstrukturen zwischen den Quarzit-Sätteln. Die Verflachungen ziehen strukturunabhängig über Sättel und Mulden hinweg.

Foto 11. At the upper Doring river near Beukesfontein, NW of Ceres (northwestern Cape Province). View to the S and SE

Am oberen Doring bei Beukesfontein, im NW von Ceres (nordwestliche Kapprovinz). Blick nach S und SO

The river is shallowly incised into the Swartruggens Plateau, in 600m, widely cutting across Witteberg shales. The plain advances into the escarpment in the way of a broad embayment, today followed by the Doring. The range to the right is part of the Hexrivier Berge (N of Touwsrivier), one of the Cape ranges.

Der Fluß ist, mit Steilrändern, flach in das Swartruggens Plateau eingelassen, das hier in Witteberg-Schiefern ausgebildet ist. Bei der Verflachung handelt es sich um jene in 600m, die hier flächenhaft weit die Landschaft prägt und buchtartig breit in die Große Randstufe hinein vorstößt, vom Doring nach aufwärts gefolgt. Die Bergkette rechts gehört jedoch noch zu den Kapketten (Hexrivier-Berge, N von Touws-rivier).

Foto 12. The Great Escarpment seen from the same situation as in foto 11. View to the E

Die Große Randstufe vom selben Standpunkt wie in Foto 11. Blick nach O

The Upper Cretaceous 600m plain cuts back right to the foot of the escarpment, dissecting it by branching-off larger and smaller embayments with inselbergs in between. Only where the dolerite sills have remained intact (as on the tafelberg in the back-ground or on the swell towards the right) a proper escarpment is formed. The higher embayment in the E (i.e. to the right of the tafelberg = S), consuming large parts of the cuesta, is part of the Lower Cretaceous 800m level.

Das breite oberkretazische Niveau in 600m greift bis zum Stufenrand zurück, ältere Niveaus ähnlich aufzehrend wie in Foto 9 anderwärts beschrieben. Die Große Randstufe ist hier buchtförmig aufgelöst, mit Inselbergen. Nur dort, wo Doleritkappen geschlossen der Abtragung widerstanden (z.B. der Tafelberg hinten links oder am Anstieg rechts = nach S) ist die sog. Randstufe deutlich. Die im E erkennbare höhere Bucht, die noch weiter in die Randstufe nach E hin eingreift, ist im unterkretazischen 800m-Niveau ausgebildet.

Foto 13. The Welwitschia Plain N of the Swakop river in the central Namib desert. View to the SW (with the morning fog still over the coastal parts)

Die Welwitschia-Ebene nördlich des Swakop in der mittleren Namib. Blick nach SW (mit den morgendlichen Küstennebeln).

The plain is part of the so-called 'oblique plain' (between Karibib in the E and Swakopmund in the W) and lies in the Upper Cretaceous 600m level. This level breaks away - along the chain of inselbergs in the centre - towards the next level below (to be seen to the left of the inselbergs and the dunes). In crystalline rocks inselbergs are typical for the scarp line between the levels. From this it follows that the 'oblique plain' is not just one single structure. The surface is covered by a fine, serir-like gravel of angular milk quartzes.

Die Ebene befindet sich im oberkretazischen 600m-Niveau als Teil der sog. 'Schiefen Ebene' (zwischen Karibib im E und Swakopmund im W). An einer Inselberg-Kette (Mittelgrund) fällt die Verebnung mit einer leichten Versteilung zum nächst tieferen Niveau ab, hinter der Inselberg-Kette und den Dünen links erkennbar. Die Inselberge sind, wie üblich, im Kristallin ausgebildet. Solche Inselbergränder oder Rampen sind typisch für die Stufen zwischen den Niveaus, sofern die Stufe im Kristallin (oder dgl.) liegt. Aus diesen Beobachtungen folgt, daß es sich bei der 'Schiefen Ebene' nicht um eine gleichmäßig durchziehende Form handelt. Die Ebene ist serir-artig von Feinkies aus eckigen Milchquarzen überkleidet.

Foto 14. View over the upper Kuiseb canyon, W of Rostock (central South West Africa). View to the SE to the Hakos highlands. Standpoint: on the Kuiseb (or Karpfenkliff) conglomerates.

Am Kuiseb-Kanon, westlich von Rostock (mittleres Südwest-Afrika). Blick nach SO auf die Ausläufer des Hakos-Berglandes. Standort: auf den Kuiseb (bzw. Karpfenkliff-) Konglomeraten.

The conglomerates, forming mesas and tafelbergs in 1000m, cover the intensely folded Damara shales below. From the right (=W) the 800m plain (centre) ingresses in the form of a large embayment (today followed by the canyon). The plain cuts smoothly over the shales, dissected from the river. In the background the plateau-like 1000m level (with rounded hilltops in the shales) borders the highlands, here also consisting of the folded Damara shales.

Die Konglomerate bilden Mesas und Tafelberge und überdecken die intensiv gefalteten Damara-Schiefer. Darunter greift im Mittelgrund die 800m-Verebnung buchtartig breit von rechts her (=W) ein (ihr folgt das heutige Tal aufwärts) und schneidet glatt über die Schiefer hinweg. Im Hintergrund grenzt das 1000m-Niveau - in den Schiefern kuppig aufgelöst - plateauartig an das höhere Bergland, das hier ebenfalls aus den gefalteten Damara-Schiefern besteht.

Foto 15. The 'oblique plain' about 90 km NE of Swakopmund. View to the S and SE

Die 'Schiefe Ebene' etwa 90 km NO Swakopmund. Blick nach S und SW

In the background, behind the incised Swakop river, the horizontally plained 800m level can be seen, from where (in the fore-ground) a very flat slope leads over to the Upper Cretaceous 600m level (on which the tracks of the Windhoek-Swakopmund railway line are visible).

Im Hintergrund, jenseits des Einschnitttes des Swakop, kann man die horizontale Verflachung des unterkretazischen 800m-Niveaus erkennen, von der (im Vordergrund sichtbar) ein sehr sanfter Hang zum oberkretazischen 600m-Niveau überleitet (hier gequert von der Eisenbahn Windhoek-Swakopmund).

Foto 16. Near Paulpietersburg (northern Natal) View to the E

Bei Paulpietersburg (nördliches Natal) Blick nach O

The view goes towards the basin of the middle Pongolo river, with two flat planation surfaces to be seen: the one below in 1000m, the other above in 1200m. Both surfaces are separated by a distinct scarp. Both planations form broad embayments to the W, today followed by the middle and upper Pongolo river. Both planations cut across granites and gneisses of different pre-Cambrian ages; they are also independent of the Jurassic faults, crossing the picture from the left (=N) to the right. The steeper and higher scarps on the right (=S) lead up to the mesas and tafelbergs developed on the Ecca sandstones and the dolerite sills. (The prominent scarp on the right is part of the Dumbe tafelberg with its top in 1536m, lying next to Paulpietersburg; the Dumbe is an outlier of the Great Escarpment). Above the crystalline basement the sandstones and sills have been widely removed in connection with the 800 and 1000m-thalassoplains. The planation, beginning at an altitude of 1200m, is not a thalassoplain, but part of a Lower Cretaceous peneplain.

Man erkennt zwei flächenhaft weite Verflachungen, einmal in 1000m (unten), ein andermal in 1200m (oben), durch eine deutliche Stufe von einander getrennt. Die Verflachungen ziehen sich buchtartig weit nach W hinein, heute vom Pongolo-Tal gefolgt. Die Verflachungen sind hier über dem Grundgebirge ausgebildet (bestehend aus Gneisen und Graniten verschiedensten Alters). Sie erstrecken sich unabhängig auch von den jurassischen Verwerfungen, die sich von links (=N) nach rechts quer durchs Bild ziehen. Die Steilanstiege rechts gehören zu Mesas und Tafelbergen, die in den Ecca-Sandsteinen bzw. in den Doleritlagen ausgebildet sind. (Der Steilanstieg rechts führt zum Dumbe in 1536m, einem Auslieger der Großen Randstufe unmittelbar im SW von Paulpietersburg.) Die flachlagernden Schichten wurden über dem Kristallin flächenhaft weit im Zusammenhang mit den Thalassoplains in 800 und 1000m aus- und abgeräumt. Die Verebnungen in 1200m gehören zur unterkretazischen P 2-Rumpffläche.

Foto 17. The Steinkopf basin (Northwestern Cape Province). View to the N

Das Becken von Steinkopf (nordwestlichste Kapprovinz).Blick nach N

The basin lies in 800m, ingressing from the west (=left) into the high-lying hills (background), here plained over by the 1000m level.

Das Becken ist in 800m ausgebildet und greift von W (=links) in das Bergland (Hintergrund), das hier im 1000m-Niveau verebnet ist.

Foto 18. Sakrivier near Brandvlei, northwest of Calvinia (northern Cape Province). View to the E

Sak-Rivier bei Brandvlei im NW von Calvinia (nördliche Kapprovinz). Blick nach O

In the foreground one sees the rest of an inselberg, formed on dolerites. The dolerite boulders consist of hard rock, covered by a very thin brown crust. The inselberg stands above the vast 1000m thalassoplain, gently undulating and gently rising to higher flat banks in 1100m. As inselberg remnants (now and then appearing on these banks) show the once higher ramps have been gradually (and almost completely) pediplained, right down into the base level of the thalassoplain.

Man blickt über die weiten Flächen der Thalassoplain in 1000m von den Resten eines Inselberges aus. Das Gestein des Inselbergs besteht aus Dolerit, ebenfalls die gerundeten Blöcke (mit sehr dünner brauner Kruste über dem festen Fels). Aus der Fläche erheben sich wellenhaft weit gespannt etwas höhere Geländezüge bis an 1100m. Diesen sitzen gelegentlich Dolerit-Restberge (wie im Vor dergrund) auf. Diese sanften höheren Geländezüge stellen somit nichts anderes dar als die von der Thalassoplain aus pediplanierten Rahmenhöhen der Fläche in 1000m als der zugehörigen Erosions- und Denudationsbasis.

Foto 19. Northern Transkei near Qumbu

Nördliche Transkei bei Qumbu

The 1000m-thalassoplain (over Beaufort sandstones and dolerite sills) dominates the landscape.

Auch hier bestimmen die flächenhaft weiten Verebnungen der 1000m-Thalassoplain (über Beaufort-Sandsteinen und Doleritlagen) das Bild.

Foto 20. Near Haarlem, to the N of Knysna (southern Cape Province). View to the SE

Bei Haarlem, im N von Knysna (südliche Kapprovinz). Blick nach SO

One looks into a part of one of the longitudinal basins (following the strike both of the Cape ranges and the Lower Cretaceous faults) which here is part of the catchment area of the Kouga river and flanked by the Kougaberge on the left (=N) and the Outweniekjaberge on the right (=S). To the SE, the Kouga basin merges into that of the Krom river near Joubertina. Following these preexisting basins, the thalassoplains appear one after the other in their specific altitudes. The foreground and the centre right is made up of the 1000m plain, into which the 800m plain ingresses, forming a scarp (near the houses). The 800m surface is itself undercut by the 600m level (in the background). (When driving from here to the SE, one crosses over the scarps from one level to the next, right down to the one in 100m.) The mesas and tafelbergs belong to the 800 and 600m surfaces. Their hard tops are made up of silicified conglomerates. As one can see on the right, the mesas slope very softly towards the flanks of the range.

Die Landschaft gehört zum Einzugsgebiet des Kouga-Flusses, der entlang dieser Längssenke (im Streichen der Kapketten wie aber auch der unterkretazischen Verwerfungen) zwischen den Kougabergen im N (links) und den Outweniekjabergen im S (rechts) in südöstlicher Richtung fließt. Nach SE geht die Kouga-Senke bei Joubertina nahtlos in jene des Krom-Flusses über. In die vorweg bestehende Längssenke hinein ziehen sich die einzelnen Thalassoplains in ihrer jeweiligen höhenbeständigen Lage. Hier bei Haarlem erblickt man die Verflachungen in 1000m (Standort, nach rechts) und in 800m, wobei die letztere mit einem Hang (bei den Häusern) deutlich in die höhere eingreift. Das gleiche ist der Fall bei der 600m-'Terrasse' (im Hintergrund) gegenüber dem 800m-Niveau. Die Verflachungen in 800 und 600m sind durch Mesas und Tafelberge charakterisiert, jeweils geknüpft an die silifizierten Konglomeratdecken. Die Oberflächen steigen hier zum gebirgswärtigen Rand der nächsten Kapkette sanft an (wie auf der rechten Seite erkennbar).- Fährt man von hier die Flucht der Längsbecken nach SE hinunter, kommt man über die jeweils zwischengeschalteten Erosionsstufen von Niveau zu Niveau herab (jeweils in deren spezifischer Höhenlage), bis hinunter zu jenem in 100m.

Foto 21. Near St. Faith's in the hinterland of Port Shepstone (Kwazulu; southern Natal)

Bei St. Faith's im Hinterland von Port Shepstone (Kwazulu; Südnatal)

The rocks are built up of upper Dwyka sandstones and harder conglomerates in between, with the layers slightly sloping to the right. The rocks - and with them the planations - are strongly dissected by the Mzimkulu river and its tributaries, by this showing that the planations are fossil forms, with no planatory processes going on any more (as in Foto 1). Two levels can be observed: one on the left in 800m, advancing embayment-like into the upper one with a distinct scarp, and the higher one in 1000m (on the right, background). The highest part in the background in 1500 to 1700m is an outlier of the Great Escarpment proper and built up of Ecca rocks and capped by dolerite sills. In between the 1000m level and the outlier a long intermediary level (on the horizon, to the left) in 1200m slopes gently up to the scarp above it. This intermediary level is part of the Lower Cretaceous P 2 peneplain here widely consumed by the younger (but equally Lower Cretaceous) thalassoplains. - Apart from the planations the landscape is characterised by slopes, with the slopes becoming more prominent the higher up they are. The lower sections of the slopes flatten out everywhere, but do not form pediments (cf. Foto 6).

Der Untergrund besteht überwiegend aus Dwyka Sandsteinen mit härteren Konglomeratlagen dazwischen. Die Landschaft ist stark vom Mzimkulu und seinen Nebenflüssen aus zerschnitten, einschließlich der Verebnungen. Diese werden somit nicht mehr weitergebildet. Zu erkennen sind zwei niveaubeständige Verebnungsbereiche: eines in 800m, das von links her buchtartig, mit deutlicher Stufe, in ein höheres in 1000m Höhe eingreift (Mittelgrund rechts, Hintergrund). Diesem oberen Niveau sitzt im Hintergrund ein Auslieger der Großen Randstufe auf, hier in 1500 bis 1700m und in Ecca-Gesteinen mit Doleritkappen ausgebildet. Das Aufsitzen erfolgt jedoch nicht unmittelbar, sondern eine lange Schleppe (am Horizont links) vermittelt mit einer geneigten Verflachung in rd. 1200m. Bei dieser Verflachung handelt es sich um einen Rest der P 2-Rumpffläche (Unterkreide), die hier durch die ebenfalls noch unterkretazischen, aber jüngeren Thalassoplains in 800 und 1000m (in das höhere Niveau eingreifend) aufgezehrt wurde. Ansonsten bestimmen Hangformen das Bild. Sie werden nach oben hin immer ausgeprägter. Nach unten hin laufen alle Hänge flach und lang aus, verschneiden sich aber nicht zu einem durchgehenden Pediment (vgl. Foto 6).

Foto 22. In the Groot Swartberge, in the neighbourhood of the Swartberg-Pass, near Oudtshoorn (southern Cape Province)

In den Groot Swartbergen, am Aufstieg zum Swartberg-Paß, in der Nähe von Oudtshoorn (southern Cape Province)

Embayment-like advance of the 800m level into the 1000m one. The higher level horizontally cuts across the ranges and interfluves made up of folded TMS quartzites.

Buchtförmiges Eingreifen des 800m-Niveaus in das von 1000m. Letzteres schneidet die Bergrücken, ausgebildet in gefalteten TMS-Quarziten, höhenkonstant ab.

Foto 23. At the Sundays river near Kirkwood, in the north of Port Elizabeth (southern Cape Province). View to the N

Am Sonntagsfluß bei Kirkwood, nördlich Port Elizabeth (südliche Kapprovinz). Blick nach N

The orange orchards in front lie on the 100m 'terrace'. Accompanying the steep rise of the Suurberg range the 200m level is horizontally cut into the quartzites as a ledge or sometimes as a 'terrace'. The higher 'terraces' above in 300, 400 and 600m follow the pre-existing valley through the range. (The 500m level can only be seen on the right behind the ravine.) The gorge of the Sundays river breaches the range from the subsequential zone north of the Suurberge towards the Algoa Bay. The breaching must be as old at least as the uppermost 'terrace' in 600m - i.e. it is at least of an Upper Cretaceous age.

Die Orangen-Huerta (Vordergrund) ist im Bereich der 100m-'Terrasse' ausgebildet. Über dieser zieht sich am Steilabfall der Suurberg-Kette höhenkonstant die 200m-'Terrasse' (oft nur als Sims) hin. Die höheren Terrassen greifen (genau wie die niederen) im Bereich der heutigen Talung durch die Kette hindurch und belegen damit, daß der Durchbruch mindestens so alt ist wie die höchste hier sichtbare Verebnung. Dies ist die oberkretazische 600m-Fläche. Darunter folgen das Niveau in 500m (rechts hinter der Kerbe schwach sichtbar) und jene in 400 und 300m. Die alte Talung bricht hier von der Subsequenzzone nördlich der Suurberge durch in Richtung auf die Algoa-Bucht.

Foto 24. The poort of the Grootrivier in the Groot Swartberge near De Rust in the neighbourhood of Oudtshoorn (southern Cape Province). View to the N

Der Durchbruch des Groot-Rivier durch die Groot Swartberge bei De Rust in der Nähe von Oudtshoorn (südliche Kapprovinz). Blick nach N

The gorge is cut through hard folded quartzites. At the outlet of the gorge (or poort) the lowest 'terrace' can be observed in 500m, though often only as a ledge or as a mere knick in the slopes. The 600m 'terrace' above is silcrete-topped. Over it follow the 'terraces' in 800 and 1000m, all of them breaching the range along the course of the poort. A predecessor must have existed therefore in the Lower Cretaceous. The valley bottom in front is formed by the 300m 'terrace' (Miocene), cutting into the Lower Cretaceous conglomerates that once filled the longitudinal intermontane basin of Oudtshoorn even higher up. The conglomerates today only reach as high up as to the line where the slope steepens above the light-coloured sediments below. This line also denotes the fault (though roughly). Along this fault the basin subsided in the Valanginian, being synchronously filled up with the sediments, later the 'terraces' were cut across the Valanginian sediments.

Der Durchbruch erfolgt durch die intensiv gefalteten Quarzite. Im Durchbruch ist ein unteres Niveau in 500m schwach zu erkennen (Simse, Hangknicke), während darüber die 'Terrassen' in 600m, 800m und 1000m kräftiger ausgebildet sind. Ein Vorläufer-Durchbruch muß somit bereits in der Unterkreide bestanden haben. Im Vordergrund vor der Kette und der 'Pforte' läuft die (miozäne) 'Terrasse' als flacher Talboden aus. Das Niveau ist hier in den unterkretazischen Konglomeraten ausgearbeitet. Die Unterkreide hört nach N ungefähr dort auf, wo die Hellfärbung des Niveaus unten gegen den sich versteilenden Hang hin aussetzt. Entlang dieser Grenze verläuft die Bruchlinie, an der das Becken von Oudtshoorn in der früheren Unterkreide (d.h. im Valanginien) relativ zum Gebirge abgesenkt und mit synchronen Sedimenten verfüllt wurde. Über diese Sedimente schneiden die Niveaus der 'Terrassen' hinweg.

Foto 25. The Fish river canyon in southern South West Africa. View to the S and SE

Der Kanon des Fischflusses im südlichen Südwestafrika. Blick nach S und SO

The canyon was worked into the flat-lying Nama layers, cutting through them, down into the crystalline basement itself, reached in the valley bottom. Because of the ingressing 300m level, a widening can be observed there. The highest parts in the background stand at 1000m. At first sight the flats to be seen there seem to be structurally controlled. This is however not the case. The layers incline to the right (=N), whereas the surface remains at the same height. The same can be observed about the next marked surface below in 800m (with flat shield-like pediments going up to 900m: on one of these the viewpoint is situated). The surface in 800m forms a 'trough' beneath the one in 1000m (which also shows pediments, up to 1100m). The interfluve between the Konkiep valley (behind the background flats) and the Fish widely carry this surface. From this it follows that the 'trough regions' of the Konkiep and Fish valleys must have existed already by Lower Cretaceous times. The 'terrace' within the canyon is that of the 600m planation (with pediments up to 700m).

Der Kanon ist im wesentlichen in den flachlagernden Sedimenten der Nama-Formation ausgearbeitet worden, schneidet im Bereich des Flusses jedoch bereits das Kristallin an. Im breiteren Talboden läuft das 300m-Niveau aus, während die Höhen im Hintergrund sich in 1000m befinden. Die Verflachungen in Hinter- und Mittelgrund scheinen strukturbedingt zu sein. Dies ist jedoch nicht der Fall: die Schichten fallen nach rechts (=N) hin ein, während die Verflachungen sich in konstanter Höhe in der Talungsregion hinziehen. Das ausgeprägte Niveau unter den Höhen im Hintergrund ist das in 800m, trogartig eingelassen in jenes in 1000m. Darunter folgt das Niveau in 600m, wiederum (verengt) trogartig in jenes in 800m eingearbeitet. Auf allen Hauptniveaus sind 'Schildpedimente' verhanden, die jeweils Höhen bis zu 100m über dem Haupt-niveau erreichen. Auf einem solchen 'Schildpediment' des 800m-Niveaus liegt in rd. 900m auch die Aufnahmestelle. Das höchste hier sichtbare Hauptniveau (das in 1000m) verebnet breit die hohen Bereiche zwischen den Taltrögen, wie hier z.B. jene zwischen dem Konkiep (das Tal verläuft hinter den Höhenzügen im Hintergrund) und dem

Fischfluß. Die Taltröge müssen somit bereits in der Unterkreide bestanden haben.

Foto 26. The 'low gradient embayment' southeast of Vioolsdrif on the Orange (northwestern Cape Province). View to E and NE

Die 'Flachböschungsbucht'von Vioolsdrif am Oranje (nordwestlichste Kapprovinz). Blick nach O und NO

The surface to be seen is the one in 600m, gently sloping down from here in the way of an 'oblique plain' towards the Orange in the NW. The embayment is surrounded by a fringe of residual hills (inselbergs), typical for crystalline or intrusive rocks, with small ragged 'bays'in between. The surface (including its extending 'bays') is completely covered by fine gravel and sand. Thalassoplanation may have followed the gently rising groundwater table and the tectonic cracks within the rocks (characterised by the small 'bays').

Im Bild erkennt man das obere Niveau dieser 'Flachböschungsbucht', hier in 600m. Die flache Böschung dacht sich von hier im Sinne einer 'schiefen Ebene' nach links (=NW) zum Oranje hin ab. Die 'Schiefe Ebene' ist stets von Rahmenhöhen und 'aufgelappten' Restberglandschaften des Grundgebirges (bzw. der Intrusiva) umgeben. Mit sehr sanftem Anstieg setzt sich das Niveau in die 'Buchtlappen' der Restberge hinein fort. Offensichtlich ist die Thalassoplanation hier einerseits einem flach ansteigenden Grundwasserspiegel gefolgt, andererseits tektonischen Zerrüttungszonen des Berglandes. In diesen sind die 'Buchtlappen' ausgebildet. Die gesamte 'Schiefe Ebene' ist bis in die 'Buchtlappen' hinein ausgekleidet von einer Decke aus Feinkies und Sand.

Foto 27. Detail of Foto 26

Ausschnitt aus Foto 26

The 'bay' developed in front of the tectonic crack behind it, was gradually widened and lowered by weathering, the widening followed from below by a developing 'bay'. The crack, itself also lowered by weathering, like a small poort, leads over to the next thalassoplained embayment behind the mountain ridge. The situation here helps to understand (and to visualise) how, during a eustatic transgression, a poort may have become a channel, serving as a new thalweg after regression. Breachings, as in the Cape ranges, can easily be explained in this way. Triangular embayments (though small here) can also easily be recognised, here in 600m. These embayments are the forerunners of pediplanation, advancing against the mountain flanks and starting from the embayment.

Man erkennt hinter der 'Bucht' die erniedrigte Zerrüttungszone sehr gut. Sie leitet paßartig hinüber zur nächsten Trogflächenbucht hinter dem Höhenrücken. Man kann sich gut vorstellen, daß bei einem Meeresspiegelanstieg der erniedrigte Paß überwältigt werden kann. Neue, auch endgültige, Talverläufe und Talsysteme können auf diese Weise nach einer Transgression entstehen, wie etwa in den Kapketten. Im Foto sind ferner die dreieckigen Pedimentbuchten gut erkennbar, eingestellt auf das 600m-Niveau. Man sieht, daß die Pedimentbuchten die Vorhut für die weitergehende Pediplanation sind, die von der Bucht aus gegen die Bergflanken vorgreift.

Foto 28. 'Low gradient embayments' on the NE margin of the Naukluftberge (central South West Africa). View to the SE

'Flachböschungsbuchten' im NE-Außenbereich der Naukluftberge (mittleres Südwestafrika). Blick nach SO

The embayments to be seen here lie at a height of 1000m; apart from that the phenomenon is principally similar to that of Foto 27, only in a far more advanced stage (wider pedimental embayments, broader pediment shields and pediments in front that already have totally consumed the mountains, especially on the right). The advanced stage can be connected to the longer timespan in which pediplanation could have been active (i.e. since the Lower Cretaceous).

Wenn man sich hier auch im 1000m-Niveau befindet, so geht es prinzipiell um dieselben Phänomene wie in Foto 27, allerdings wegen des längeren Andauerns der Vorgänge (seit der Unterkreide) weiter fortgeschritten (verbreiterte Pedimentbuchten, breitere Pedimentschilde, voll durchgreifende Pedimente im bereits völlig aufgezehrten Bergland, wie besonders im rechten Hintergrund).

Foto 29. Kneersvlakte and Lepel se Vlakte, S of Kliprand (northwestern Cape Province). View to the NE and E

Kneersvlakte und Lepel se Vlakte, S von Kliprand (nordwestliche Kapprovinz). Blick nach NO und O

The 'vlaktes' (=flats) form the southernmost 'low gradient embayment' or 'oblique plain'. The levels in 500, 600 and 800m are here gently merged into each other, though distinct flattenings can be observed in the heights of the levels. The hills in 1000m around the 'vlaktes' frame the embayment and, are flattened over by pediplanation. The 'oblique plain' is covered by serir-like fine angular quartzes and sand.

Die ineinanderübergehenden 'Vlaktes' stellen das südlichste Vorkommen einer 'Flachböschungsbucht' bzw. einer 'Schiefen Ebene' dar. In dieser 'Schiefen Ebene' verschmelzen gewissermaßen die Niveaus in 500, 600 und 800m sacht miteinander, jedoch nicht völlig unmerklich, da in den Höhen der Niveaus sich Verflachungen einstellen. Die Rahmenhöhen sind in 1000m ausgebildet und durch Pediplanation ebenfalls verflacht. Die 'Schiefe Ebene' wird durch eine Sand- und Feinkiesdecke serirartig über- und verkleidet.

Foto 30. A 'vega' along the Gamka river (S of Calitzdorp, W of Oudtshoorn; southern Cape Province). View to the E and NE

Eine 'Vega' am Gamka (S Calitzdorp, W Oudtshoorn, südliche Kapprovinz). Blick nach O und NO

The Gamka river here crosses the longitudinal Lower Cretaceous basin of Oudtshoorn from the N (=left) to the S. The basin is the catchment area of the Olifants river system. The Gamka flows into the Olifants in the S (=right), outside the foto. The lowest level here is the one in 200m (Plio-Pleistocene), stretching into the basin through the Cape ranges (to the right, not to be seen) and followed by both rivers. The level is intensely cultivated because of irrigation. The level broadens very much once it has reached the basin floor, but narrows in the gorges (or poorts) in the S, forming a 'terrace' there on both sides of the valley. Upstream, both 'terraces' close together, forming the lowest plain in the basin - immediately when the valley reaches the basin. The level narrows again towards Calitzdorp (=N), finding its limit where the Gamka, on entering the next gorge or poort in the N, passes the scarp between this level and the next one higher up (in 300m). The scarp between the two levels runs on into the basin, clearly separating the two levels there. The foto was taken from a standpoint on the 300m 'terrace', looking across the 200m level towards the scarp on its other side. The 300m level is accentuated by silcretes (on both sides). The mesa in the centre background is formed by the conglomerates of the 500m level (not much silicified), surrounded by the (narrow) 400m level, here strongly consumed by pediments. The mesa is continued to the left, though more consumed there. Behind it, in the farther background, the Groot Swartberge (behind Oudtshoorn) rise up, running parallel to the strike of the basin (from left = W to right = E).

Der Gamka quert hier von links (=N) nach rechts das im Streichen verlaufende unterkretazische Becken von Oudtshoorn. Dieses wird durch den Olifants-Fluß entwässert (der auch das Grootrivier aufnimmt: Foto 24). Olifants und Gamka fließen etwas weiter im S (= rechts) zusammen. An beiden Flüssen ist das 200m-Niveau durch Intensivkulturen inwertgesetzt. Das Niveau verbreitert sich im Längsbecken, während es sich im Durchbruch durch die Kapkette nach S hin (= rechts, nicht im Bild) wieder sehr verengt und auf beiden Seiten des Durchbruches eine 'Terrasse' bildet. Diese 'Terrassen' schließen sich umlaufend dort zu einem breiten Talboden zusammen (eben dem 200m-Niveau), wo der heutige Flußlauf in das Niveau dieser ehemaligen Thalassoplain im Längsbecken eintritt. Das Niveau verengt sich wieder langsam auf Calitzdorp zu; es findet am Ausgang des nördlichen Gamka-Durchbruches, unterhalb einer umlaufenden Versteilung, sein Ende.
Das 200m-Niveau, das hier den unteren Boden des Beckens bildet, ist mit einer deutlichen Stufe eingesenkt in das nächsthöhere Niveau in 300m, an dessen westlicher Kante der Betrachter steht. Die Kanten des Niveaus sind auf beiden Seiten in Silcretes ausgebildet und daher verfestigt. Auf der gegenüberliegenden Seite zieht diese Kante in einem Oval um die 200m-Verflachung herum, dabei eine umlaufende Versteilung bildend.

Die durch die 300m-Thalassoplain herausgeschnittene Mesa im Bildhintergrund (Mitte) befindet sich im Bereich der 500m-Verebnung, umsäumt vom schmalen, pedimentartig aufgelösten 400m-Niveau. Die Oberkante der Mesa wird von hier wenig verfestigten Schottern gebildet. Die Fortsetzung der Mesa ist nach links hinten im selben Niveau sichtbar, wenn auch randlich stärker zerschnitten. Dahinter ragen die Groot Swartberge nördlich von Oudtshoorn auf. Diese Kette verläuft genau im Streichen des Beckens von W (=links) nach E (=rechts).

Foto 31. Part of the Buffels river landscape. View to the E and SE

Landschaft am Buffelsrivier. Blick nach O und SO

The Buffels river is the upper part of the Groot river within the longitudinal basin between the Klein Swartberge in the south (= right) and Surklcos se Berg in the north (=left). The deepest part of the landscape can be found in the background to the left. It consists of the Upper Cretaceous 600m level, immediately along the Buffels. Within the basin, it widens broadly and narrows again in the poorts through the ranges. From the widening level pediments slope up rather flatly , but break away rather abruptly at the foot of the scarp below the next higher thalassoplain (i.e. the T 2 in 800m of Lower Cretaceous age). Within the basin, this T 2 level is also widely developed, forming mesas and tafelbergs because of its being topped by conglomerates and silcretes (typical for a warm and semi-arid climate). The T 2 is strongly dissected from below, i. e. the 600m level as its base level. In the foreground, folded Witteberg quartzites have become bared as a consequence of denudation. Altogether, the T 2 level rises softly towards heights or ramps framing this level. On top of these framing heights the T 1 planation (1000m, Lower Cretaceous) can be seen, though only as remnants (background, right and left). The reason for this is that pediments branch off from the lower level, thus dissecting the higher one. On the higher level no mesas and no tafelbergs can be observed because silcretes are missing here totally.

Das Buffelsrivier bildet den Oberlauf des Grootsriviers im Bereich der Längssenke zwischen den Klein-Swartbergen im S (=rechts) und dem Suurklos se Berg im N (=links). Das niedrigste Gelände im vorderen Hintergrund erstreckt sich direkt am Rivier entlang als breite Talweitung im (oberkretazischen) 600m Niveau. In den beiden Durchbrüchen durch diese Kapketten ist das Niveau zwar verengt, zieht sich aber jeweils voll hindurch. Von seiner ausraumartigen Verbreiterung im Längsbecken zwischen den beiden Ketten, ziehen flache Pedimente zum nächsthöheren Niveau hinauf, das jedoch seinerseits mit deutlichem Steilrand von den Pedimenten abgesetzt ist. Bei dem höheren Niveau handelt es sich um die (unterkretazische) T 2-Fläche, die, wie man sieht, im Becken ebenfalls weit ausgreift. Wegen der Silcret-Verbackung der durch die Plain abgeschnittenen Schichten bzw. wegen der auflagernden Konglomerate ist die Thalassoplain mesa- und tafelbergartig ausgebildet. Die Silcret-Bildung ist die Folge eines warmen, semi-ariden Klimas. Die Zerschneidung der einst geschlossenen Plain in Mesas (Mittelgrund) und Tafelberge (rechts) ist die Folge einer jüngeren Zerschneidung, die bis heute auf das 600m-Niveau als Erosionsbasis eingestellt ist. Im Vordergrund sind die mesabildenden Horizonte völlig abgetragen worden, so daß dort die gefalteten Witteberg-Quarzite freigelegt wurden. Die T 2 steigt sanft zu den Rahmenhöhen hin an. Über diese Rahmenhöhen erstrecken sich die Verflachungen des noch älteren (aber ebenfalls unterkretazischen) T 1-Niveaus in 1000m. Im Bildausschnitt ist diese Plain jedoch nur in der Form von Restbergen zu sehen, weil der Pediplanationsvorgang, vom 800m-Niveau ausgehend, stark mit Pedimenten eingegriffen hat, wodurch die Restberglandschaft entstand (Hintergrund rechts und links). Wegen des Fehlens von Kieselhartkrusten auf dem 1000m-Niveau sind dort keine Mesa- und Tafelbergformen vorhanden.

Foto 32. Landscape in central Natal, NW of Richmond. View to the SW

Landschaft in Mittelnatal, im NW von Richmond. Blick nach SW

The planation softly rising to the left is the P 2 peneplain (setting in at altitudes of 1200m). The plain has become dissected from the left (= E). In the foreground, a scarp leads down, step-like, to the level in 1000m.

Die nach links hin sanft ansteigende Verebnung ist die P 2-Rumpffläche, die in rund 1200m Höhe einsetzt. Sie ist von links her (=E) zerschnitten. Sie wird nicht mehr weitergebildet. Stattdessen greifen die Hangbildungsprozesse des humiden Klimas ein. Im Vordergrund erkennt man den Abfall zum nächstniederen Niveau, dem in 1000m.

Foto 33. A peneplain landscape south of Otjiwarongo in northern South West Africa. View to the N and NW

Rumpfflächenlandschaft südlich Otjiwarongo im nördlichen Südwestafrika. Blick nach N und NW

The highway leads to Okahandja in the south. Beyond the road the P 2 peneplain of Lower Cretaceous age softly rises towards the W (=left). It ingresses into the higher mountains in the background by way of larger and smaller embayments and flattish pediments. When the peneplain began to be formed and to be extended, the Etjo beds (sediments and volcanites of an end-Jurassic/early Cretaceous age), originally covering the whole stretch shown in the foto, became gradually and almost totally denuded. Only some relics of the beds were preserved. One remnant of them is the hill from which the foto was taken.

Die Straße verläuft nach S nach Okahandja. Jenseits der Straße dehnt sich weithin die P 2-Rumpffläche, die von E (=rechts), hier in etwas über 1400m, nach W (links) sanft ansteigt und buchtartig in das höhere Bergland eingreift. Bei der Ausbildung der Fläche sind im Bereich des Bildes die Etjo-Schichten (spät-jurassisch bis früh-kretazisch, bestehend aus Sedimenten und Vulkaniten) entfernt worden. Sie stehen heute nur noch in Resten an, so z.B. im Bereich der Kuppe, von der aus die Aufnahme gemacht wurde.

Foto 34. Embayment of the P 2 peneplain N of Rehoboth in central South West Africa. View to the N and NW

Bucht der P 2-Rumpffläche nördlich Rehoboth im mittleren Südwestafrika. Blick nach N und NW

The embayment is situated to the south of the Windhoek basin (lying behind the mountain ridge in the background). The P 2 plain extends into the embayment, rising with a distinctly steeper gradient from an altitude of 1550m in the front towards 1650m in the background. In the farthest corners of the embayment the altitudes reach 1700m. The P 2 plain clearly dissects and destroys the highest level in roughly 2000m, still widely to be seen on top of the range. The 2000m level is the P 1 pene-plain - the true Gondwana level. The P 1 level is no longer being formed on, as its almost complete dissection shows, whereas the forming of the P 2 still does. The forming of the P 2 level must have begun after the break-up of Gondwanaland. It is therefore the first and proper 'African' level, widely distributed in southern Africa, as the preceding fotos will have revealed. The savanna bushes form strips along the groundwater streams, extending with these streams into the smaller embayments that cut into the scarp as branches of the P 2 level. - In the foreground: calcretes.

Die Rumpfflächenbucht von Rehoboth liegt südlich des Beckens von Windhoek, das hinter der Bergkette im Hintergrund beginnt. Die Rumpffläche (es handelt sich um die P 2-Fläche) steigt von 1550m im Vordergrund zwar sanft, aber deutlich sich versteilend an, und zwar auf 1650m. In den hintersten Winkeln der Seitenbuchten werden knapp 1700m erreicht. Die P 2-Fläche greift somit zerstörend in das höchste Niveau ein, das sich flach in rd. 2000m über die Rücken des Hochlandes von Windhoek erstreckt (Hintergrund). Dieses höchste Niveau muß als das wahre Gondwana-Niveau bezeichnet werden. Dieses ist die P-Rumpffläche, die heute nicht mehr weitergebildet, sondern zerstört wird. Dagegen wird die P 2-Fläche an den hinteren Rändern durchaus noch weitergebildet. Ihre weite Verbreitung rings um Südafrika herum (vgl. auch die vorhergehenden Fotos) läßt sie zum eigentlichen und ersten 'Afrika'-Niveau werden, wegen ihres Umlaufens offensichtlich ausgebildet nach der beginnenden Trennung des Gondwana-Kontinentes.
Links hinten sieht man den Savannenbusch streifenartig in schmale Buchten hineinziehen, dem Grundwasserstrom folgend. Ähnlich greift hier die P 2-Fläche ein. - Im Vordergrund: Kalkkruste.

Foto 35. The 'African' plain (P 2 level) W of Usakos in central South West Africa. View to the N

Die 'Afrika'-Fläche (P 2-Niveau) westlich von Usakos im mittleren Südwestafrika. Blick nach N

The Kleine and the Große Spitskoppe, 'sitting' on the 'African' surface of the P 2-level, here in ca. 1200m. On all sides, the P 2 level is surrounded by the pediplanation level of 1100m, itself belonging to the younger 1000m level farther down (not in the foto).

Die Kleine und die Große Spitskoppe sitzen gewissermaßen dem 'Afrika'-Niveau der P 2-Fläche auf, die sich hier in rd. 1200m Höhe befindet. Dieses Niveau wird ringsum vom niedrigeren Pedimentationsniveau in rd. 1100m abgelöst, das seinerseits zeitlich zum nächstniederen Niveau in 1000m gehört (nicht mehr im Bild).

Foto 36. The Great Escarpment in the Sani pass region at the frontier between Natal and Lesotho. View towards the S and SE

Die Große Randstufe an der Auffahrt zum Sani-Paß an der Grenze von Natal und Lesotho. Blick nach S und SO

The edge of the very thick dolerite sills here stands at ca. 3000m. The sills form walls that tower for several hundred metres above the ledges below. The wall was forced back by denudation for about 1 km. The ledges below are formed on softer Stormberg sandstones and conglomerates, again and again under-layered by dolerite sheets of different thicknesses. Each sill forms a smaller or larger cuesta of its own. Thus the ledges are nothing but structurally controlled forms and do not show a continuous (though strongly dissected) surface, as is KING's opinion.

Die Oberkante der mehre hundert m mächtigen Doleritpakete liegt hier in rd. 3000m Höhe. Die Wände sind denudativ rd. 1km zurückverlegt worden. Die entblößten Riedel darunter sind in weicheren Stormbergschichten (Sandsteine und Konglomerate) ausgebildet, in die jedoch immer wieder unterschiedlich mächtige Doleritbänke eingedrungen sind. Jede dieser Bänke bildet eine eigene Schichtstufe, mal höher, mal niedriger. Bei den Riedeln handelt es sich nicht um eine zerschnittene, einst durchlaufende Fläche, wie KING meint, sondern um nichts als strukturangepaßte Hangformen.

Foto 37. The Great Escarpment near Aberdeen in the southern Cape Province. View to the NW (Cf. foto 20)

Die Große Randstufe bei Aberdeen in der südlichen Kapprovinz. Blick nach NW (Vgl. Foto 20)

The steep escarpment consists of several thick dolerite sills (with vertical cracks) inter-layered by softer Stormberg sandstones and shales. The highest edge is also formed by the dolerites (cf. Foto 38). It is these dolerites that shape the Great Escarpment into a true cuesta (or cuesta chain), effectively desisting denudation at the front. It is only along the triangular embayments that the levels ingress into the escarpment. In the foreground one can see the 900m-pediplanation level, belonging to the younger of the two Lower Cretaceous thalassoplains, viz. that in 800m. The escarpment - or cuesta rather - has therefore to be dated here as having an age of ca. 120my.

Mehrere Lagen mächtiger, senkrecht geklüfteter Dolerite bauen den Steilabfall mit zwischenlagernden weicheren Stormberg-Sandsteinen und -Schiefern auf. Die oberste Kante wird, wie in Foto 38, eindeutig von einem mächtigen Dolerit-Paket gebildet. Dieses ist hier der grundsätzliche Anlaß für das Herausarbeiten der Großen Randstufe als riesiger Schichtungsstufe. Sie ist nicht nur der Anlaß und Ansatzpunkt, sondern bewahrt die Stufe vor weitgehender Abtragung an der Stirn. Nur entlang der Dreiecksbuchten greifen die jüngeren Niveaus in die Schichtungsstufe zurück, diese auf diese Weise aufschlitzend. Im Vordergrund reicht in einer solchen Bucht das 900m-Pedimentationsniveau bis an die Große Randstufe zurück. Da das 900m-Niveau der jüngeren unterkretazischen Thalassoplain in 800m zugeordnet werden muß, besitzt die Große Randstufe im Sinne einer Schichtungsstufe hier ein Alter von sicherlich 120 Mill. J. und mehr.

Foto 38. On the Sani Pass road, in 1700m. View to the NE (Cf. Foto 35)

An der Straße zum Sani-Paß, in 1700m. Blick nach NO (Vgl. Foto 35)

The forming of slopes in the humid subtropical region can well be observed here. The flat slopes on the left are on shales and, as that, typical of slopes on all similar rocks, including the crystalline ones. Steeper and flatter sections interchange, forming a kind of 'vertical garland'. Very characteristic are the flat lower slope sections, often several hundred metres long. Nowhere in Natal do these flat lower sections intermerge with each other in the way of true pediments. Each lower section is completely independent of either the one to the right or to the left, being genetically connected only with the middle and upper slope sections immediately above. The 'surfaces' on the right are structurally controlled. The layers show a flat fall towards the background.
In an altitude of 2400m the next dolerite buttress higher up follows, underlain by Cave Sandstone (Stormberg), rising, to the right, in the form of a plateau. Originally, this plateau was also covered by the thick dolerite sills to be seen in the background, but having been completely denuded here because of the very efficient 'amphitheatre effect'. The walls of the dolerite buttress are covered by periglacial cones of falling stones.

Das Bild vermittelt eine Vorstellung von der Hangformung im humiden subtropischen Bereich. Die flachen Hänge auf der linken Bildseite sind in Schiefern ausgebildet und damit typisch für alle ähnlichen Gesteine einschließlich des Kristallins. Versteilungen und Verflachungen wechseln girlandenartig-rhythmisch miteinander ab. Die flachen Unterhänge sind immer wieder gut ausgeprägt, häufig bis zu 500m lang. Doch nirgendwo in Natal (auch nicht in den niederen Regionen) bilden sie zusammenhängende und durchlaufende Pedimente. Jeder flache Unterhang bildet eine selbständige Form, die mit den benachbarten Flachhängen nicht in Verbindung steht. Vielmehr steht jeder Unterhang genetisch allein nur im Zusammenhang mit den mittleren und oberen Hangabschnitten, die ihn nach oben direkt fortsetzen. Die auf der rechten Bildseite vermeintlichen Verflachungen sind allein strukturbedingt. Die Schichten fallen nach hinten hin flach ein. Im Hintergrund erscheint das nächste mächtige Dolerit-bollwerk, bis in 2400m Höhe. Unmittelbar unter dem Bollwerk streicht der Höhlensandstein (Stormberg) aus. Im Schichtsteigen setzt sich dieser nach rechts vorn hin fort, nach vorne hin somit eine Art Plattform bildend. Über dieser Plattform ist das auch hier ehemals befindliche Doleritpaket völlig abgetragen worden, und zwar aufgrund des sehr wirksamen sog. 'Amphitheater-Effekts'. Unter den steilen Hängen des Dolerit-Bollwerks befinden sich periglaziale Steinschutthalden.

Foto 39. Ecca shales in central Natal

Ecca Schiefer im mittleren Natal

Typical weathering and denudation of shales in the form of small loamy wafers.

Typische Verwitterung und Abtragung an Schiefern mit Plättchenzerfall.

Foto 40. The hilly countryside of the lower Pongolo river near the frontiers of Natal, Transvaal and Swaziland. View to the SW

Das Pongolo-Hügelland im Grenzbereich von Natal, Transvaal und Swasiland. Blick nach SW

In the foreground the Pongolo river, here flowing within the 100m-'terrace'. The 'terrace' ingresses far inland first breaching the Lebombo range (to the left, not in the foto), as similarly do the other 'terraces' in 200, 300 and 400m. Behind the valley, the tops of the dolerite hills, rounded from denudation, stand in the heights of these 'terraces'. Roundedness clearly increases from the lower and younger (Tertiary) 'terraces' to the highest one in 400m (with an Upper Cretaceous age). Roundedness is thus a function of the time-span in which disgranulation could work on the volcanites. Similar climatic conditions must have prevailed within the time-span in question, i.e. for the last 65my, since disgranulation stands in need both of warmth and of humidity.

Im Vordergrund verläuft das Bett des Pongolo im Niveau der 100m-'Terrasse', die von links her durch die Lebombo-Kette (nicht mehr im Bild) binnenwärts weit zurückgreift, wie im übrigen auch die darüberfolgenden höheren 'Terrassen' in 200, 300 und 400m. Die Kuppen der Vulkanit-Hügel im Hintergrund spielen auf diese Niveaus ein. Die Abgerundetheit der Kuppen nimmt von unten nach oben deutlich zu, d. h. von den jüngeren (tertiären) Terrassen zu der älteren (oberkretazischen) in 400m. Die abgerundeten Hänge verdanken ihre Entstehung der Abgrusung, die somit ebenfalls von unten nach oben zunimmt. Die aus der Abgrusung resultierende Abrundung und Hangentwicklung ist, hier deutlich erkennbar, eine Funktion der Zeit über 65 Mill. J. hinweg, innerhalb einer Zeitspanne, in der vergleichbare klimatische Bedingungen obwaltet haben müssen. Je länger die für die Abgrusung notwendigen Bedingungen (d.h. entsprechende Wärme und Humidität) angedauert haben, umso stärker ist - von unten nach oben - die Abrundung fortgeschritten.

Foto 41. Near Wakkerstrom in Transvaal

Bei Wakkerstrom in Transvaal

Slope development by undercutting in layered rocks, with undercutting bound to horizons of springs and wet bags. The rocks are Ecca sandstones and quartzites topped by dolerite sills (with their vertical cracks).

Hangentwicklung durch Unterschneidung an Quellhorizonten im Schichtgestein. Es handelt sich um Ecca-Sandsteine und -Quarzite, die oben von Doleritpaketen (mit senkrechter Klüftung) bedeckt werden.

Foto 42. In the Sani-Pass region, central Natal (cf. Fotos 36 and 38)

In der Sani Pass Region in Mittelnatal (vgl. Fotos 36 und 38)

Quartzitic Beaufort conglomerates break away in thick boulders over a groove formed in connection with a spring horizon on top of the wetter shales beneath the conglomerates. A small 'amphi-theatre' has developed.

Abbrechen mächtiger quarzitisierter Beaufort-Konglomerate über einer Hohlkehle, die im Horizont von Quellen und Naßgallen ausgebildet ist, wobei dieser Horizont seinerseits geknüpft ist an die feuchteren Schiefer unmittelbar unter den harten Bänken. Es ist ein kleines 'Amphitheater' entstanden.

Foto 43. Near Marienthal in southern South West Africa. View to the E and SE

Bei Marienthal im südlichen Südwestafrika. Blick nach O und SO

The foto demonstrates the forming of grooves and the breaking away of boulders at cuestas under semi-arid conditions. In the foreground the flat 1100m-pediplanation level (itself belonging to the 1000m-thalassoplain of the upper Fish river region) can be seen advancing to the foot of the cuesta. The resistant rocks on top of the cuesta represent the intact upper slope section. Above it, the P 2 plain sets in, immediately in 1200m. The cuesta slope is formed within the softer Dwyka layers (below) and the hard Ecca sandstones and greywackes (above). Dwyka and Ecca both belong to the Karoo Super Group that transgresses over the basement along a flat unconformity which itself falls to the E i.e. towards the Kalahari basin. We here stand at the western margin of the basin.

Auch an Schichtstufen des semi-ariden Bereichs kann Hohlkehlenbildung und nachfolgendes Abbrechen als Prozeß der Hangformung beobachtet werden. Das 1100m-Pediplanationsniveau (das seinerseits auf die 1000m-Thalassoplain des oberen Fischflusses eingestellt ist) greift bis an den Rand der Schichtstufe zurück. Die Stufe selber bildet die intakte Arbeitskante zur P 2-Rumpffläche, die unmittelbar über der Kante in 1200m einsetzt. Die Stufe ist hier an Ecca-Sandsteinen und -Grauwacken geknüpft (Stufenbild-ner), die die weniger widerständigen Dwyka-Schiefer überlagern. Diese Schichtglieder der Karu-Formation liegen dem Untergrund an einer flachen Diskordanz auf. Diskordanz und Schichten fallen insgesamt nach E zum uralten Kalahari-Becken hin ein. Die Schichtstufe bezeichnet hier den westlichen Außenrand des Beckens.

Foto 44. North of Willowmore, southern Cape Province. View to the N and NE

Im N von Willowmore in der südlichen Kapprovinz. Blick nach N und NO

Standing on the northern flank of the northernmost Cape range the view extends across the subsequential zone between the Cape ranges and the escarpment. The escarpment itself lies to the north and cannot be seen here, because it is too far away. The 'plain' forming the surface here is the 900m pediplanation level with its gentle undulations. The road track makes visible the flat 'valleys' between the equally flat backs of the pediments. The foto provides a good notion of the vast nature of the level.

Der Blick geht vom nördlichen Abhang der nördlichsten Kapkette hinweg über die sehr breite Subsequenzzone zwischen den Kapketten und der Randstufe. Diese ist hier allerdings zu weit entfernt, um noch sichtbar zu sein. Die 'Ebene', die die Oberfläche der Subsequenzzone hier bildet, gehört zum 900m-Pediplanationsniveau. Man erhält einen guten Eindruck von der enormen Verbreitung des Niveaus. Zwischen den einzelnen flachen Pedimentrücken, in Wellen von der Straße gequert, sind Flachmuldentäler eingeschaltet.

Foto 45. North of Willowmore in the southern Cape Province. View to the SE

Nördlich Willowmore in der südlichen Kapprovinz. Blick nach SO

The standpoint of this foto is very near to the one in Foto 44. In the background one can see the northernmost Cape range and in front of it the 900m pediplanation level. The level rises softly towards the range, in accordance with the groundwater table. In the foreground and in the centre, the level cuts across folded Dwyka shales and greywackes. The inselberg on the left in the background is also built up of these rocks, itself 'sitting' on top of the pediments by which it is surrounded on all sides. The pediments form a flat shield-like pediment dome. Denudation is carried on by way of small embayments cutting back into the flanks of the inselberg and eventually doing completely away with it. All stages leading towards such a complete destroyal can be observed in the immediate neighbourhood of this standpoint. The shales and greywackes are weathered into small wafers and still smaller particles. They are sheet-flooded into the shallow valleys and transported away within these channels as loamy sands.

Vor der nördlichsten Kapkette erstreckt sich nach N die Pediplanations-Verebnung in 900m. Zur Kapkette hin steigt das Niveau ganz sanft an, in Übereinstimmung mit dem Grundwasserstrom. Die Dwyka-Schiefer und -Grauwacken (die hier noch leicht in die Faltung einbezogen wurden) werden durch die Pediplanation glatt gekappt, wie man im Steinbruch sehr schön beobachten kann.
Aus dem Vordergrund heraus erstreckt sich das Pediment zum Inselberg hinten links. Dieser wird auf allen Seiten vom Pediment (in der Form eines insgesamt flachen Schild-Pedimentes) umgeben. Vom Pediment ausgehende kleine Buchten treiben die Abtragung des 'Bergknopfes' in der Mitte des Schildpedimentes voran. Diese Buchten bewirken auf Dauer das völlige Verschwinden des Inselbergs. Sämtliche Stadien dazu können im unmittelbaren Umland der Aufnahmesituation vorzüglich beobachtet werden. Die zerfallenen Schieferplättchen und Grauwackenpartikel, die ständig noch weiter zerfallen, werden von Spülfluten in die Flachmuldentäler hinein abgewaschen und dort als Sand und Lehm weitertransportiert. Auch die Spuren dieses Prozesses sind in der Umgebung vorzüglich zu beobachten.

Foto 46. North of Vanrhynsdorp, northwestern Cape Province. Cf. Foto 9

Nördlich Vanrhynsdorf in der nordwestlichen Kapprovinz. Vgl. Foto 9

The bottom of the 'valley' to be seen in the foreground and stretching towards the background is used by a tributary of the Soutrivier, flowing towards the Olifants in the SW. The bottom itself lies in the level of the 100m 'terrace', which here ingresses - by way of a larger embayment - into the 200m 'terrace'in the background. Both levels are separated from each other by a distinct erosional scarp. After larger rainfalls the bottom is sheet-flooded and lateral erosion is enabled to work into the scarp (here consisting of easily disgranulating rocks). In this way the 100m level is still being enlarged at the cost of the next level higher up.

Der 'Talboden', den man hier erblickt, wird benützt von einem episodischen Rinnsal, das dem Soutrivier tributär ist, der seinerseits nach SW zum Olifants hin entwässert. Der 'Talboden' gehört der Verebnung der 100m-'Terrasse' an, die hier in einer größeren Bucht in das 200m-Niveau im Hintergrund eingreift. Nach größeren Regenfällen (wie kurz vor der Aufnahme) laufen auf der ganzen Breite des 'Talbodens' Spülfluten ab, so daß am Abfall (der hier aus leicht vergrusendem Gestein besteht) die Verbreiterung des unteren Niveaus auf Kosten des höheren auch heute noch weitergehen kann.

Foto 47. Silcrete in the neighbourhood of Oudtshoorn in the southern Cape Province

Kieselhartkruste im SO von Oudtshoorn in der südlichen Kapprovinz

The silcrete here forms the top of a mesa. The surface of the mesa is part of the Upper Cretaceous level in 600m. The crust has a thickness of 6m, comprising a top layer of conglomerates. It is here formed across Bokkeveld shales. A sharp groove is extant where the hard crust ends and the whitish kaolinisation horizon begins. This horizon has a thickness of 2m, part of it covered by slope débris. The kaolinite - illite ratio is 60 : 40. The crust to be seen is an example of the silcrete type I. It became developed in connection with the dammed-up groundwater table of the 600m thalassoplain.

Das Foto gibt einen Hangausschnitt wieder, der unterhalb einer Mesakante (oben gerade sichtbar) entwickelt ist. Die Mesakante selbst liegt im Bereich der 600m-Thalassoplain. Die Kieselhartkruste ist hier über anstehendem Bokkeveld-Schiefer entwickelt, der z.T. völlig intakt in die Kruste einbezogen worden ist (unten rechts am vorderen Bildrand z.B.). In den oberen Horizont der Kruste sind Konglomerate eingebacken. Mit scharfer Hohlkehle grenzt die hier insgesamt 6m mächtige Hartkruste an den Weißverwitterungshorizont darunter (im Bild 2m mächtig, im unteren Teil durch Hangschutt verdeckt). Im Bereich dieses Weißverwitterungshorizontes wurden die Schiefer kaolinisiert. Das Kaolinit-Illit-Verhältnis beträgt 60 : 40. Es handelt sich im Bild um die Kieselhartkruste vom Typ I. Verbackung und Kaolinisierung liefen im Grundwasserstaubereich der 600m-Thalassoplain ab.

Foto 48. In the basin of the Swartrivier NW of Port Elizabeth, southern Cape Province

Im Becken des Swart Riviers im NW von Port Elizabeth, südliche Kapprovinz

The foto gives a view of a slope beneath the Miocene 300m level. Two horizons can be seen: a darker one on top with silicification and a whitish one below with kaolinisation, both processes being due to a fossile groundwater table dammed up in connection with the 300m thalassoplain. Both horizons are not as thick as the ones shown in Foto 47. The upper horizon measures 1.0 to 1.5m, the lower one 0.5 to 1.0m. Silicification is not as intensive as in Foto 47 and a crust has not been formed. The layers of the rocks are better preserved than in Foto 47. The rocks here consist of TMS quartzites and sandstones on top and Bokkeveld shales below. The foto represents a good example of the silcrete type II.

Das Foto gibt einen Hangausschnitt unterhalb des miozänen 300m-Niveaus wieder. Die beiden Horizonte (Verkieselung oben, Weißverwitterung aufgrund von Vergleyung unten) sind deutlich weniger mächtig als in Foto 47. Der obere Horizont erreicht im Aussschnitt 1.0 - 1.5m, der untere 0.5 - 1.0m. Silifizierung und Vergleyung sind zudem weniger intensiv: keine Verbackung im Silifizierungshorizont (nur kleinere, isolierte 'Scherben': s. im Foto rechts oben), bessere Erhaltung des Schichtgefüges der Gesteine (hier TMS-Sandsteine und -Quarzite an der Grenze zu Bokkeveld-Schiefern darunter). Es handelt sich hier um eine Hartkruste vom Typ II.

Foto 49. Red dolerite cover on top of Ecca shales, northern Natal

Rotlehm aus Dolerit über Ecca Schiefern, Nord-Natal

The cover consists of rubefied and disgranulated dolerite particles. The shales below are partly turned into clays (local gley horizon). The red cover has obviously been spread from above (left) on to and over the lower slope section (right). Rubefaction does not advance from the cover into the shales below as the sharp line between the two clearly shows. Rubefaction is obviously bound (1) to the lithology and the geo-chemism of the dolerite particles and (2) to the topogenic situation of the lower slope section. Equivalent and very similar examples (and on top of other rocks as well) can be observed all over Natal. (The vertical cracks in the red cover are the result of a drying-up effect along the frest road cutting.)

Man erkennt, daß die rote Decke sekundär vom Hanghöheren (links) ins Hangniedere (rechts) gebreitet worden ist - zugleich mit Abnahme der Mächtigkeit von links nach rechts. Die zum größeren Teil hier vergleyten Ecca-Schiefer (lokaler Stauhorizont) werden an scharfer Linie glatt überlagert. Unter rezenten wie aber auch unter subrezenten Klimabedingungen dringt die Rubifizierungsfront nicht aus der Doleritdecke in das Gestein darunter mit dessen anderem Chemismus ein. Die Rubifizierung ist somit an den Chemismus des Ausgangsgesteines allein gebunden. Sie kommt nur im Zusammenhang mit den Doleritpartikeln - in flacher Lagerung als Hangdecke - vor. Entsprechende Aufschlüsse können in ganz Natal, auch über anderen Gesteinen, beobachtet werden. - Die senkrechte Klüftung der Rotdecke ist eine Folge der Austrocknung am frischen Anschnitt.

Foto 50. Region of the lower Kneersvlakte (between Bitterfontein and Gareis, northwestern Cape Province). View to the E

Bereich der unteren Kneersvlakte (zwischen Bitterfontein und Gareis, nordwestliche Kapprovinz). Blick nach O

The foto shows the serir-like cover of fine angular milk quartzes above the weathered Nama shales (foreground) with their typical brown colour. The shales are not completely capped everywhere by the quartz cover. The cover has an average thickness of 25 to 50cm, filling up all depressions.

Das Bild zeigt die serir-artige Feinkiesdecke aus Milchquarzen. An der Kante im Vordergrund steht der braun verwitterte Nama-Schiefer an, der gelegentlich auch die Feinkiesdecke durchragt. Die Feinkiesdecke überkleidet mit Mächtigkeiten von 25 bis 50cm die flacheren und steileren Hänge und gleicht Unebenheiten aus.

Foto 51. In the neighbourhood of Mt. Frère (northern Transkei)

Bei Mt. Frère (nördliche Transkei)

The foto shows a very thick fossile latosol with complete profile over deeply weathered Beaufort sandstones (below, in the foreground). The total height is 3m. The altitude of this roadcutting is 1250m, i.e. the level of the (Lower Cretaceous) P 2-peneplain.

Das Foto zeigt einen mächtigen, fossilen Latosol mit vollständigem Bodenprofil über Beaufort-Sandstein. Der Sandstein steht unten (Vordergrund) an, ist stark zermürbt und entfärbt (weißlich-grau). Der Aufschluß ist 3m hoch. Der Straßenanschnitt liegt in 1250 m Höhe, d.h. im Niveau der unterkretazischen P 2-Rumpffläche.

Foto 52. In the middle catchment area of the Great Fish river (eastern Cape Province)

Im mittleren Einzugsgebiet des Großen Fischflusses (östliche Kappprovinz)

The foto shows the flat 'landscape of removal' with the complete dissection of the Great Escarpment into a 'landscape of residual hills'. The layers are Stormberg sandstones with dolerite sills in between or on top. The layers have nearly no fall. They could easily be removed in connection with the groundwater table being dammed up by a risen sea level.

Das Foto zeigt eine flache Ausraumlandschaft mit Auflösung der Großen Randstufe zu einer Restberglandschaft im Bereich flachlagernder Schichten im Stormberg. Mehr oder weniger mächtige Doleritbänke sind in die Sandsteine eingeschaltet oder überlagern sie. Der flächenhaft weite Ausraum entwickelte sich im Bereich des Grundwasserhorizontes im Aufstaubereich eines angestiegenen Meeresspiegels.

Foto 53. Barberton basin (eastern Transvaal). View to the NW

Becken von Barberton (Osttransvaal). Blick nach NW

Though 'landscapes of wide removal' are especially well developed on layered rocks and in subsequential zones, they can also occur on crystalline rocks, as in the Barberton basin, when and where a persistent groundwater table could be formed in connection with a thalassoplain (here the one in 600m). - Cf. fotos 47 and 48.

Zwar sind Ausraumlandschaften in geschichteten Gesteinen besser entwickelt (und dort besonders in Subsequenzzonen), doch kommen sie durchaus auch im kristallinen Grundgebirge vor, wie hier im Becken von Barberton, wo sich - hier im Zusammenhang mit der Thalassoplain in 600m - ein Grundwasserstau herausbilden konnte.

Foto 54. In the middle catchment area of the Sabie river (eastern Transvaal). View to the W and NW

Am mittleren Sabiefluß (Osttransvaal). Blick nach W und NW

The 'terraces' of the thalassoplains reach inland, forming deeply ingressing embayments, today followed by rivers, as here, for instance, by the Sabie. By later erosion, the 'terraces' became shaped into a kind of 'corrugated iron landscape' with numerous gullies, especially where the 'terraces' were formed over the crystalline basement in front of the giant cuestas (one of them to be seen in the background). As the gullies show, the 'terraces' here are fossile forms. No forming-on processes are visible today, in spite of the marginally tropical climate. The Lowveld peneplain reaches up into the Sabie valley (and into the valleys of some tributaries), but it has not been able so far to consume the scarps between one 'terrace' and the next.

Die Niveaus der Thalassoplains sind hier terrassen- bzw. trogflächenartig ausgebildet. Dem somit älteren Trog folgt der Sabie-Fluß. Die 'Terrassen' (höhenkonstant) sind durch die junge Rillenerosion in ein wellblechartiges Relief umgestaltet worden, besonders im kristallinen Grundgebirge vor den Riesen-Schichtstufen (eine im Hintergrund sichtbar).

Eine rumpfflächenhafte Verbreiterung greift vom Lowveld aus in den Talraum des Sabie hinein sowie in dessen untere Seitentäler, vermag aber nicht die Stufen zwischen den einzelnen Trogniveaus (Thalassoplains) zu überspringen. Die intensive Gullybildung zeigt, daß die Trogniveaus heute zerschnitten und nicht mehr weitergebildet werden, trotz des marginal tropischen Klimas.

Foto 55. In the hinterland of Margate and Ramsgate (southern Natal). View to the N and NE

Im Hinterland von Margate und Ramsgate (Südnatal). Blick nach N und NO

The foto shows the crystalline basement in front of a TMS scarp (cf. Foto 4). The Indian ocean will be seen in the background. Coastal 'terraces' were planed across the basement in altitudes of 200m and 300m. They have become intensely dissected by the many streams and their meanders. Along the streams the flat lower sections of the slopes do not merge into each other and thus do not form a fluvial terrace.

Vor der TMS-Stufe (vgl. Foto 4) erstreckt sich das kristalline Grundgebirge zum Indischen Ozean (Hintergrund). Über das Grundgebirge wurden die 'Terrassen' in 200 und 300m Höhe eingeschnitten, sind inzwischen allerdings von den vielen Gerinnen und ihren Mäandern intensiv zerkerbt worden. Die unteren Hangabschnitte der Kerben verschneiden sich in unterschiedlichen Reliefsituationen. Fluviale Terrassen sind nicht ausgebildet.

BIBLIOGRAPHY

Abel, H. (1954) Geomorphologische Untersuchungen im Kaokoveld (Südwestafrika). Verh. d. Dt. Geogr.tages Essen 1953, 115 - 125. Wiesbaden

Abel, H. (1954) Beiträge zur Landeskunde des Kaokoveldes. Dt. Geogr. Blätter, 47, H. 1 - 2. Bremen

Abel, H. (1955) Beiträge zur Landeskunde des Rehobother Westens. Mitt. d. Geogr. Ges. Hamburg, 51, 55 - 97. Hamburg

Abel, H. (1959) Beiträge zur Morphologie der Großen Randstufe im südwestlichen Afrika. Dt. Geogr. Blätter, 48, 131 - 268. Bremen

Ahnert, F. (1970) Functional relationships between denudation, relief anduplift in large mid-latitude drainage basins. Am. J. Sci, 168, 243 - 263

Beater, B.E., Maud, R.R. (1960) The occurrence of an extensive fault system in S. E. Zululand and its possible relationship to the evolution of part of the coastline in South Africa. Trans. Geol. Soc. S-Afr, 63, 51 - 60

Beetz, P.F.W. (1934) Geology of South West Angola, between Cunene and Lunda axis. Trans. Geol. Soc. S-Afr, 35, 137 ff

Beetz, P.F.W. (1950) Die Geheimnisse des Kunenelaufes in Südangola und Südwestafrika. Erdkunde, 4, 43 - 54

Besler, H. (1980) Die Dünen - Namib. Stutt. Geogr. Stud., Vol. 96. Stuttgart

Bibus, E. (1984) Reliefentwicklung im Rheinischen Schiefergebirge - neue Befunde, neue Probleme. Jungtertiär und Quartär.Verh. d. 44. Dt. Geogr.tages Münster, 1983, 74 - 76. Wiesbaden

Biesiot, B.G. (1957) Miocene foraminifery from the Uloa sandstone, Zululand. Trans. Geol. Soc. S-Afr., 60, 61 - 80

Birch, G.F. (1978) The distribution of clay minerals on the continental margin of the west coast of South Africa. Trans. Geol. Soc. S-Afr., 81, 23 - 34

Bird, E.C.F. (1984) Coasts. Oxford

Birkenhauer, J. (1970a) Der Klimagang im Rheinischen Schiefergebirge und in seinem näheren und weiteren Umland zwischen dem Mitteltertiär und dem Beginn des Pleistozäns. Erdkunde, 24, 268 - 274

Birkenhauer, J. (1970b) Zur Talgeschichte des unteren und mittleren Nahegebiets. Decheniana, 123, 1 - 18. Bonn

Birkenhauer, J. (1970c) Vergleichende Betrachtung der Hauptterrassen in der rheinischen Hochscholle. Festschr.f. K. Kayser = Kölner Geogr. Arb., Sonderband, 99 - 140. Wiesbaden

Birkenhauer, J. (1973) Die Entwicklung des Talsystems und des Stockwerkbaus im zentralen Rheinischen Schiefergebirge zwischen dem Mitteltertiär und dem Altpleistozän. Arb. z. Rhein. Landeskunde, H. 34, Bonn

Birkenhauer, J. (1983) Tal- und Höhenrelief der deutschen Mittelgebirge. Geogr. Rundschau, Vol. 35, 27 - 34

Birkenhauer, J. (1987) Höhere, im marinen Niveau entstandene Verflachungen in Mitteleuropa und in Südafrika. In: Aktuelle geomorphologische Feldforschung = Gött. Geogr. Abh. 84, 19 - 29

Bleich, K.E. (1960) Über das Alter des Albtraufs. Jh. Ver. Vaterländ. Naturkunde Württ., 115, 39 - 92. Stuttgart

Blümel, W.P. (1981) Pedologische und geomorphologische Aspekte der Kalkkrustenbildung in Südwestafrika und Südostspanien. Karlsr. Geogr. Hefte, 10

Bourdon, M.; Magnier, P. (1969) Notes on the Tertiary fossils at Birbury, Cape Province. Trans. Geol. Soc. S-Afr., 72, 123 - 125

Bremer, H. (1965) Ayers Rock - ein Beispiel für klimagenetische Morphologie. Zeitschr. f. Geom., N.F. 9, 249 - 284

Bremer, H. (1975) Intramontane Ebenen - Prozesse der Flächenbildung. Zeitschr. f. Geom., Suppl.-bd. 25, 26 - 48

Bremer, H. (1978) Zur tertiären Reliefgenese der Eifel. In: Bremer, H., Pfeffer, K.H. (Ed., 1978) Zur Landschaftsentwicklung der Eifel. Kölner Geogr. Arb., 36, 195 - 255

Bremer, H. (1981)Reliefformen und reliefbildende Prozesse in Sri Lanka. In: Bremer, H. et al. (Ed., 1981) Zur Morphogenese der feuchten Tropen = Relief - Böden - Paläoklima, 1, 7 - 184

Büdel, J. (1957) Grundzüge der klimamorphologischen Entwicklung Frankens. Würzb. Geogr. Arb., 415, 5 - 46

Büdel, J. (1958) Die Flächenbildung in den feuchten Tropen und die Rolle fossiler solcher Flächen in anderen Klimazonen. Verh. d. 31. Dt. Geogr.tages Würzburg, 1957, 89 - 121. Wiesbaden

Büdel, J. (1977) Klima-Geomorphologie. Berlin

Bubnoff, S. von (1948) Einführung in die Erdgeschichte. 2 vol. Halle

Butzer, K.W., Helgren, D.M. (1972) Late cenozoic evolution of the Cape Coast between Knysna and Cape St. Francis, South Africa. Quaternary research, 2, 143 - 169

Climate of South Africa. (1951) Pretoria

Cloos, H. (1919) Der Erongo. Beitr. z. Erforsch. d. Dt. Schutzgebiete. Berlin

Cloos, H. (1937) Zur Großtektonik Hochafrikas. Geol. Rundsch., 28, 333

Cooks, J., De Villiers, G. du T. (1979) The polycyclic nature of the Umbilo river drainage basin tested by hypsometric analysis. Journ. Univ. Durban-Westville, vol. 3, 347 - 353

Credner, W. (1931) Das Kräfteverhältnis morphogenetischer Faktoren und ihr Ausdruck im Formenbild Südost-Asiens. Bull. Geol. Soc. China, 11, 138 - 145. Peiping

De Swardt, A.M.H., Bennet, G. (1974) Structural and physiographic development of Natal since the Late Jurassic. Trans. Geol. Soc. S-Afr., 77, 309 - 322

Dingle, R.V., Siesser, W.G., Newton, A.R. (1983) Mesozoic and Tertiary geology of Southern Africa. Rotterdam

Dingle, R.V., Hendey, Q.B. (1984) Late Mesozoic and Tertiary sediment supply to the eastern Cape Basin (SE Atlantic) and paleo-drainage systems in south western Africa. Marine Geology, 56, 13 - 26

Dixey, F. (1942) Erosion cycles in central and southern Africa. Trans. Geol. Soc. S-Afr., 45, 151 - 180

Dixey, F. (1944) Contribution to the geomorphology of the Natal Drakensberg. Trans. Geol. Soc. S-Afr., 47, 125 - 133

Dongus, H. (1972) Schichtflächenalb, Kuppenalb, Flächenalb (Schwäbische Alb). Zeitschr. f. Geomorph., N.F. 16, 374 - 392

Douglas, J. (1980) Climatic geomorphology. Zeitschr. f. Geomorph., Suppl.bd., 36, 27 - 47

Drury, G.H. (1971) The face of the earth. 2nd edition. Hammondsworth

Du Toit, A.L. (1920) The geology of Pondoland. Pretoria

Du Toit, A.L. (1922) The evolution of the South African coast line. Afr. Geogr. Journ., 1923, 5 - 13

Du Toit, A.L. (1933) Crustal movement as a factor of the geographical evolution of South Africa. S-Afr. Geogr. Journ., 1933

Du Toit, A.L. (1954) The geology of South Africa. 3rd. edition. Edinburgh

Dürr, E. (1970) Kalkalpine Sturzhalden und Sturzschuttbildung in den westlichen Dolomiten. Tüb. Geogr. Stud., H.37. Tübingen

Faber, F.J. (1926) Bijdrage tot the Geologie van Zuid-Angola (Afrika). Delft

Fabry, R. (1924) Die Entstehung der Durchbruchstäler im Gouritz-Flußsystem in Südafrika. Mitt. d. Geogr. Ges. München, 17, 82 - 96

Fair, T.J.D. (1948) Slope formation and development in the interior of Natal, South-Africa. Trans. Geol. Soc. S-Afr., 50, 105 - 108

Fair, T.J.D., King, L.C. (1954) Erosional land-surfaces in the eastern marginal areas of South Africa. Trans. Geol. Soc. S-Afr., 57, 19 - 26

Flemming, N.C., Roberts, D.G. (1973) Tectono-eustatic changes in sea level and sea floor spreading. Nature, 243, 19 - 22

Frankel, J.J. (1960) The geology along the Umfolozi River, south of Mtubatuba, Zululand. Trans. Geol. Soc. S-Afr., 63, 231 - 252

Frankel, J.J. (1968) Tertiary sediments in the lower Umfolozi River valley, Zululand. Trans. Geol. Soc. S-Afr., 71, 135 - 145

Frankel, J.J., Kent, L.E. (1937) Grahamstown surface quartzites (silcretes). Trans. Geol. Soc. S-Afr., 10, 1 - 42

Frankenberg, P., Lauer, W. (1981) Eine Karte der hygrothermischen Klimatypen Afrikas. Erdkunde, 35, 245 - 248

German, R. (1965) Morphologie des Traufes der schwäbischen Alb. = Festschr. f. G. Wagner. = Jh. geol. LA B-W, 7, 463 - 474. Freiburg

Gevers, T.W. (1933) Zur Gliederung des Grundgebirges im Windhoeker Bezirk Südwestafrikas. Geol. Rdsch., 24, 285 - 297

Gevers, T.W. (1933) Zur Tektonik des mittleren Südwestafrika. Geol. Rdsch., 24, 237 - 248

Gevers, T.W. (1934) Untersuchungen des Grundgebirges im westlichen Damaraland. N. Jb. f. Min., Geol. Pal., Part

I: 72. Beil. bd. Abt. B, 283 - 325; Part II: 73. Beil. bd., Abt. B, 27 - 41

Gevers, T.W. (1936) The morphology of western Damaraland and the adjoining Namib Desert of South West Africa. Geogr. Journ. 19, 61 - 79

Gevers, T.W. (1961) Geology along the northwestern margin of the Khomas Highlands between Otjimbingwa-Karibib and Okahandja. Trans. Geol. Soc. S-Afr., 66, 199 - 251

Gevers, T.W., Frommurze, H.F. (1929) The geology of northwestern Damaraland in South West Africa. Trans. Geol. Soc. S-Afr., 32, 31 - 55

Gierloff-Emden, H.G. (1979) Geographie des Meeres. 2 vol. Berlin

Grees, H. (1983) Die Schwäbische Alb. In: Borcherdt, C. (Ed., 1983) Geographische Landeskunde von Baden-Württemberg. 310 - 339. Stuttgart

Hagedorn, J., Brunotte, E. (1983) Flächen- und Talentwicklung im südöstlichen Kapland/Südafrika und ihre Faktoren. Zeitschr. f. Geom., Suppl.bd., 48, 235 - 246

Hancock, J.M., Kauffman, E.G. (1979) The great transgressions of the Late Cretaceous. Journ. Geol. Soc., 136, 175 - 186

Handbuch ausgewählter Klimastationen der Erde. (1979). Trier

Haughton, S.H. (1925) The Tertiary deposits of the south eastern districts of the Cape Province. Trans. Geol. Soc. S-Afr., 28, 27 - 32

Haughton, S.H. (1930) Note on the occurrence of Upper Cretaceous marine beds in South West Africa. Trans. Geol. Soc. S-Afr., 33, 61 - 63

Haughton, S.H. (1969) Geological history of southern Africa. Cape Town

Haughton, S.H., Fromurze, H.F. (1928) The Karoo beds of the Warmbad district, South West Africa. Trans. Geol. Soc. S-Afr., 30, 133 ff.

Hays, J.D., Pitman III, W.C. (1973) Lithospheric plate motion, sea level changes and climatic and ecological consequences. Nature, 246, 18 - 22

Helgren, D.M., Butzer, K.W. (1977) Paleosols of the southern Cape coast, South Africa. Geogr. Review, 67, 430 - 445

Helgren, D.M. (1979) Rivers of diamonds. Chicago

Herz, N. (1977) Timing of spreading in the South Atlantic: information from Brazilian alkali rocks. Geol. Soc. Am. Bull., 88, 101 - 112

Hövermann, J. (1978) Formen und Formung in der Pränamib (Flächennamib). Zeitschr. f. Geom., Suppl. Bd. 30, 55-73

Hüser, K. (1976) Namibrand und Erongo. Karlsruher Geogr. Hefte 9, Karlsruhe

Jaeger, F. (1923) Die Grundzüge der Oberflächengestaltung von Südwestafrika. Zeitschr. Gesellsch. Erdk. Berlin, 1923, 14 - 24

Jaeger, F. (1930) Probleme der Großformen Afrikas. = G. Wagner Gedächtnisschr. = Pet. Geogr. Mitteil., Erg.h. 209, 136 - 146. Gotha

Jaeger, F. (1932) Zur Morphologie der Fischflußsenke in Südwestafrika. Festschr. f. C. Uhlig, 30 - 37. Öhringen

Jessen, O. (1932) Berichte über seine Forschungsreise nach Angola. Mitt. d. Geogr. Ges. München, 25, 82 - 90

Jessen, O. (1936) Reisen und Forschungen in Angola. Berlin

Jessen, O. (1943) Die Randschwellen der Kontinente. Pet. Geogr. Mitt., Erg.h. 241. Gotha

Jutta's Magister Atlas. 2nd edition. London

Kalpagè, F.S.C.P. (1974) Tropical soils. London

Kayser, K. (1970) Namib-Studien. In: Wilhelmy, H. (ed., 1970) Deutsche geographische Forschung in der Welt von heute. = Festschr. f. E. Gentz., 181 - 186. Kiel

Kayser, K. (1986) Geomorphologie - Südafrika = Afrika Kartenwerk, Serie S, H. 2. Berlin

Kent, L.E. (1938) The geology of a portion of Victoria County, Natal. Trans. Geol. Soc. S-Afr., 41, 1 ff.

Kennedy, W.J., Cooper, M. (1975) Cretaceous ammonite distribution and the opening of the South Atlantic. Journ. Geol. Soc. London, 131, 283 - 288

King, L.C. (1962, 1967) Morphology of the earth. London

King, L.C. (1963) South African scenery. 3rd edition. Edinburgh

King, L.C. (1970) Uloa revisited. Trans. Geol. Soc. S-Afr., 73, 151 - 158

King, L.C. (1972) Pliocene marine fossils from the Alexandria Formation in the Paterson district, eastern Cape Province, and their geomorphic significance. Trans. Geol. Soc. S-Afr., 72, 159 - 160

King, L.C. (1972) The Natal monocline. Durban

King, L.C. (1982) The Natal monocline. 2nd revised edition. Pietermaritzburg

Klebelsberg, R. von (1922) Die Hauptflächensysteme der Ostalpen. Wien

Knetsch, G. (1940) Zur Frage der Küstenbildung und der Entwicklung des Oranje-Tals in Südwestafrika. Sonderveröffentl. III, Geogr. Ges. Hannover. Hannover

Koeppen, W., Geiger, F. (1961) Weltklimakarte. Darmstadt

Korn, H., Martin, H. (1937) Die jüngere und klimatische Geschichte Südwestafrikas. Zentr.bl. f. Min., Geol., Pal., 1937, Abt. B, 456 ff.

Krenkel, F. (1928) Geologie Südafrikas. Vol. II. Stuttgart

Krige, A.J. (1926) An examination of the Tertiary and Quaternary changes of sea level in South Africa. Ann. Univ. Stellenbosch, vol. V, A. 1

Kuntz, J. (1912) Erläuterungen zur Karte des Kaokoveldes. Mitt. aus d. dt. Schutzgebieten, vol. 25, 233 - 235

Kuntze, H., Niemann, J., Roeschmann, G. (1981) Bodenkunde. Stuttgart

Larson, R.L., Ladd, J.W. (1973) Evidence for the opening of the South Atlantic in the early Cretaceous. Nature, 246, 209 - 211

Lock, B.E. (1973) Tertiary limestones at Needs Camp, near East London. Trans. Geol. Soc. S-Afr., 76, 1 - 5

Louis, H. (1964) Über Rumpfflächen- und Talbildung in den wechselfeuchten Tropen, besonders nach Studien in Tansania. Zeitschr. f. Geom., N.F. 8, Sonderheft 1964, 43 - 70

Louis, H., Fischer, K. (1979) Allgemeine Geomorphologie. 4th edition. Berlin

Maack, R. (1923) Der Brandberg. Zeitschr. Ges. Erdk. Berlin, 1923, 1 - 14

Mabbut, J.A. (1955) Erosion surfaces in Namaqualand. Trans. Geol. Soc. S-Afr., 58, 1 - 30

Mabbut, J.A. (1957) Pediment landforms in Little Namaqualand. S-Afr. Geogr. Journ., 121, 77 - 82

McCarthey, T.S., Moon, B.P., Levin, M. (1985) Geomorphology of the western Bushmanland plateau, Namaqualand, South Africa. S-Afr. Geogr. Journ., 67, 160 - 178

McCarthy, M.J. (1969) Stratigraphical and sedimentological evidence from the Durban region of major sea level movements since the Later Tertiary. Trans. Geol. Soc. S-Afr., 70, 135 - 165

Machatschek, F. (1955) Das Relief der Erde. 2nd vol. Berlin

McLachlen, J.R., Brenner, P.W., McMillan, J.K. (1976) Micropalaeontological study of Cretaceous beds of Mbotyi and

Muganzane, Transkei, South Africa. Trans. Geol. Soc. S-Afr., 79, 321 - 340

McLachlen, J.R., Brenner, P.W., McMillan, J.K. (1976) The stratigraphy and micropalaeontology of the Cretaceous Brenton Formation and the PB-A/1 well near Knysna, Cape Province. Trans. Geol. Soc. S-Afr., 79, 341 - 370

Marker, M., Sweeting, M.M. (1983) Karst development in the Alexandria Limestone Formation, Eastern Cape Province, South Africa. Zeitschr. f. Geom., N.F. 27, 21 - 38

Martin, H. (1968) Paläomorphologische Formenelemente in den Landschaften Südwestafrikas. Geol. Rdsch., 58, 121 - 128

Martin, H. (1972) The Atlantic margin of southern Africa between latitude 17° south and the Cape of Good Hope. In: Nairn, A.E.M., Stehli, F.G. (1972) The ocean basins and margins. vol. 1, 277 - 300

Martin, H. (1983) Der Große Brukkaros. Journal, vol. 36/37, SWA Wiss. Gessellschaft, 7 - 10, Windhoek

Martin, A.K., Goodlad, S.W., Salmon, D.A. (1982) Sedimentary basin infill in the northernmost Natal Valley, hiatus development and Agulhas Current paleooceanography. Journ. Geol. Soc., 139, 183 201

Maud, R.R. (1961) A preliminary review of the structure of coastal Natal. Trans. Geol. Soc. SAfr., 64, 247 256

Maud, R.R. (1968) Quarternary geomorphology and soil formation in coastal Natal. Zeitschr. f. Geom., Suppl.bd 7, 155 ff.

Maud, R.R. (1968) Further observations on the laterites of coastal Natal, South Africa. Intern. Congr. Soil Sci., 4, 151 158

Maud, R.R. (1971) The occurrence and properties of ferricretes in Natal. Soil Mech. Found. Engl. Reg. Conf. Afr., Procedings, 5, vol. 1, 3 5

Maud, R.R., Orr, W.N. (1975) Aspects of post Karoo geology in the Richards Bay area. Trans. Geol. Soc. SAfr., 78, 101 - 190

Meyer, R. (1967) Studien über Inselberge und Rumpfflächen in Nordtransvaal. Münchn. Geogr. Hefte, 31. Kallmünz

Moon, B.P., Selby, M.J. (1983) Rock mass strength and scarp form in southern Africa. Geogr. Annaler, 65, A, 135 145

Nance, R., Worsley, R., Moody, B. (1988) Der Superkontinent-Zyklus. Spektrum der Wissenschaft, 9/88, 80 - 87

Nance,R., Worsley, R., Moody, B. (1988) The super-continent cycle. Scientific American, 1988, July, 44 - 51

Obst, E., Kayser, K. (1949) Die Große Randstufe auf der Ostseite Südafrikas. Hannover

Olbert, G. (1975) Talentwicklung und Schichtstufenmorphogenese am Südrand des Odenwaldes. Tüb. Geogr. Stud., H. 31, Tübingen

Ollier, C.D. (1981) Tectonics and landforms. London

Ollier, C.D. (1985) Morphotectonics of passive continental margins: introduction. Zeitschr. f. Geom., Supplbd. 54, 1 9, Berlin

Ollier, C.D., Marker, M.E. (1985) The Great Escarpment of southern Africa. Zeitschr. f. Geom., Supplbd. 54, 37 56

Penck, A. (1908) Der Drakensberg und der Quathlambabruch. Sitz.ber. d. Kgl. Preuß. Akademie d. Wiss., 9, 230 258

Pfeffer, K.H. (1984) Die präquartäre Entwicklung des Reliefs der westlichen Eifel. Verh. d. 44. Dt. Geogr.tages Münster 1983, 79 83. Wiesbaden

Pitman, W.V., Middleton, B.J., Midgley, D.C. (1981) Surface water resources of South Africa. vol. VI: The eastern escarpment. Parts 1 a, 2. Report No. 9/81 of the Water Research Commission. Johannesburg

Renk, W. (1977) Die räumliche Struktur und Genese der Bodendecke im Bereich der Großen Randstufe Transvaals und Swasilands. Diss. Kiel

Rhodes, R.C., Leith, M.J. (1967) Lithostratigraphic zones in the TMS of Natal. Trans. Geol. Soc. SAfr., 70, 15 28

Rogers, A.W. (1903) The geological history of the Gouritz river system. Trans. Geol. Soc. S-Afr., 14

Rogers, A.W. (1920) Anniversary address. Trans. Geol. Soc. S-Afr., 23, XIX - XXXIII

Rogers, A.W. (1928) The geology of the Union of South Africa. = Handbuch der regional. Geol., vol. 7, 1 - 8

Rust, U. (1970) Beiträge zum Problem der Inselberglandschaften aus dem mittleren Südwestafrika. Hamb. Geogr. Stud., 23, Hamburg

Rust, U. (1987) Geomorphologische Forschungen im südwestafrikanischen Kaokoveld zum angeblichen vollariden quartären Kernraum der Namibwüste. Erdkunde 41, 118 - 133

Rust, U., Wieneke, F. (1973) Grundzüge der quartären Reliefentwicklung der zentralen Namib, Südwestafrika. Journ. SWA Wiss. Gesellsch., 27, 5 - 30

Rust, U., Wieneke, F. (1980) A reinvestigation of some aspects of the evolution of the Kuiseb river valley upstream of Gobabeb, South West Africa. Madoqua, vol. 12, 163 - 173

Rust, J., Reddering, K. (1983) Sedimentology of Keurbooms estuary and Robbery Formation. Sedplett 83. Port Elizabeth

Schieber, M. (1983) Bodenerosion in Südafrika. Gieß. Geogr. Schr., 51. Gießen

Schofield, J.C. (1968) Palaeogeogr., Palaeoclim., Palaeoecol. vol. 5, 141. Amsterdam

Schupp, H. (1962) Zur Morphologie des mittleren Westrichs. Mitteil. d. Polichia, 123, 75 - 196

Schwarz, E.H.C. (1906) Coast ledges in the south west of the Cape Colony. Quart. Journ. Geol. Soc.

Schwarz, E.H.C. (1916) Post Jurassic earth movements in South Africa. Geol. Mag. Decade V, vol. IX

Schwarzbach, M. (1974) Das Klima der Vorzeit. 3rd. edition. Stuttgart

Semmel, A. (1966) Zur Entstehung von Flächen und Schichtstufen im nördlichen Rhönvorland. Verh. Dt. Geogr.-tag Bochum 1965, 340 - 350. Wiesbaden

Semmel, A. (1984) Reliefentwicklung im Rheinischen Schiefergebirge. Neue Befunde, neue Probleme. Zur präquartären Entwicklung. Verh. d. 44. Dt. Geogr.-tages, Münster 1983, 71 - 74. Wiesbaden

Shone, R.W. (1978) A case for lateral gradation between the Kirkwood and the Sundays River Formation. Algoa Basin. Trans. Geol. Soc. S-Afr., 81, 319 - 326

Siesser, W.G. (1980) Late Miocene origin of the Benguela upwelling system off northern Namibia. Nature, 208, 283 - 285

Siesser, W.G., Dingle, R.V. (1981) Tertiary sea level movements around South Africa. Journ. Geol. Soc., 89, 83 - 96

Siever, R. (1983) Die dynamische Erde (The dynamic earth), Spektrum der Wissenschaft, 11, 1983, 22 - 32 (= Scientific American, German Edition)

Spreitzer, H. (1963) Die zentrale Namib. Mitt. d. Österreich. Geogr. Ges., 105, 340 - 356

Spreitzer, H. (1966) Landschaft und Landformung in der zentralen Namib. Nova Acta Leopoldina, N.F. 31, 131 ff.

Stein, W. (1981) Berechnung des quartären Abraums unterhalb der 300ft-Isohypse im immerfeuchten Südwesten und wechselfeuchten Südosten der Insel Sri Lanka. In: Bremer, H., Schnütgen, A., Späth, H. (1981; ed.) Zur Morphogenese der feuchten Tropen: Verwitterung, Boden- und Reliefbildung am Beispiel von Sri Lanka. = Relief, Boden, Paläoklima, H. 1. Berlin

Sueß, E. (1885) Das Antlitz der Erde, Vol. I, Wien

Tankard, A.J. (1975) The late Cenozoic history and palaeoenvironment of the coastal margin of the south-western Cape

Province, South Africa. Ph. D. thesis unpubl., Rhodes University, Grahamstown

Tankard, A.J. (1975) The marine neogene Saldanha Formation. Trans. Geol. Soc. S-Afr., 78, 257 - 264

Tankard, A.J. (1976) Pleistocene history and coastal morphology of the Ysterfontein-Elandsbay area, Cape Province. Ann. S-Afr. Mus., 69, 73 - 119

Tankard, A.J. (1976) Cenozoic sea level changes: a discussion. Ann. S-Afr. Mus., 71, 1 - 17

Tankard, A.J., Jackson, M.P.A., Eriksson, K.A., Hobday, D.H., Hunter, D.R., Minter, W.E.L. (1982, ed.) Crustal evolution of southern Africa. Berlin

Thompson, A.O. (1942) The coastal terraces of eastern Pondoland. Trans. Geol. Soc. S-Afr., 45, 37 - 53

Tooley, M.J. (1978) Sea level changes. Oxford

Toy, T. (1977) Hillslope form and climate. Geol. Soc. Am. Bull., 88, 16 - 22

Turner, J.L. (1967) The mapping and study of certain geomorphological features of Natal and its adjacent region. Mag. thesis, unpubl., University of Durban, Dept. of Geography. Durban

Twidale, C.R. (1978) On the origin of Ayers Rock, Central Australia. Zeitschr. f. Geom., Suppl.bd., 31, 177 - 206

Tyson, P.D. (1987) Climatic change and variability in southern Africa. Cape Town

Vageler, P. (1920) Beobachtungen in Südwestangola und im Ambolande. Zeitschr. d. Gesellsch. f. Erdk., Berlin, 1920, 179 - 193

Vail, P.R. et al. (1977) Seismic stratigraphy and global changes of sea level. In: Payton, C.E. (ed. 1977) Sesmic stratigraphy. Mem. Amer. Assoc. Pet. Geol., 26, 49 -212

Vail, P.R., Hardenbol, J. (1977) Sea level changes during the Tertiary. Oceanus. 71 - 79

Valeton, I. (1983) Klimaperioden lateritischer Verwitterung und ihr Abbild in den synchronen Sedimentationsräumen. Zeitschr. d. dt. geol. Gesellsch., 134, 413 - 452

Valeton, I. (1983) Palaeoenvironment of lateritic bauxites with vertical and lateral differentiation. Wilson C. (Ed.), (1983). Residual deposits. Geol. Soc. Spec. Publ., 11, 77 - 90. London

Van der Eyck, J.J., McIvor, C.N., De Villiers, J.M. (1969) soils of the Tugela basin. The Town and Regional Planning Commission, Pietermaritzburg

Van der Merwe, D.H. (1962) Weathering of various types of Karoo dolerite. Trans. Geol. Soc. S-Afr., 65, 119 - 132

Van der Merwe, C.R. (1962) Soil groups and sub-groups of South Africa. Sci. Bull., 2nd edition. Pretoria

Vogel, J.C. (ed., 1983) Late Cainozoic palaeoclimates of the southern hemisphere. Amsterdam

Wagner, G. (1924) Über das Zurückweichen der Stufenränder in Schwaben und Franken. Jb. u. Mitteil. d. Oberrh. geol. Ver., NF 13, 170 - 175. Freiburg

Wagner, G. (1960) Einführung in die Erd- und Landschaftsgeschichte. 2nd edition. Öhringen

Wagner, G. (1958) Die schwäbische Alb. Essen

Ward, J.D. (1982) Aspects of a suite of Quaternary conglomeraticsediments in the Kuiseb valley, Namibia. In: Coetzee, J.A., Van Zinderen Bakker, E.N., (ed., 1982) Paleoecology of Africa and the surrounding islands. Vol. 15, 211 - 216, Rotterdam

Ward, J.D. (1983) On the antiquity of the Namib. S. Afr. Journ. of Sci., 79, 175 - 183

Ward, J.D. (1984) A reappraisal of the Cenozoic stratigraphy in the Kuiseb valley, of the central Namib desert. In: Vogel, J.C. (ed., 1984) Late Cainozoic palaeoclimates of the southern hemisphere, 455 - 463. Sasqua Intern. Symposium. Proceedings. Rotterdam

Ward, J.D., Von Brunn, V. (1985) Geological history of the Kuiseb valley west of the escarpment. In: Huntley, B.J. (ed., 1985): The Kuiseb environment: The development of a monitoring base-line. S. Afr. Nat. Sient. Progr., Report No. 106, 21 - 25

Watson, A., Williams, P.D. (1985) Early Pleistocene river gravels in Swaziland and their geomorphological and structural significance. Zeitschr. f. Geom. N.F., 29, 71 - 87

Wellington, J.H. (1955) Southern Africa. A geographical study. vol.1, Cambridge

Windley, B.F. (1984) The evolving continents. 2nd edition, Chichester

Wirthmann, A. (1981) Täler, Hänge und Flächen in den Tropen. Geoökodynamik, 2, 165 - 204

Wirthmann, A. (1983) Lösungsabtrag von Silikatgesteinen und Tropengeomorphologie. Geoökodynamik, 4, 149 - 172

Wißmann, H. von (1950) Über seitliche Erosion. Colloqu. Geogr., vol. 1, Bonn

Yair, A., Gerson, R. (1974) Mode and rate of escarpment retreat in an extremly arid environment. Zeitschr. f. Geom., Suppl. bd., 21, 202 - 215

Young, A. (1975) Slopes. London (2nd edition)

Fotos

For explanations see page 305 ff.

Erläuterungen zu den Fotos S. 305 ff.

2

3

5

8

10

11

12

13

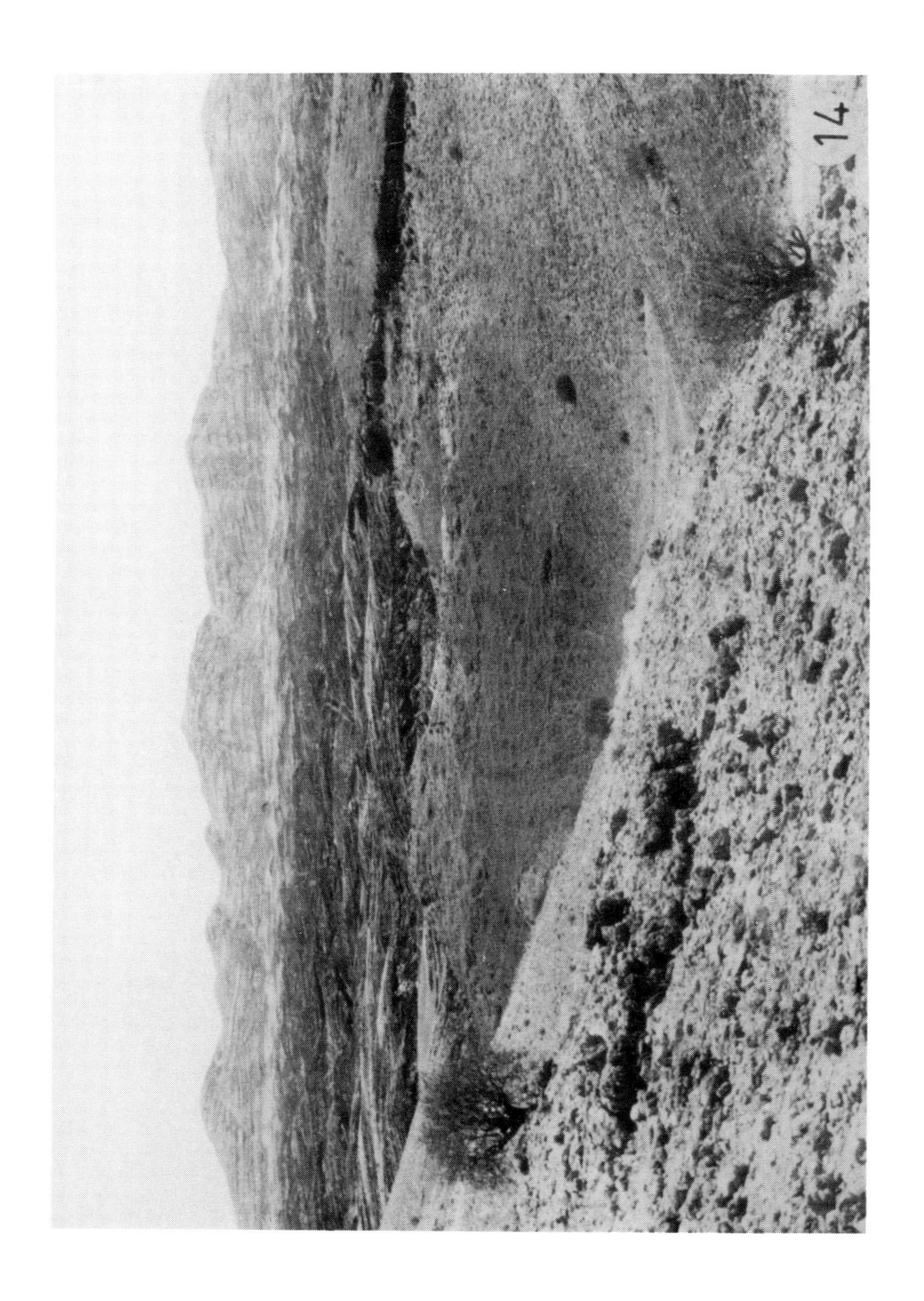
14

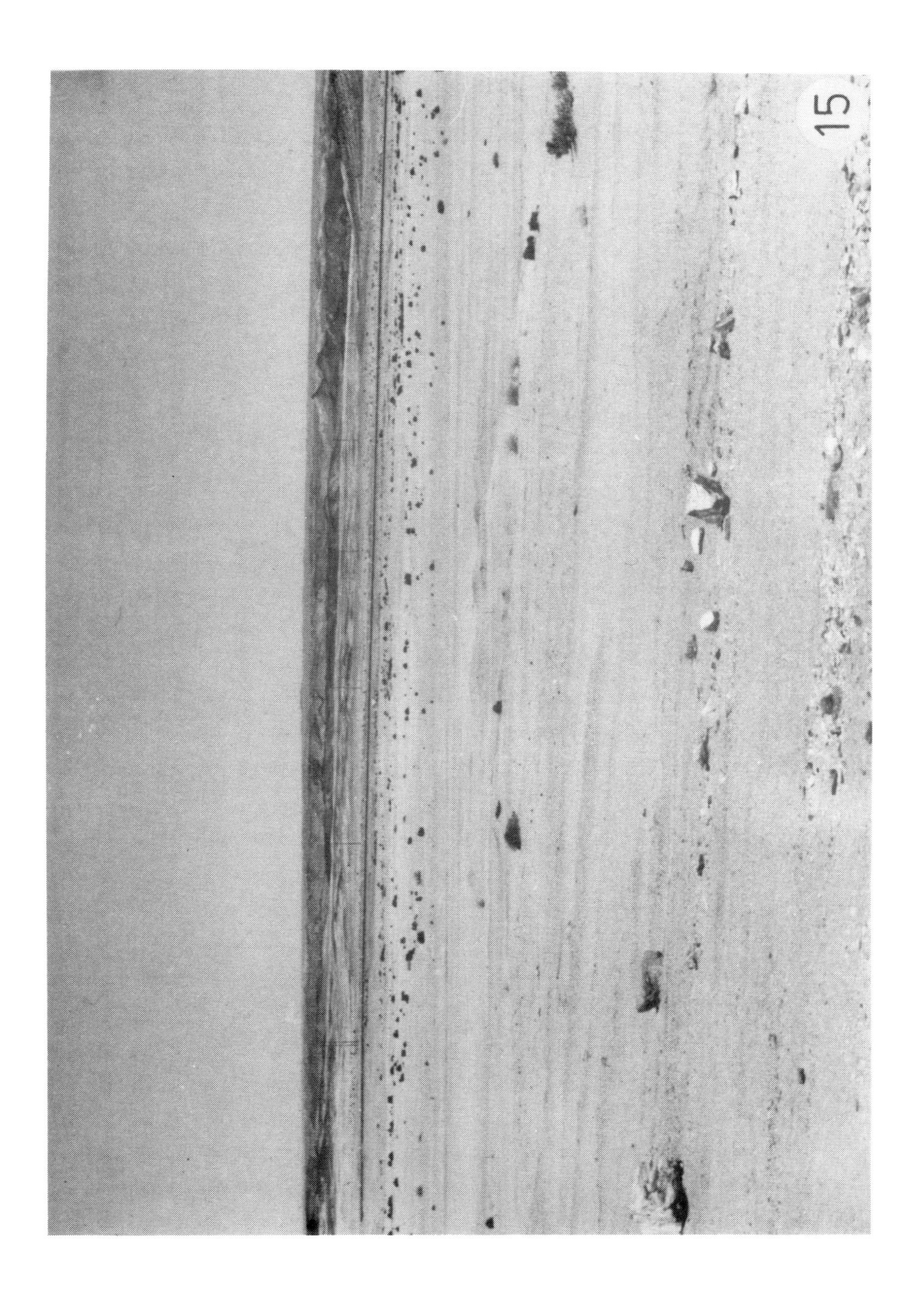
15

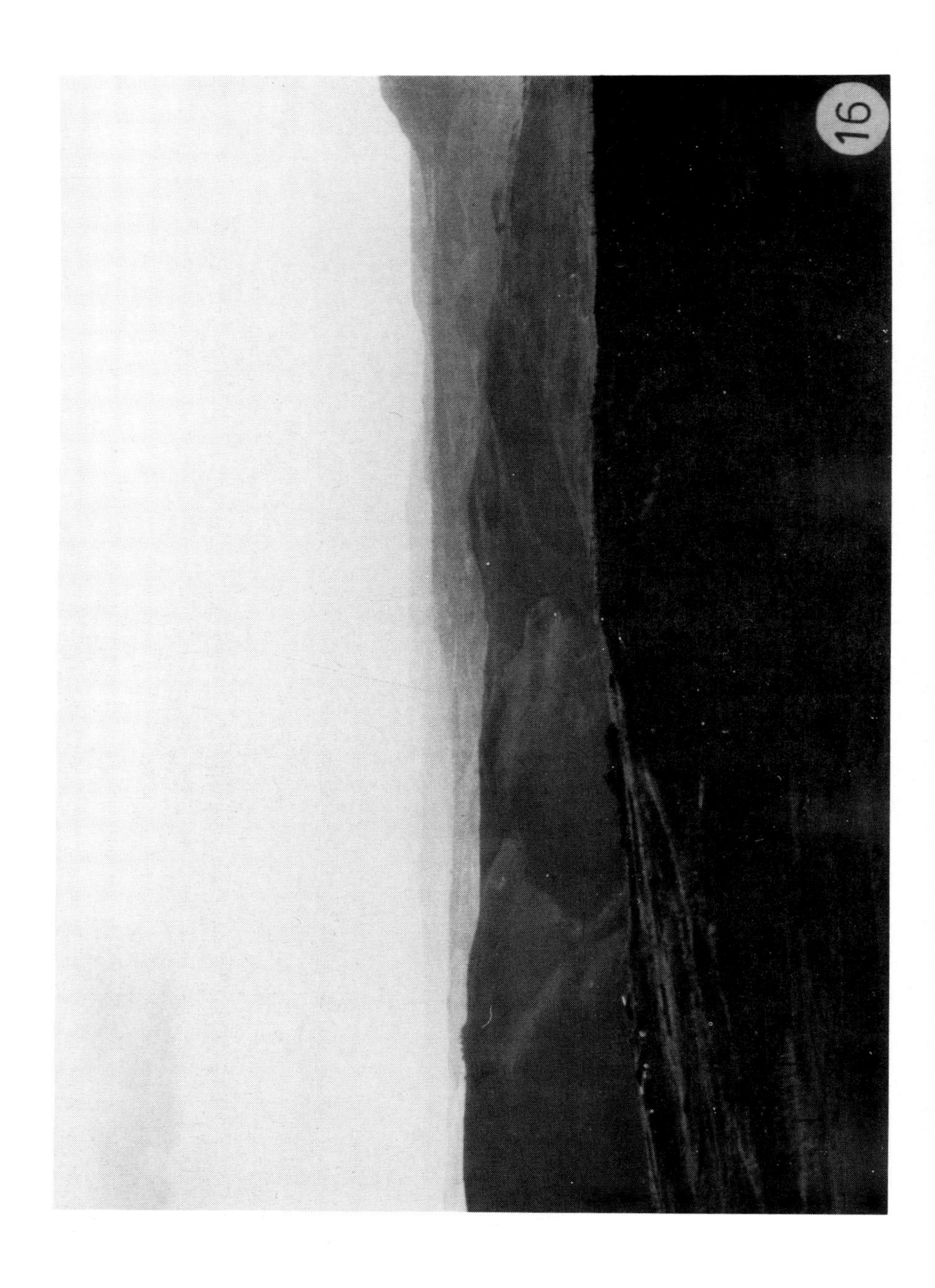
16

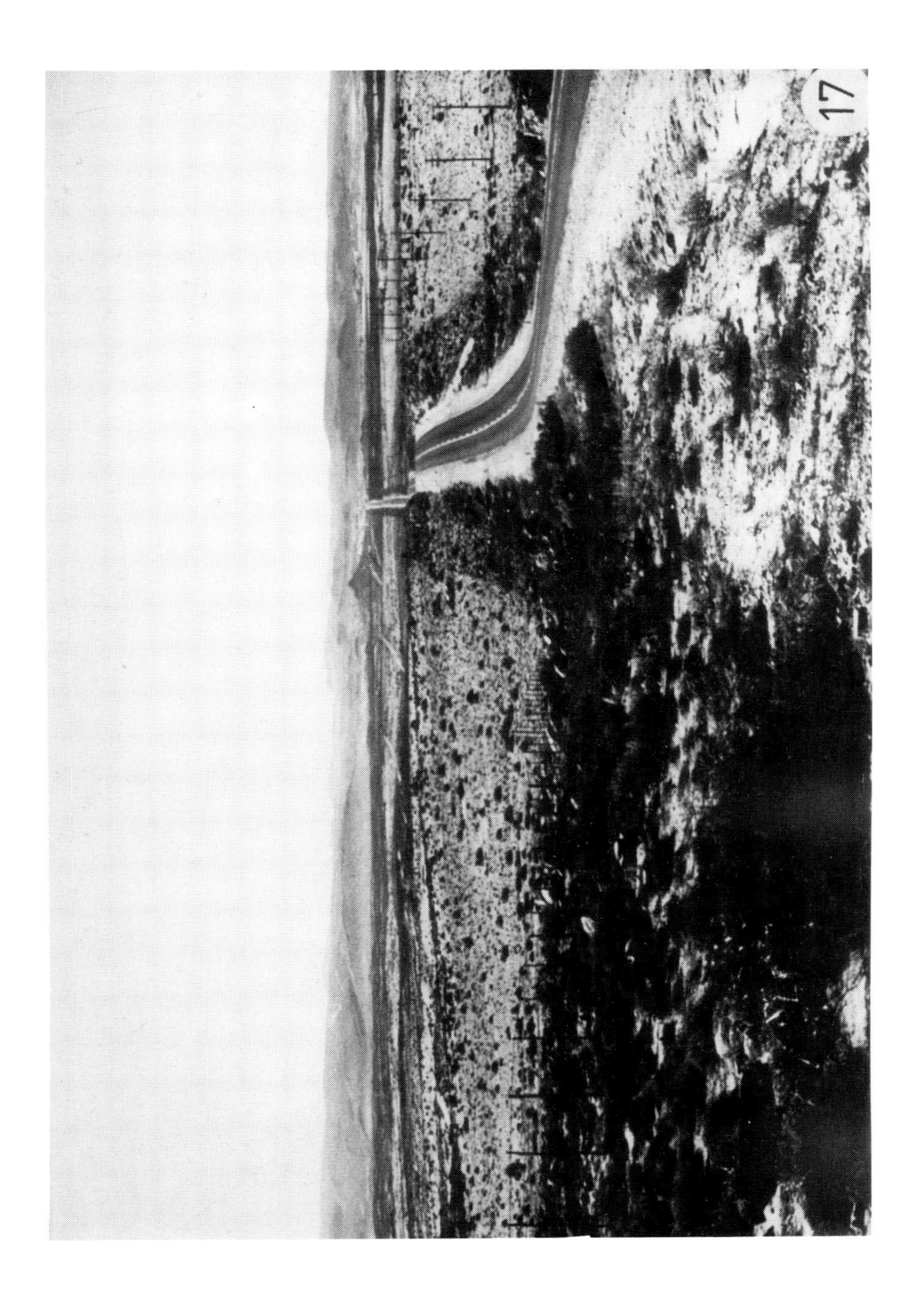
17

18

19

20

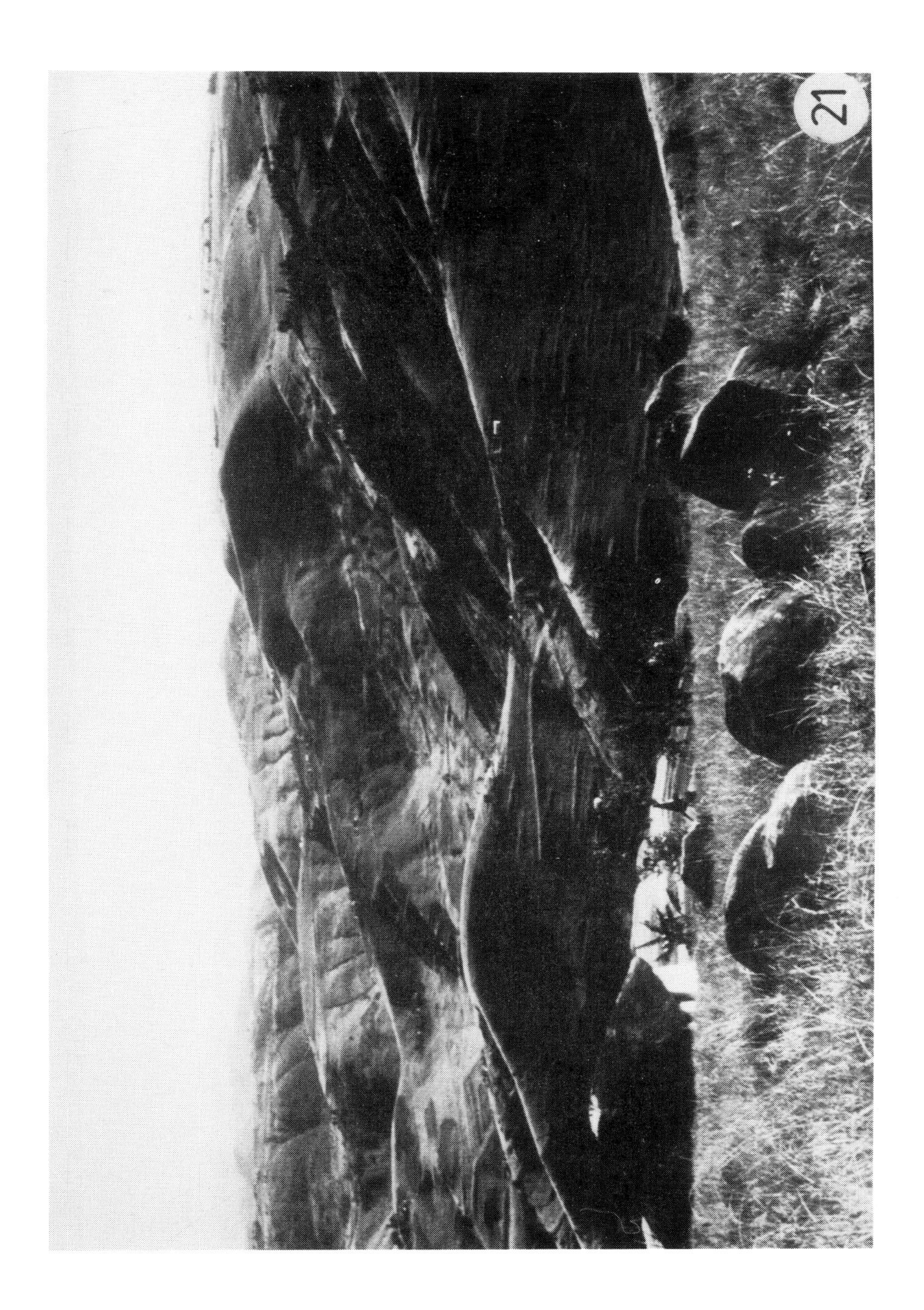
21

22

23

24

25

26

27

28

29

30

31

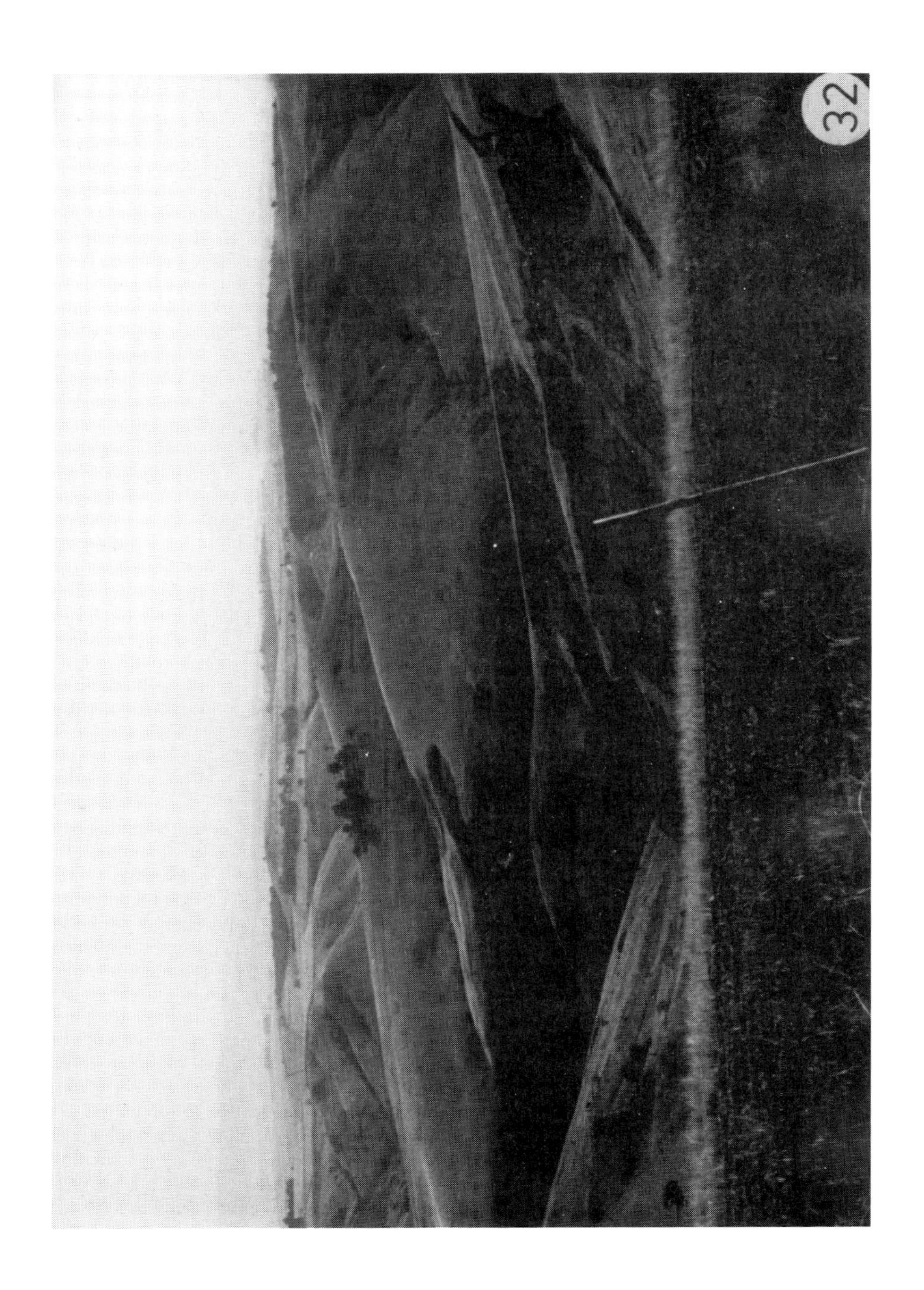
32

33

34

35

36

37

38

39

41

43

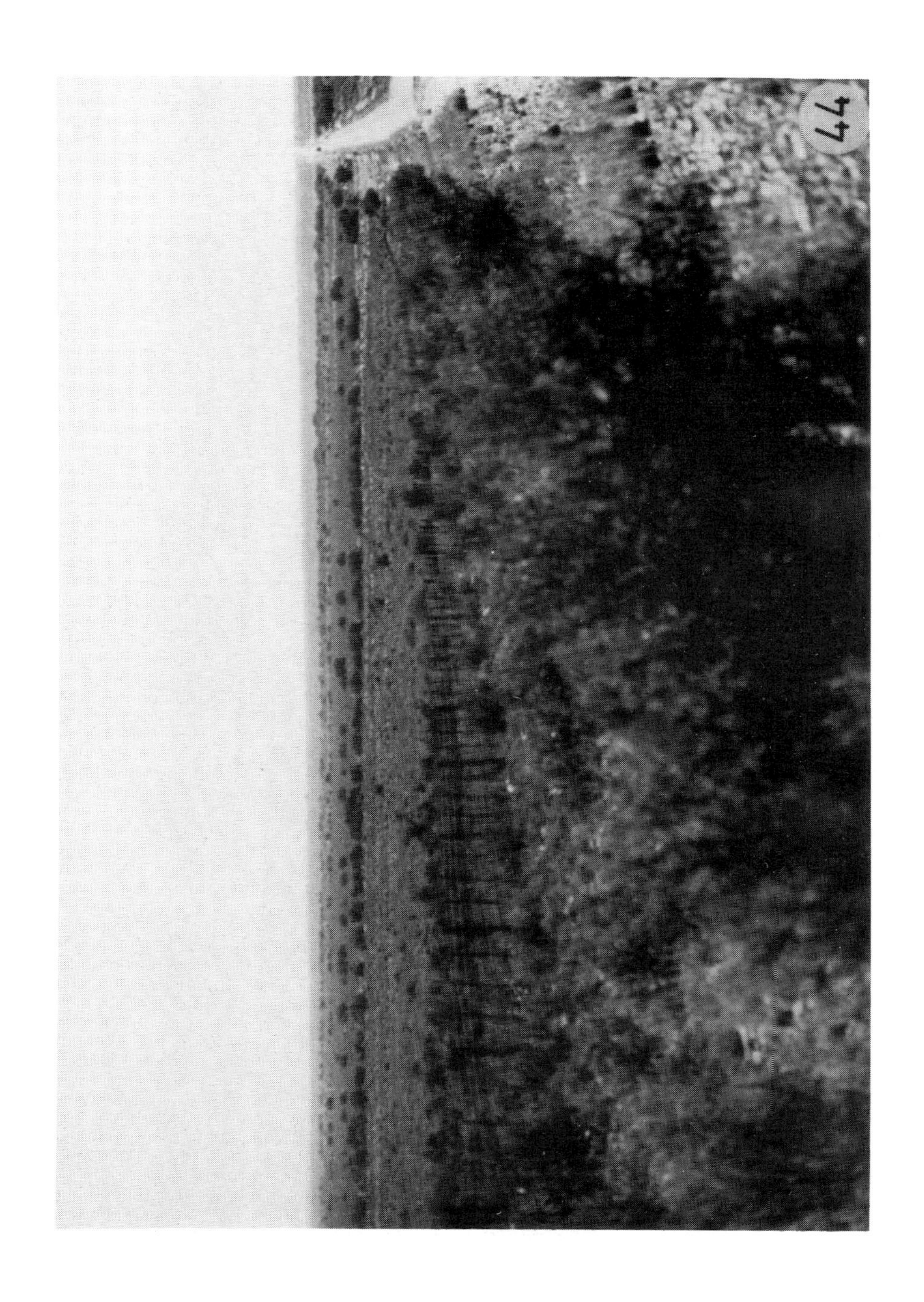
44

47

97

47

67

50

51

52

53

54

55

MÜNCHENER GEOGRAPHISCHE ABHANDLUNGEN

Institut für Geographie der Universität
Fakultät für Geowissenschaften
8 München 2, Luisenstr. 37

Herausgeber: Prof. Dr. H.-G. Gierloff-Emden und Prof. Dr. F. Wilhelm
Schriftleitung: Dr. K. R. Dietz und Dr. H.-J. Mette

Band 1 — Das Geographische Institut der Universität München, Fakultät für Geowissenschaften, in Forschung, Lehre und Organisation. 1972, 101 S., 3 Abb., 13 Fotos, 1 Luftb., DM 10,- — ISBN 3 920397 60 6

Band 2 — KREMLING, H.: Die Beziehungsgrundlage in thematischen Karten in ihrem Verhältnis zum Kartengegenstand. 1970, 128 S., 7 Abb., 32 Tab., DM 18,- — ISBN 3 920397 61 4

Band 3 — WIENEKE, F.: Kurzfristige Umgestaltungen an der Alentejoküste nördlich Sines am Beispiel der Lagoa de Melides, Portugal (Schwallbedingter Transport an der Küste). 1971, 151 S., 34 Abb., 15 Fotos, 3 Luftb., 10 Tab., DM 18,- — ISBN 3 9203997 62 2

Band 4 — PONGRATZ, E.: Historische Bauwerke als Indikatoren für küstenmorphologische Veränderungen (Abrasion und Meeresspiegelschwankungen in Latium). 1972, 144 S., 56 Abb., 59 Fotos, 8 Luftb., 4 Tab., 16 Karten, DM 24,- — ISBN 3 920397 63 0

Band 5 — GIERLOFF-EMDEN, H.-G. und RUST, U.: Verwertbarkeit von Satellitenbildern für geomorphologische Kartierungen in Trockenräumen (Chihuahua, New Mexico, Baja California) - Bildinformation und Geländetest. 1971, 97 S., 9 Abb., 17 Fotos, 2 Satellitenb., 5 Tab., 6 Karten, DM 10,- — ISBN 3 920397 64 9

Band 6 — VORNDRAN, G.: Kryopedologische Untersuchungen mit Hilfe von Bodentemperaturmessungen (an einem zonalen Strukturbodenvorkommen in der Silvrettagruppe). 1972, 70 S., 15 Abb., 5 Fotos, 12 Tab., DM 10,- — ISBN 3 920397 65 7

Band 6 — WIECZOREK, U.: Der Einsatz von Äquidensiten in der Luftbildinterpretation und bei der quantitativen Analyse von Texturen. 1972, 195 S., 20 Abb., 27 Tafeln, 10 Tab., 2 Karten, 50 Diagr., DM 42,- — ISBN 3 920397 66 5

Band 8 — MAHNCKE, K.-J.: Methodische Untersuchungen zur Kartierung von Brandrodungsflächen im Regenwaldgebiet von Liberia mit Hilfe von Luftbildern. 1973, 73 S., 13 Abb., 7 Fotos, 1 Luftb., 1 Karte, vergriffen. — ISBN 3 920397 67 3

Band 9 — Arbeiten zur Geographie der Meere. Hans Günter Gierloff-Emden zum 50. Geburtstag. 1973, 84 S., 27 Abb., 20 Fotos, 3 Luftb., 7 Tab., 3 Karten, DM 25,- — ISBN 3 920397 68 1

Band 10 — HERRMANN, A.: Entwicklung der winterlichen Schneedecke in einem nordalpinen Niederschlagsgebiet. Schneedeckenparameter in Abhängigkeit von Höhe über NN, Exposition und Vegetation im Hirschbachtal bei Lenggries im Winter 1970/71. 1973, 84 S., 23 Abb., 18 Tab., DM 18,- — ISBN 3 920397 69 X

Band 11 — GUSTAFSON, G.C.: Quantitative Investigation of the Morphology of Drainage Basins using Orthophotography - Quantitative Untersuchung zur Morphologie von Flußbecken unter Verwendung von Orthophotomaterial. 1973, 155 S., 48 Abb., DM 27,- — ISBN 3 920397 70 3

Band 12 MICHLER, G.: Der Wärmehaushalt des Sylvensteinspeichers. 1974, 255 S., 82 + 7 Abb., 7 Photos, 23 Tab., DM 28,- ISBN 3 920397 71 1

Band 13 PIEHLER, H.: Die Entwicklung der Nahtstelle von Lech-, Loisach- und Ammergletscher vom Hoch- bis Spätglazial der letzten Vereisung. 1974, 105 S., 16 Abb., 13 Fotos, 14 Tab., 1 Karte, DM 20,- ISBN 3 920397 72 X

Band 14 SCHLESINGER, B.: Über die Schutteinfüllung im Wimbach-Gries und ihre Veränderung. Studie zur Schuttumlagerung in den östlichen Kalkalpen. 1974, 74 S., 9 Abb., 12 Tab., 7 Karten, DM 18,- ISBN 3 920397 73 8

Band 15 WILHELM, F.: Niederschlagsstrukturen im Einzugsgebiet des Lainbaches bei Benediktbeuren, Obb. 1975, 85 S., 40 Fig., 19 Tab., DM 19,- ISBN 3 920397 74 6

Band 16 GUMTAU, M.: Das Ringbecken Korolev in der Bildanalyse. Untersuchungen zur Morphologie der Mondrückseite unter Benutzung fotografischer Äquidensitometrie und optischer Ortsfrequenzfilterung. 1974, 145 S., 82 Abb., 8 Tab., DM 38,- ISBN 3 920397 75 4

Band 17 LOUIS, H.: Abtragungshohlformen mit konvergierend-linearem Abflußsystem. Zur Theorie des fluvialen Abtragungreliefs. 1975, 45 S., 1 Fig., DM 14,- ISBN 3 920397 76 2

Band 18 OSTHEIDER, M.: Möglichkeiten der Erkennung und Erfassung von Meereis mit Hilfe von Satellitenbildern (NOAA-2, VHRR). 1975, 159 S., 65 Abb., 10 Tab., DM 36,- ISBN 3 920397 77 0

Band 19 RUST, U. und WIENEKE, F.: Geomorphologie der küstennahen Zentralen Namib (Südwestafrika). 1976, 74 S., Appendices (50 Abb., 23 Fotos, 17 Tab.,), DM 60,- ISBN 3 920397 78 9

Band 20 GIERLOFF-EMDEN, H.-G. und WIENEKE, F. (Hrsg.): Anwendung von Satelliten- und Luftbildern zur Geländedarstellung in topographischen Karten und zur bodengeographischen Kartierung. 1978, 69 S., 6 Abb., 6 Luftb., 6 Tab., 2 Karten, 4 Tafeln, DM 44,- ISBN 3 920397 79 7

Band 21 PIETRUSKY, U.: Raumdiffereenzierende bevölkerungs- und sozialgeopgraphische Strukturen und Prozesse im ländlichen Raum Ostniederbayerns seit dem frühen 19. Jahrhundert. 1977, 174 S., 25 Abb., 32 Tab., 9 Karten, Kartenband (12 Planbeilagen), DM 46,- ISBN 3 920397 40 1

Band 22 HERRMANN, A.: Schneehydrologische Untersuchungen in einem randalpinen Niederschlagsgebiet (Lainbachtal bei Benediktbeuern/Oberbayern). 1978, 126 S., 68 Abb., 14 Tab., DM 32,- ISBN 3 920397 41 X

Band 23 DREXLER, O.: Einfluß von Petrographie und Tektonik auf die Gestaltung des Talnetzes im oberen Rißbachgebiet (Karwendelgebiet, Tirol). 1979, 124 S., 23 Abb., 16 Tab., 2 Karten, DM 60,- ISBN 3 920397 47 9

Band 24 GIERLOFF-EMDEN, H.-G.: Geographische Exkursion: Bretagne und Nord-Vendée. 1981, 50 S., 19 Abb., 9 Tab., 50 Karten, DM 18,- ISBN 3 88618 090 5

Band 25 DIETZ, K.R.: Grundlagen und Methoden geographischer Luftbildinterpretation. 1981, 110 S., 51 Abb., 9 Tafeln, 9 Karten, DM 40,- ISBN 3 88618 091 3

Band 26 STÖCKLHUBER, K.: Erfassung von Ökotopen und ihren zeitlichen Veränderungen am Beispiel des Tegernseer Tales - Eine Untersuchung mit Hilfe von Luftbildern und terrestrischer Fotografie. 1982, 113 S., 72 Abb., 6 Tab., 8 Tafeln, DM 56,- ISBN 3 88618 092 1

Band 27 WIECZOREK, U.: Methodische Untersuchungen zur Analyse der Wattmorphologie aus Luftbildern mit Hilfe eines Verfahrens der digitalen Bildstrukturanalyse. 1982, 208 S., 20 Abb., 6 Tab., 4 Tafeln, 3 Karten, DM 103,- ISBN 3 88618 093 X

Band 28 SOMMERHOFF, G.: Untersuchungen zur Geomorphologie des Meeresbodens in der Labrador- und Irmingersee. 1983, 86 S., 39 Abb., 2 Tab., 7 Beilagen, DM 25,- ISBN 3 88618 094 8

Band 29 GIERLOFF-EMDEN, H.-G.: Geographische Exkursion: Niederlande. 1982, 36 S., 13 Abb., 2 Tab., 44 Karten, DM 18,- ISBN 3 88618 095 6

Band 30 GIERLOFF-EMDEN, H.G und WILHELM, F. (Hrsg.): Forschung und Lehre am Institut für Geographie der Universität München. 1982, 50 S., 21 Abb., 14 Bilder, DM 20,- ISBN 3 88618 096 4

Band 31 JACKSON, M.: Contributions to the Geology and Hydrology of Southeastern Uruguay Based on Visual Satellite Remote Sensing Interpretation. 1984, 72 S., 36 Abb., 7 Tab., DM 22,- ISBN 3 88618 097 2

Band 32 GIERLOFF-EMDEN, H.-G. und DIETZ, K.R.: Auswertung und Verwendung von High Altitude Photography (HAP) (Hochbefliegungen aus Höhen von 12 - 20 km). Kleinmaßstäbige Luftbildaufnahmen von 1 : 125000 bis 1 : 30000 mit Beispielen von UHAP und NHAP aus den USA. 1983, 106 S., 66 Abb., 23 Tab., DM 50,- ISBN 3 88618 098 0

Band 33 GIERLOFF-EMDEN, H.-G., DIETZ, K.R. und HALM, K. (Hrsg.): Geographische Bildanalysen von Metric-Camera-Aufnahmen des Space-Shuttle-Fluges STS-9. Beiträge zur Fernerkundungskartographie. 1985, 164 S., 130 Abb., DM 60,- ISBN 3 88618 099 9

Band 34 STRATHMANN, F.-W.: Multitemporale Luftbildinterpretation in der Stadtforschung und Stadtentwicklungsplanung. Methodische Grundlagen und Fallstudie München-Obermenzing. 1985, 132 S., 31 Abb., 8 Tab., 39 Luftbilder/Bildtafeln, DM 33,50 ISBN 3 88618 100 6

Band 35 HALM, K.: Photographische Weltraumaufnahmen und ihre Eignung zur thematischen und topographischen Kartierung, zur Umweltverträglichkeitsprüfung (UVP) und zur wasserwirtschaftlichen Rahmenplanung (WRP) - dargestellt am Beispiel der Metric-Camera-Aufnahmen des Rhône-Deltas. 1986, 123 S., 127 Fig., DM 30,- ISBN 3 88618 101 4